MATHEMATICAL AND COMPUTATIONAL CONCEPTS IN CHEMISTRY

Based on the invited and special lectures presented at the International Symposium on Applications of Mathematical Concepts to Chemistry, Dubrovnik, Croatia,

September 2–5, 1985.

Symposium was sponsored by IUPAC, Union of Chemical Societies of Yugoslavia, The Croatian Chemical Society, and Department of Physical Chemistry in the Rugjer Bošković Institute, Zagreb.

Scientific Committee

N.L. Allinger, Athens, GA, USA
F.DeLos DeTar, Tallahassee, FL, USA
F.E. Harris, Salt Lake City, UT, USA
R.B. King, Athens, GA, USA
Z.B. Maksić, Zagreb, Croatia, Yugoslavia
P.G. Mezey, Saskatoon, Saskatchewan, Canada
O.E. Polansky, Mülheim/Ruhr, BRD
L.J. Schaad, Nashville, TN, USA
N. Trinajstić, Zagreb, Croatia, Yugoslavia

Organizing Committee

S. Bosanac
T. Cvitaš
A. Graovac
K. Kovačević
Z. Majerski
Z.B. Maksić
Z. Meić
N. Nekić
Lj. Ruščić
A. Sabljić
B. Špoljar
N. Trinajstić
T. Živković

This book is dedicated to great Croatian scientists

Rugjer Bošković (1711–1787)
Andrija Mohorovičić (1857–1936)
Lavoslav Ružička (1887–1976)
Božo Težak (1907–1980)

MATHEMATICAL AND COMPUTATIONAL CONCEPTS IN CHEMISTRY

Editor:
NENAD TRINAJSTIĆ
The Rugjer Bošković Institute,
Zagreb, Croatia, Yugoslavia.

Published for the
INTERNATIONAL UNION OF PURE AND APPLIED CHEMISTRY

ELLIS HORWOOD LIMITED
Publishers · Chichester

Halsted Press: a division of
JOHN WILEY & SONS
New York · Chichester · Brisbane · Toronto

First published in 1986 by

ELLIS HORWOOD LIMITED
Market Cross House, Cooper Street, Chichester, West Sussex, PO19 1EB, England

The publisher's colophon is reproduced from James Gillison's drawing of the ancient Market Cross, Chichester.

Distributors:

Australia, New Zealand, South-east Asia:
Jacaranda-Wiley Ltd., Jacaranda Press,
JOHN WILEY & SONS INC.,
G.P.O. Box 859, Brisbane, Queensland 4001, Australia

Canada:
JOHN WILEY & SONS CANADA LIMITED
22 Worcester Road, Rexdale, Ontario, Canada.

Europe, Africa:
JOHN WILEY & SONS LIMITED
Baffins Lane, Chichester, West Sussex, England.

North and South America and the rest of the world:
Halsted Press: a division of
JOHN WILEY & SONS
605 Third Avenue, New York, N.Y. 10158 U.S.A.

British Library Cataloguing in Publication Data

Applications of mathematical concepts to chemistry.
1. Chemistry- Mathematics
2. Numerical calculations
I. Trinajastic Nenad
II. International Union of Pure and Applied Chemistry
III. Internationial Symposium on Applications of Mathematical
   Concepts to Chemistry (1985–Dubrovnik)
540'.1'51   QD39.3.M3

ISBN 0-85312-934-7 (Ellis Horwood Limited)
ISBN 0-470-20289-0 (Halsted Press)

Printed in Great Britain by Buttler and Tanner, Frome, Somerset.

Contents

Contents

Contents

PREFACE

This book contains the invited and special lectures presented at the INTERNATIONAL SYMPOSIUM ON THE APPLICATIONS OF MATHEMATICAL CONCEPTS TO CHEMISTRY held in Dubrovnik, Croatia, from September 2 to September 5, 1985. The Symposium was sponsored by the International Union of Pure and Applied Chemistry in conjunction with the Union of Chemical Societies of Yugoslavia, the Croatian Chemical Society, and the Department of Physical Chemistry of the Rugjer Bošković Institute in Zagreb.

Not long ago, several researchers from different countries working in various fields of theoretical chemistry, suggested that an international symposium on some aspects of Mathematical Chemistry be organized. These included Professor B.M. Gimarc (Columbia, SC), Professor M. Randić (Ames), Professor D.J. Klein (Galveston), Dr. R.B. Mallion (Canterbury), Dr. D. Bonchev (Burgas), Professor P.G. Mezey (Saskatoon), Professor J.V. Knop (Düsseldorf), Dr. P. Křivka (Pardubice), Professor B.A. Hess, Jr. (Nashville), Professor A.T. Balaban (Bucharest). The suggestion specifically indicated the Theoretical Chemistry Group at the Rugjer Bošković Institute in Zagreb as organizer, primarily because of our past active role in developing and promoting Mathematical

Preface

Chemistry in general and Chemical Graph Theory in particular. We accepted the responsibility, both as an honour and a privilege. We are especially happy that the long standing tradition of Mathematical Chemistry in this country, and in Croatia in particular, has been recognized in this way. Hence, this, the first symposium dealing exclusively with topics that constitute Mathematical Chemistry has been organized by our Group with great fervour and enthusiasm.

We selected Dubrovnik as the site of the Symposium because it is one of the most interesting Croatian cities with a distinguished historical, cultural, and scientific heritage. The famous Croatian scientist Rugjer Bošković, fellow of the Royal Society (London), Professor of Mathematics at Collegium Romanum in Rome, and founder of the Observatory at Brera, was born there (May 18, 1711). Besides Dubrovnik, known for its mild Mediterranean climate, is a frequent site for scientific meetings, and is easily accessible by land, sea and air.

The International Symposium on the Applications of Mathematical Concepts to Chemistry was attended by 121 participants from 21 countries. The Symposium brought together mathematical chemists, computer chemists, theoretical chemists, mathematically and theoretically minded experimental chemists, mathematical physicists and mathematicians, for discussions about the status of Mathematical Chemistry, its perspectives, and its influence on various aspects of Chemistry.

Last day of the Symposium has been dedicated to the memory of the late Professor Andrej Ažman (Celje 1937 - Ljubljana 1980) who in his time was the leading chemical theoretician in Yugoslavia and inspiring supporter of Mathematical Chemistry. Last, but not least, during the Symposium important preliminary steps have been made to establish an International Society of Mathematical Chemistry. We expect that the combination of exciting science, beautiful scenery, warm weather and the hospitality of the local people made the Symposium a memorable experience for all participants.

The lectures and posters covered a significant part of the broad spectrum of problems in Mathematical Chemistry and its interactions with other areas of Chemistry. The Symposium also revealed the permanent need for Mathematics in all branches of Chemistry. In addition, it became evident that Mathematical Chemistry will continue to grow as a part of

Theoretical Chemistry; other overlapping but distinguishable parts include Quantum Chemistry, Statistical Mechanics and Computational Chemistry.

The contributed papers presented at the Symposium will appear in a special issue of Croatica Chemica Acta, a chemistry journal published by the Croatian Chemical Society.

Zagreb, September 1985.

Nenad Trinajstić

Chapter 1

INTRODUCTORY REMARKS

V. Prelog
Organic Chemistry Laboratory of the Swiss Federal Institute of Technology,
CH-8092 Zürich, Switzerland

About 200 years ago Immanuel Kant wrote in his "Kleinere Schriften zur Naturphilosopie": "Ich behaupte, dass in jeder besonderen Naturlehre nur so viel Wissenschaft angetroffen werden könne, als darin Mathematik anzutreffen ist". He wrote then: "Solange also noch für die chemischen Wirkungen der Materien aufeinander kein Begriff aufgefunden wird, der sich konstruieren lässt ....eine Forderung, die schwerlich jemals erfüllt werden wird - so kann Chemie nichts mehr als Kunst oder Experimentallehre, niemals aber eigentliche Wissenschaft werden". Just about the same time Lavoisier was introducing mathematics into chemistry through his use of the balance. The law of definite proportions became the basis of stoichiometry and every chemist-artist had to learn some arithmetic.

The next great step in development of chemistry was the structural theory, but the pioneers in this field, Butlerov, Couper and Kekulé did not realize that the structural formulae are actually mathematical objects, graphs. It was the mathematician Arthur Cayley who first became aware of that and he developed the theory of tree-graphs by trying to calculate the number of possible constitutional isomers of paraffin hydrocarbons C_nH_{2n+2}. Later, Georg Polya, in an encounter with problems of isomerism discovered his famous theorem which became one of the fundaments of combinatorics. These mathematical achievements, however, had very little impact on the daily praxis of chemists, because their everyday problems were usually so simple that they could mostly be solved by trial and error.

Soon after the structural theory had been developed Jacobus Hendricus van't Hoff and Achilles Le Bel founded organic stereochemistry and Alfred Werner the stereochemistry of inorganic complexes. Symmetry, disymmetry and asymmetry became recognized as important features of molecules, and although group theory was clearly the branch of mathematics most appropriate for treating problems in this area very few chemists made use of it.

The gate of entrance for more sophisticated mathematics into chemistry was and still is physical chemistry. Step by step, methods of mathematical physics penetrated into chemistry: first calculus, to deal with problems of chemical thermodynamics and kinetics, then linear algebra and group theory, as additional tools of quantum chemistry and molecular spectroscopy.

The more recent general impact of computers on chemistry and on chemists cannot be overestimated. Without computers theoretical chemistry, structural analysis by diffraction methods, molecular mechanics, etc. could not have developed far, and the collection, retrieval and classification of the terrifying multiplicity of chemical data would be almost impossible . The design of syntheses and analyses of reaction pathways are other promising applications.

Last but not least topological aspects of chemical concepts such as e.g. aromaticity, should be mentioned. Thus the progress of chemistry depends more than ever on applications of mathematics.

The important aims of such conferences as this one are to bring together scientists who use the many different branches of mathematics in chemistry and to bridge existing gaps between them, to learn about the new work of pioneers and about the progress in established fields. The titles in the programme indicate that almost all mentioned topics will be covered.

In summary one can say that chemistry today includes quite enough mathematics to be considered a respectable science from the Kantian point of view, but let us also not forget what Lord Kelvin said about physics: "It is as dangerous to let mathematics take charge of physics as to let an army run a government."

Chapter 2

THE MATHEMATICAL TRAINING OF CHEMISTS

J.N. Murrell
School of Molecular Sciences, The University of Sussex,
Brighton BN1 9QJ, England, UK

I was recently a member of an appointing committee for a university lectureship in mechanical engineering and was struck by the fact that none of those interviewed had dirty fingernails. I should put it less glibly; everyone that was seriously considered for the post was involved in the computer simulation of structures rather than the building and testing of rigs. Of course, one can understand why this is so; it is far cheaper to do a computer simulation of an oil rig than to build one and test it to destruction. I assume that somewhere in the background there must be engineers with oily hands who are still building and testing, because one can only confidently use mathematical models to extend empirical knowledge by small amounts. Nevertheless, the balance between the mathematical model builders, and the rig builders and testers, seems to have swung far towards the clean handed people in recent years.

I do not think this situation has occurred yet in chemistry which is still predominantly an empirical subject. In fact, I would go as far as to say that the mathematical models used in chemistry still have rather poor predictive performance. How often does an experimental chemist ask for the results of a calculation before deciding whether an experiment is likely to prove fruitful? In part this is because chemical experiments are still generally cheap and not too time consuming. It is quite often quicker to do a good experiment than a poor calculation.

However, the situation is bound to change in the future and one can see already that this is beginning in some of the more expensive fields such as the pharmaceutical industry. It costs several million pounds to bring a new drug to the market so that any model which can narrow the field of likely candidates is very useful.

The point of this preamble is to stress that what has in the past been found adequate by the chemical community for their models may not be adequate for the future. I am not saying that our chemistry courses should ignore the Friedel-Craft's reaction in favour of more time for the Schrödinger equation. Rather that we should move our theoretical studies a little away from conceptual models towards more quantitative models.

I have for many years been struck by the fact that chemists and physicists seem to have different approaches to theoretical problems. Physicists like to solve approximate models exactly and chemists prefer to solve good models approximately. Think how far the physicist has gone with the particle-in-a-box. In fact, I believe that most physicists are by training or inclination even less inclined to seek the help of quantitative mathematical models than are chemists.

You might justly ask whether I practice what I preach. Well, with my close colleagues Sydney Kettle and John Tedder I have written two undergraduate textbooks on the theory of the chemical bond. The first of these, Valence theory, was written in 1965 and is quite firmly based on mathematical skills. For example, the book shows how to derive the ligand field matrix for p^2 in the intermediate (weak-strong) coupling regime, which I think is quite tough for an undergraduate text. We assumed that the reader had a prior knowledge of vector, operator and matrix algebra and possessed a spirit which was not overawed by mathematical manipulations. However, Valence theory is essentially a book that deals with concepts. In a later edition we derived the Hartree-Fock-Roothaan equations but there were few references to the results which could be obtained from them.

Our second book, The Chemical Bond, was published in 1978 and is much less mathematically demanding; nothing more is required than the ability to expand a secular determinant.

However. the principles underlying the ab-initio SCF approach are explained and the results of such calculation are used to establish the validity of the qualitative MO approach. For example, the concept of a molecular orbital is developed not only by arguments based on the LCAO approximation but also by examining the actual ab-initio results for H_2^+. Our view in 1978 was that students could appreciate the basis of quantitative calculations without having the mathematical ability to follow the problems encountered in, say, integral evaluation. In a new edition of the book, soon to appear, we are more expansive on the mathematical basis of the ab-initio method and on the general structure of the commonly available black-box SCF programs. I stress my belief that one can get students to understand the basis of the ab-initio method and to appreciate the results of such calculations even though their mathematical background is poorer than is required to follow many of the conceptual models of valence theory.

We have been struggling for many years to produce students who are literate and numerate. We should now also require them to be computerate. I do not much like the word but I am sure it will not be long before it appears in the Oxford English Dictionary.

I had originally intended to start this essay with a list of the mathematical tools that all chemists should know. On further reflection I decided that was not very useful. I am sure that most of us would take the view that almost any mathematical skill can be valuable. Putting this the other way round; I would be most reluctant to pick out any mathematical topic as being unworthy of study or as being of no use to chemists. The reason that we are at this conference rather than the Burgenstock conference on Natural Products (I know a few of us go to both) is not that we know a great deal of mathematics but that we have an appreciation for the mathematical approach. I would even say that most of us would find some joy in the proof that the square root of two is not a rational number - and I cannot think that has much relevance to theoretical chemistry.

However. I will finish by telling you what mathematics is known by one of our most eminent organic chemists. I told Professor John Cornforth that I was to give this talk and in order to see what was the minimum requirement for a good organic chemist I asked for the things he knew. This was his list:

   (i)   Elementary arithmetic
  (ii)   Two-dimensional geometry including analytical

        geometry
(iii)   Plane trigonometry and trigonometric functions
 (iv)   Conic sections
  (v)   Elementary algebra
 (vi)   Simultaneous equations
(vii)   Permutations and combinations
(viii) Differential calculus
 (ix)   Integral calculus
  (x)   Differential equations (including partial)
 (xi)   Theory of probability
(xii)   Theory of errors
(xiii) Exponential functions
(xiv)   Binomial theorem

He regretted not knowing:

  (i)   Prime numbers
 (ii)   Diophantine equations
(iii)   Topology
 (iv)   Symmetry

Well, I suspect we do not have a typical case here.   I am
certain that if he had sought my opinion on what organic
chemistry a theoretician needed to know my list would have
been much less impressive.

Chapter 3

THE EFFECT OF ELECTRO-NEGATIVITY ON BOND LENGTHS IN MOLECULAR MECHANICS CALCULATIONS

Norman L. Allinger, Mita R. Imam, Manton R. Frierson, and Young Yuh
Department of Chemistry, University of Georgia, Athens, GA 30602
Lothar Schäfer
Department of Chemistry, University of Arkansas, Fayetteville, AR 72701

ABSTRACT

The attachment of an electronegative atom or group to a carbon-carbon bond causes that bond to have a reduced length. The magnitude of the reduction is roughly proportional to the electronegativity of the attached atom. The substitution of multiple electronegative atoms on the same bond leads to a further reduction of the bond length, unless the size and number of the groups is sufficient so that steric effects outweigh the electronegativity effect. Carbon-hydrogen bonds and carbon-halogen bonds behave similarly, except that the degree of shrinkage differs, depending upon the particular bond. Attachment of an electropositive atom causes a bond lengthening. Within the context of molecular mechanics, these changes in bond length occur in the "natural bond length", or l_o value for the bond, and whatever other effects may be present in the bond due to its particular environment also occur as usual, giving a resulting bond length which may be smaller or larger than the usual value by as much as 0.050 Å or so.

INTRODUCTION

The energy of an electron in a 2s orbital on carbon is much lower than that of an electron in the corresponding 2p orbital. Hence, if a hydrogen atom in an alkane is replaced by an electronegative atom, the carbon bonded to that electronegative atom responds by donating electron density to the electronegative atom, which means the bond to that atom contains more p character than it did in the alkane. If one bond to a carbon has the amount of p character in it increased, the other bonds to that carbon must between them have an increased amount of s character. Considering carbon-carbon bonds in general, an increase in s character (from an sp^3 hybridization) will give a shorter bond, while increased p character will give a longer bond. Thus the attachment of an electronegative atom in, say, ethane, to give fluoroethane, yields a molecule with a shorter carbon-carbon bond than in ethane itself. And attachment of an electropositve atom would yield a longer carbon-carbon bond, for analogous reasons. Experimentally these changes in bond length have long been known, as has the interpretation given.[1-5]

The molecular mechanics treatment of hydrocarbons has been worked out in considerable detail. While it cannot be said that all problems have been solved in this area, most of them have been, at least to a reasonably good approximation. Thus for ordinary saturated hydrocarbons, one can calculate good structures by what are now standard methods.[6]

When one adds other atoms onto the hydrocarbon framework, then the situation is somewhat different. Depending on the type and number of the substituents on the hydrocarbon, the geometries as now calculated by molecular mechanics may be appreciably in error. Since most organic molecules contain various kinds of functionalized substituents, it is also important to be able to deal with the effects of these substituents accurately.

We have been interested in refining the MM2 force field[7] so as to obtain more accurate structures for functionalized molecules. As a prelude to this, we gathered a large amount of representative experimental data so as to quantitatively study this electronegativity effect. We have also carried out ab initio calculations on this effect, and in particular on the torsional dependence of the effect. Some of these calculations have been published previously.[8]

As far as molecular ground states, theory and experiment show substantially the same thing, namely that a given bond will be shortened if an electronegative atom or substituent is attached to it, and lengthened if an electropositive atom is attached to it.[1-5,8] The amount of shortening or lengthening is roughly proportional to the electronegativity of the attached substituent[9], but also depends upon steric effects and other interactions of the substituent with the rest of the molecule. Theory indicates that these electronegativity effects should be general, that is, they should occur for all substituents and all bonds, the only requirement being a difference in electronegativity between the substituent and the atom to which it is attached. But quantitatively, the effects can be small or large, and important or otherwise, depending upon the case. The torsional dependence of these bond length effects upon the substituent is a somewhat more complicated matter, and that will not be discussed further here.[8] There is also some dependence of the bond angles upon the electronegativity of the attached substituent, and in the direction predicted by the model outlined. Thus, if we substitute a fluorine atom into ethane, the C-C-F bond angle, because of the extra p character in the bond to the fluorine, will be somewhat smaller than tetrahedral, whereas the other bond angles at the central carbon will be somewhat larger. The effect has also been studied experimentally with respect to benzene rings having attached substituents, where a great deal of information is available.[10]

RESULTS

In the present work we have examined a few typical substituents, and we have noted how carbon-carbon bonds are affected by their attachment. Many of the substituents examined were halogens, and if one attaches two or more halogens to the same carbon, then one may also note the effect of one halogen on the other carbon-halogen bond. To a lesser extent we have looked also at carbon-hydrogen bonds.

Our formulation is that the natural bond length, l_o, is changed by an amount δl_e, to give a new natural bond length (l_o') such that;

$$l_o' = l_o + \delta l_e$$

where δl_e is a negative quantity for an atom more electronegative than the atom to which it is attached, and is a positive quantity for an atom more electropositive than the atom to which it is attached. The quantities δl_e are given for various atoms and bonds in Table 1. If we consider a given bond, for example C-C, which has multiple electronegative (or electropositive) substituents, A, B, C, D,... etc;

$$l_o' = l_o + \delta l_{e(A)} + 0.62\delta l_{e(B)} + 0.62^2\delta l_{e(C)} + 0.62^3\delta l_{e(D)} + ..$$

where the substituents are ordered A, B, C, D,... in order of their decreasing values of δl_e (absolute values). The factor of 0.62 in the above equation was arrived at by studying the data for a number of compounds belonging to six different classes, but it was mainly chosen to fit the available data on fluorides. Because there is a relatively large amount of gas phase data for alkyl fluorides, and because the fluorine, being most electronegative, has the largest effect, this was taken as the case to fit.

In Table 1 are given the electronegativity correction parameters to l_o. For a carbon-carbon bond the amount of shrinkage is largest in the case of fluorine (.022 Å), but other electronegative atoms such as halogens and oxygen show effects of the order of 5-10 thousandths of an Å. The only electropositive atoms studied were silicon[9], which in silanes yields a lengthening of .015 Å, and hydrogen when attached to oxygen or nitrogen.

The effects tend to be smaller, usually unobservably small, for a C-H bond. (They can, however, be seen in ab initio calculations). But for carbon-halogen bonds, the effects can be even larger than they are for carbon-carbon bonds. Thus, fluorine attached to the carbon of a C-F bond yields a shortening of .034 Å.

It is also of interest that a hydrogen attached to an oxygen (alcohol) or to a nitrogen (amine) yields a substantial bond lengthening, relative to the analogous molecule in which there is a carbon attached instead of a hydrogen. (The hydrogen may be thought of as being more electropositive than the carbon.) Thus, the C-O bond in an alcohol is in general longer than that in an ether, and the C-N bond in methylamine is longer than that in dimethylamine, which in turn is longer than that in trimethylamine. These bond length relationships are the reverse of those which result from simple steric effects, as were applied in the earlier versions of the MM2 program. With those versions of the program, the best that could be done was to average out the bond lengths, so that we calculated the C-N bond length in methylamine as being too short, and that in trimethylamine as being too long. But with this electronegativity correction, we can get this order correct.

We might mention the anomeric effect, which occurs when one has two atoms, each of which contain a lone pair of electrons, attached to a common atom. If these atoms are both halogens, no special treatment is needed, because the collective orientation of lone pairs can always be viewed as being optimal. If one or more of these atoms is an oxygen, nitrogen, or sulfur, then a special treatment of the anomeric effect is required, depending on the orientations of the lone pairs. We presented a formulation of this treatment elsewhere for the case where the two atoms are both oxygen.[11] One could similarly formulate an analogous treatment for other combinations of these electronegative atoms.

Table 1. Electronegativity Correction Parameters to l_o[a]

BOND TYPE	END OF BOND	ATOM TYPE	CORRECTION TO l_o (Å)
1-1	1	6	-.009
1-1	1	8	-.001
1-1	1	11	-.022
1-1	1	12	-.008
1-1	1	13	-.012
1-1	1	14	-.005
1-1	1	15	-.001
1-1	1	19	.015
1-5	1	6	-.002
1-5	1	8	-.001
1-5	1	11	-.010
1-5	1	12	-.005
1-5	1	13	-.002
1-5	1	14	-.001
1-11	1	11	-.034
1-11	1	12	-.020
1-11	1	13	-.004
1-12	1	11	-.030
1-12	1	12	-.020
1-12	1	13	-.003
1-13	1	11	-.002
1-13	1	12	-.003
1-13	1	13	-.002
1-14	1	11	-.001
1-14	1	12	-.001
1-14	1	13	-.001
1-6	6	21	.019
1-8	8	23	.015

[a]The atom type numbers have the usual meaning: 1 is carbon, 5 hydrogen, etc.[7]

In Table 2 are given parameters that should be used to replace those in earlier versions of MM2, which need to be changed because of the introduction of the electronegativity effect.

In a few cases they are just updates, because better information is available now than was available when the original formulation was carried out.

In Table 3 are summarized some experimental bond length data for a group of fluorides, together with the corresponding values as given now by the MM2 program. Table 4 gives a similar summary for chlorides, and for a few compounds which contain both chlorine and fluorine.

Table 2. Revised MM2 Parameters

Torsional Parameters (kcal/mol)

Atom Type Nos.				V1	V2	V3
11	1	1	11	-.100	-2.000	.200
6	1	1	12	.000	-1.400	.180
6	1	1	13	.000	-1.400	.180
6	1	1	14	.000	-1.400	.180

Stretching Parameters

Bond Type	K_s(mdyn/Å)	l_o (Å)
1 - 6	5.36	1.402
1 - 8	5.10	1.4380
1 - 11	5.10	1.3920
1 - 12	3.23	1.7950
1 - 13	2.30	1.9490
1 - 14	2.20	2.1490

Bending Parameters

Atom Types			K_θ(mdyn Å/rad^2)	θ_o (deg)	Ed. Type
5	1	11	.490	110.500	
11	1	11	1.070	107.100	
1	1	11	.650	109.500	1
1	1	11	.650	107.500	2
1	1	11	.650	109.500	3

Table 3. C–C and C–F Bond Lengths in Some Fluorides[12]

Compound	C-C bond length		C-F bond length	
	Exper.	MM2	Exper.	MM2
Ethyl Fluoride[14]	1.504(5)	1.511	1.399(4)	1.394
1,2-Difluoro-ethane[a,15]	1.505(3)	1.507	1.391(2)	1.395
1,1-Difluoroethane[16]	1.500(4)	1.496	1.366(2)	1.360
1,1,1-Trifluoro-ethane[17]	1.496(3)	1.486	1.342(2)	1.338
1,1,2-Trifluoro-ethane[a,18]	1.502(5)	1.505	1.355(4)CF_2H 1.389(8)CH_2F	1.361 1.396
1,1,1,2-Tetra-fluoroethane[19]	1.503(4)	1.507	1.391(6)CH_2F 1.336(2)CF_3	1.397 1.339
1,1,2,2-Tetra-fluoroethane[a,20]	1.520(5)	1.518	1.352(2)	1.361
Pentafluoroethane[14]	1.527(4)	1.532	1.349CHF_2 1.329CF_3 1.337(2)av.	1.361 1.340 1.348
Hexafluoroethane[21]	1.545(6)	1.560	1.326(2)	1.340
2-Fluoropropane[22]	1.514(4)	1.512	1.405(5)	1.396
1,3-Difluoro-propane[a,b,23]	1.515(3)	1.516	1.393(2)	1.394
t-Butyl Fluoride[24]	1.522(24)	1.517	1.427(24)	1.399

[a]The MM2 values are for the conformation which is calculated to be of lower enthalpy.

[b]The experimental numbers are average values for the conformers.

Table 4. C–C and C–Cl Bond Lengths in Some Chlorides[12]

Compound	C–C bond length		C–Cl bond length	
	Exper.	MM2	Exper.	MM2
Ethyl Chloride[25]	1.528(4)	1.523	1.802(3)	1.799
1,2-Dichloro-ethane[a,26]	1.530(4)	1.523	1.795(1)	1.800
1,1-Dichloro-ethane(r_o)[27]	1.540(15)	1.521	1.766(15)	1.780
1,1,1-Trichloro-ethane(r_o)[28]	1.541(15)	1.524	1.7712(100)	1.772
1,1,2-Trichloro-ethane[a,b,29]		1.530	1.778(5)	1.789
Hexachloroethane[30]	1.566(20)	1.623	1.771(6)	1.780
1-Chloro-propane[a,b,31]	1.525(2)	1.531	1.796(2)	1.799
2-Chloropropane[32]	1.529(1)	1.527	1.814(1)	1.805
1,1-Dichloro-propane[a,b,33]	1.522(6)	1.531	1.781(6)	1.781
2,2-Dichloro-propane[34]	1.523(4)	1.526	1.799(3)	1.789
1,3-Dichloro-propane[a,b,35]	1.531(8)	1.529	1.798(6)	1.799
Octachloro-propane[36]	1.657(60)	1.640	1.764(24)(CCl_3) 1.812(80)(CCl_2)	1.778 1.812
t-Butyl Chloride[37]	1.532(6)	1.531	1.828(10)	1.814
1-Chloro-1,1-di-fluoroethane(r_o)[38]	1.490(20)	1.496	1.736(15)(C–Cl) 1.328(20)(C–F)	1.753 1.348
1-Chloro-2-fluoro-ethane[a](r_o)[39]	1.530(20)	1.511	1.787(20)(C–Cl) 1.365(20)(C–F)	1.801 1.395
1,1,1-Trichloro-2-2-2-trifluoro-ethane(r_o)[40]	1.539(15)	1.573	1.771(10)(C–Cl) 1.330(20)(C–F)	1.776 1.341

[a]The MM2 values are for the conformation which is calculated to be of lower enthalpy. If there is more than one bond of a particular kind, the MM2 value is an average bond length.

[b]The experimental numbers are average values for the conformers.

Alkyl bromides were studied in a manner similar to that described for the chlorides and fluorides. The number of compounds for which data are available is much smaller. Compounds actually used were ethyl bromide[41], 1,2-dibromoethane[42], 2-bromopropane[43], 1,3-dibromopropane[44], n-butyl bromide[45], and 2-bromo-1-chloro-2-methylpropane[46]. The last three of these were studied by electron diffraction, the others by spectroscopic methods. Constants chosen to fit the data are included in Table 1.

The gas phase structural data available for ethers are limited, and for alcohols extremely so. Many of the data which are available are from microwave studies with large uncertainties. In general, the effect of an oxygen attached to a C-C bond is not very great, so corrections are not very large. On the other hand, the change in the C-O bond length when a hydrogen is attached to an oxygen (rather than a carbon as in an ether) is considerably larger. The numbers that were arrived at were -0.009 Å for the shrinkage of a C-C bond due to the attachment of an alcohol or an ether oxygen, and $+0.019$ for the lengthening of the C-O bond from attachment of a hydrogen. The electronegativity correction caused by oxygen on the C-H bond was taken to be -0.002 Å. The natural bond length for a C-O bond was reduced from 1.407 Å to 1.402 Å to better fit ethers and alcohols simultaneously. (The earlier bond length had been chosen to do as well as could be done with a single value for both alcohols and ethers, but weighting the fit towards the ethers). The data used were taken from the following compounds: methanol[47], ethanol[48], dimethyl ether[49], ethyl methyl ether[50], diethyl ether[51], methyl propyl ether[51], tetrahydropyran[52] dimethoxy-methane[11,53], and 1,3-dioxane[54]. The last two compounds also involve the anomeric effect, which has been treated elsewhere.[11]

Amines have also been examined (Table 5). Since the electronegativity difference between nitrogen and carbon is quite small, the correction needed is also quite small (-0.001 Å). But the correction needed for a hydrogen attached to nitrogen is sizeable ($+0.015$ Å). Amine C-N bond lengths in fact get shorter along the series primary, secondary, tertiary. When only steric interactions are considered, the calculated C-N bond is found to be too short in methylamine, and too long in trimethylamine, a necessary result from the steric effects. However, in the current procedure when the effects of electronegativity are taken into account, these bond lengths turn out to be in the correct order.

Table 5. C–N and C–C Bond Lengths in Some Amines[12]

Compound	C-N bond length		C-C bond length	
	Exper.	MM2	Exper.	MM2
Methylamine[55]	1.467(2)	1.466		
Dimethylamine[56]	1.457(2)	1.463		
Trimethylamine[56]	1.456(2) 1.458(r_α)	1.455		
Ethylenediamine[57]	1.469	1.469	1.545(8)	1.534
Dimethyl-ethylamine[58]	1.454(6)	1.458	1.541(24)	1.536
Piperidine[59]	1.474(11)	1.464	1.533(6)	1.534

CONCLUSIONS

In general we feel that the introduction of the "electro-negativity effect" including both variable natural bond lengths and angles will bring considerable improvement to molecular mechanics results for functionalized molecules without a significant increase in computation time. Geometries in molecular mechanics should take into account the most significant effects of the environment on bond distances and angles. These effects include the electronegativity effect discussed here and the conformational effects discussed previously.[8,11] Systematic studies of these effects are underway in our laboratory. Results will be reported elsewhere.

REFERENCES

(1) Bent, H. A. Chem. Rev. **1961**, 61, 275.
(2) Moffitt, W. Proc. Roy, Soc. (London) **1949**, A196, 524.
(3) Mulliken, R. S. J. Phys. Chem. **1937**, 41, 318.
(4) Walsh, A. D. Discussions Faraday Soc. **1947**, 2, 18.
(5) Ingold, C. K. Structure and Mechanism in Organic Chemistry, Cornell University Press, Ithaca, New York, 1953, p. 70.
(6) Burkert, U. and Allinger, N. L., Molecular Mechanics, American Chem. Soc., Washington, D.C., 1982.
(7) The original version of this program (MM2(77)) is available from the Quantum Chemistry Program Exchange, Department of Chemistry, Bloomington, Indiana 47405. Program #395. A later version (MM2(82)) is available from Molecular Design, Ltd., 2132 Farallon Drive, San Leandro, CA 94577. The refinements discussed herein are included in MM2(85), which will be distributed in due course. The details of MM2 were described in an

earlier paper (Allinger, N. L. J. Am. Chem. Soc., **1977**, <u>99</u>, 8127). Applications to functionalized molecules are described in later papers, and these are summarized in ref. 6.

(8) Allinger, N. L., Schäfer, L., Siam, K., Klimkowski, V. J. and Van Alsenoy, C. J. Comput. Chem., **1985**, <u>6</u>, in press.

(9) Frierson, III, M. R., dissertation presented to the University of Georgia in partial fulfillment of the Ph.D. requirements, 1984.

(10) Domenicano, A., Murray-Rust, P. and Vaciago, A. <u>Acta Cryst.</u>, **1983**, <u>B39</u>, 457, and other papers in this series.

(11) Nørskov-Lauritsen, L. and Allinger, N. L. <u>J. Comput. Chem.</u>, **1984**, <u>5</u>, 326.

(12) The bond lengths calculated by MM2 are r_g values. The electron diffraction values given in the literature for these compounds are sometimes r_g, but sometimes r_a. To facilitate comparison, r_a values have been converted[13] to approximate r_g values by: $r_g = r_a + .002$ Å.

(13) Molecular Structure by Diffraction Methods, Vol. 1, The Chemical Society London, 1973, p. 18.

(14) Beagley, B.; Jones, M. O.; Yavari, P. <u>J. Mol. Struct.</u> **1981**, <u>71</u>, 203.

(15) Friesen, D.; Hedberg, K. <u>J. Am. Chem. Soc.</u> **1980**, <u>102</u>, 3987.

(16) Beagley, B.; Jones, M. O.; Houldsworth, N. <u>J. Mol. Struct.</u> **1980**, <u>62</u>, 105.

(17) Beagley, B.; Jones, M. O.; Zanjanchi, M. A. <u>J. Mol. Struct.</u> **1979**, <u>56</u>, 215.

(18) Beagley, B.; Brown, D. E. <u>J. Mol. Struct.</u> **1979**, <u>54</u>, 175.

(19) Al-Ajdah, G. N. D.; Beagley, B.; Jones, M. O. <u>J. Mol. Struct.</u> **1980**, <u>65</u>, 271.

(20) Brown, D. E.; Beagley, B. <u>J. Mol. Struct.</u> **1977**, <u>38</u>, 167.

(21) Gallaher, K. L.; Yokozeki, A.; Bauer, S. H. <u>J. Phys. Chem.</u> **1974**, <u>78</u>, 2389.

(22) Kakubari, H.; Iijima, T.; Kimura, M. <u>Bull. Chem. Soc. Jpn.</u> **1975**, <u>48(7)</u>, 1984.

(23) Klaeboe, P.; Powell, D. L.; Stølevik, R.; Vorren, Ø. Acta Chem. Scand. **1982**, <u>A36</u>, 471.

(24) Haas, B.; Haase, J.; Zeil, W. <u>Z. Naturforsch.</u> **1967**, <u>22a</u>, 1646.

(25) Hirota, M.; Iijima, T.; Kimura, M. <u>Bull. Chem. Soc. Jpn.</u> **1978**, <u>51(6)</u>, 1594.

(26) Kveseth, K. <u>Acta Chem. Scand.</u> **1974**, <u>A28</u>, 482.

(27) Flygare, W. H. <u>J. Mol. Spectry.</u> **1964**, <u>14</u>, 145.

(28) Holm, R.; Mitzlaff, M.; Hartmann, H. <u>Z. Naturforsch.</u> **1968**, <u>23a</u>, 307.

(29) Huisman, P.; Mijlhoff, F. C. <u>J. Mol. Struct.</u> **1974**, <u>21</u>, 23.

(30) Almenningen, A.; Andersen, B.; Traetteberg, M. <u>Acta Chem. Scand.</u> **1964**, <u>18</u>, 603.

(31) Yamanouchi, K.; Sugue, M.; Takeo, H.; Matsumura, C.; Kuchitsu, K. <u>J. Phys. Chem.</u> **1984**, <u>88</u>, 2315.

(32) Iijima, T.; Seki, S. Kimura, M. Bull. Chem. Soc. Jpn. **1977**, 50(10), 2568.

(33) Rydland, T.; Seip, R.; Stølevik, R., Vorren, Ø. Acta. Chem. Scand. **1983**, A37, 41.

(34) Hirota, M.; Iijima, T.; Kimura, M. Bull. Chem. Soc. Jpn. **1978**, 51(6), 1589.

(35) Grindheim, S.; Stølevik, R. Acta Chem. Scand. **1976**, A30, 625.

(36) Fernholt, L.; Stølevik, R. Acta Chem. Scand. **1974**, A28, 963.

(37) Momany, F. A.; Bonham, R. A.; Druelinger, M. L. J. Am. Chem. Soc. **1963**, 85, 3075.

(38) Graner, G; Thomas, C. J. Chem. Phys. **1968**, 49, 4160.

(39) Mukhtarov, I. A.; Mukhtarov, E. I.; Akhundova, L. A. Zh. Strukt. Khim. **1966**, 7, 607; J. Struct. Chem. (USSR) (English Transl.) **1966**, 7, 565.

(40) Holm, R.; Mitzlaff, M.; Hartmann, H. Z. Naturforsch. **1968**, 23a, 1040.

(41) Flanagan, C.; Pierce, L. J. Chem. Phys. **1963**, 38, 2963.

(42) Fernholt, L.; Kveseth, K. Acta Chem. Scand. **1978**, A32, 63.

(43) Schwendeman, R. H.; Tobiason, F. L. J. Chem. Phys. **1965**, 43 201.

(44) Farup, P. E.; Stolevik, R. Acta Chem. Scand. **1974**, A28, 680.

(45) Momany, F. A.; Bonham, R. A.; McCoy, W. H. J. Am. Chem. Soc. **1963**, 85, 3077.

(46) Shen, Q. J. Mol. Struct. **1984**, 112, 169.

(47) Kimura, K.; Kubo, M. J. Chem. Phys. **1959**, 30, 151.

(48) Sasada, Y. J. Mol. Spectros. **1971**, 38, 33.

(49) Tamagawa, K.; Takemura, M.; Konaka, S.; Kimura, M. J. Mol. Struct. **1984**, 125, 131.

(50) Oyanagi, K.; Kuchitsu, K. Bull. Chem. Soc. Jpn. **1978**, 51(8), 2237.

(51) Hayashi, M.; Adachi, M. J. Mol. Struct. **1982**, 78, 53.

(52) Breed, H. E.; Gundersen, G.; Seip, R. Acta Chem. Scand. **1979**, A33, 225.

(53) Astrup, E. E. Acta Chem. Scand. **1973**, 27, 3271.

(54) Schultz, G.; Hargittai, I. Acta Chim. Acad. Sci. Hung. **1974**, 83, 331.

(55) Higginbotham, H. K.; Bartell, L. S. J. Chem. Phys. **1965**, 42, 1131.

(56) Beagley, B.; Hewitt, T. G. Trans. Faraday Soc. **1968**, 64, 2561.

(57) Yokozeki, A.; Kuchitsu, K. Bull. Chem. Soc. Jpn. **1971**, 44, 2926.

(58) Ter Brake, J. H. M.; Mom, V.; Mijlhoff, F. C. J. Mol. Struct. **1980**, 65, 303.

(59) Gundersen, G.; Rankin, D. W. H. Acta Chem. Scand. **1983**, A37, 865.

Chapter 4

ON GRAPH-THEORETICAL POLYNOMIALS IN CHEMISTRY

K. Balasubramanian*
Department of Chemistry, Arizona State University, Tempe, AZ 85287 USA

Recent developments in the area of applications of graph-
theoretical polynomials to several branches of Chemistry are
outlined. In particular, the use of characteristic
polynomials, matching polynomials, king and color polynomials
is considered. The developments of important computational
techniques such as Frame's method (LeVerrier-Faddeev method),
recursive Pascal programs for matching polynomials etc. are
reviewed. Applications to several areas of chemistry such as
statistical mechanics, quantum chemistry, random walks on
graphs and lattices, electronic structure of organic polymers
and periodic lattices, exact finite lattice statistics etc.
are considered.

INTRODUCTION

Graph theory and combinatorics have made significant impact on
several areas of chemistry such as quantum chemistry,
spectroscopy, stereochemistry, chemical kinetics, statistical
mechanics etc. The applications of graph theory and
combinatorics to spectroscopy and quantum chemistry were
recently reviewed by the present author [1]. Graphs are
useful as representation of molecules, chemical reactions,

---

* Alfred P. Sloan fellow.

isomerizations, quantum mechanical and statistical-mechanical interactions, NMR spin hamiltonians etc.

Graphs are useful in characterizing carcinogenic benzenoid hydrocarbons and identification of potentially carcinogenic bay regions [2,3].

A number of polynomials can be associated with graphs. A few such polynomials are characteristic polynomials, matching polynomials, king polynomials, color polynomials, sextet polynomials, chromatic polynomials, cyclic polynomials etc.

The characteristic polynomial of a graph is defined as the secular determinant of the adjacency matrix of a graph. The ijth matrix element of the adjacency matrix is 1 if the vertices i and j are connected; otherwise, it is zero.

In recent years characteristic polynomials and related polynomials of graphs and other applications of graph theory to chemistry have been the subjects of a large number of investigations [4-53]. Characteristic polynomials play an important role in several branches of chemistry. These polynomials are structural invariants and are thus useful in coding chemical structures. They are generating functions for dimer statistics on trees (such as Bethe lattices) and thus they play an important role in statistical mechanics [35,44,46]. Characteristic polynomials of graphs have applications in quantum chemistry, chemical kinetics, dynamics of oscillatory reactions etc. They are also useful in estimating the stability of conjugated systems.

The present author [47,48] showed the use of Frame's method for evaluating characteristic polynomials of graphs containing large numbers of vertices and further developed a computer program based on this method. Křivka, Jeričević and Trinajstić [49] have recently shown that the Frame's method outlined in the present author's paper is similar to Le Verrier-Faddeev's method. Other versions of the Frame's method could also be found in the literature [50].

Characteristic polynomials of organic polymers and periodic networks have been evaluated recently by extending the computer program developed by the author to complex hermetian matrices [51].

Matching polynomials of graphs generate the number of ways a given number of disjoint dimers can be placed on graphs and lattices. They are generators of Kekule structures and dimer coverings on Ising lattices. These polynomials are also useful in the calculation of the grand canonical partition function of a lattice gas. Ramaraj and present author [45] recently developed a computer program in Pascal to compute the

matching polynomials of a number of graphs and lattices. The
use of matching polynomials in the chemical literature can be
found in the papers of Hosoya [26,54], Mohar and Trinajstić
[55] and Gutman and Hosoya [56].

The king polynomial was first defined by Motoyama and Hosoya
[57] who also showed the potential applications of these
polynomials. These polynomials are useful in a number of
applications such as enumeration of Kekulé structures,
adsorption of molecules on surfaces, aromaticity, exact finite
lattices statistics etc. The present author and Ramaraj [44]
developed a computer program to generate the king polynomials
and color polynomial and demonstrated the usefulness of these
polynomials.

In the next section we review the construction and
applications of characteristic polynomials. In the third
section the uses of matching polynomials and king or color
polynomials are outlined.

CHARACTERISTIC POLYNOMIALS

Frame [41] developed a very elegant method discussed, for
example, in the book by Dwyer [44]. Křivka, Jeričević and
Trinajstić [49] have recently shown that this is the same as
Le Verrier-Faddeev's method. This method provides an
excellent algorithm for the computer generation of the
characteristic polynomials of graphs of interest in
chemistry. We outline here first, the essential steps of the
Frame method.

Let A be the adjacency matrix of a graph. Define the set of
matrices B_k's recursively by the following recipe.

$$C_1 = \text{Trace } A \tag{1}$$

$$B_1 = A(A - C_1 I)$$

$$C_2 = 1/2 \text{ Trace } B_1 \tag{2}$$

$$B_2 = A(B_1 - C_2 I)$$

$$C_3 = 1/3 \text{ Trace } B_2 \tag{3}$$

$$\cdot$$
$$\cdot$$
$$\cdot$$

$$B_{n-1} = A(B_{n-2} - C_{n-1} I)$$

$$C_n = 1/n \text{ Trace } B_{n-1} \tag{4}$$

The characteristic polynomial of the graph whose adjacency matrix is A, is given by

$$\lambda^n - C_1\lambda^{n-1} - C_2\lambda^{n-2} \ldots - C_{n-1}\lambda - C_n \qquad (5)$$

Thus the coefficients $C_1, C_2, \ldots$ are generated as traces of matrices obtained in the above recursive matrix product. Hence the Frame method provides a very efficient algorithm for the generation of coefficients C_1, C_2, etc., and consequently, the characteristic polynomial. The above algorithm can be coded into FORTRAN. The present author [47] developed such a program which is applicable to a number of graphs.

We now give a few examples. Table I lists a few graphs containing 10 vertices and their characteristic polynomials. These polynomials were generated using our computer program [47].

Table 1.

Graph	Characteristic polynomial
	$\lambda^{10} - 9\lambda^8 + 28\lambda^6 - 35\lambda^4 + 15\lambda^2 - 1$
	$\lambda^{10} - 9\lambda^8 + 27\lambda^6 - 30\lambda^4 + 9\lambda^2$
	$\lambda^{10} - 13\lambda^8 + 48\lambda^6 - 52\lambda^4 + 16\lambda^2$
	$\lambda^{10} - 11\lambda^8 + 41\lambda^6 - 65\lambda^4 + 43\lambda^2 - 9$
	$\lambda^{10} - 17\lambda^8 - 16\lambda^7 + 78\lambda^6 + 132\lambda^5 - 29\lambda^4 - 168\lambda^3 - 100\lambda^2 - 16\lambda$
	$\lambda^{10} - 14\lambda^8 - 4\lambda^7 + 59\lambda^6 + 18\lambda^5 - 91\lambda^4 - 22\lambda^3 + 37\lambda^2 + 10\lambda - 1$

We next consider two lattice graphs that are of interest in lattice statistics. In Figure 1 we show a square lattice graph and in Figure 2 we show a honeycomb lattice containing 54 vertices. The characteristic polynomials of both these graphs were obtained using our program. The characteristic polynomial of the square lattice graph in Figure 1 is given below.

$$\lambda^{16} - 24\lambda^{14} + 206\lambda^{12} - 804\lambda^{10} + 1481\lambda^8 - 1260\lambda^6 + 400\lambda^4$$
$$(6)$$

The characteristic polynomial of the hexagonal lattice graph in Fig. 2 is given below.

$\lambda^{54} - 72\lambda^{52} + 2430\lambda^{50} - 51,152\lambda^{48} + 753,867\lambda^{46}$

$- 8,227,552\lambda^{44} + 70,356,380\lambda^{42} - 474,823,692\lambda^{40}$

$+ 2,589,615,333\lambda^{38} - 11,556,300,564\lambda^{36}$

$+ 42,569,538,372\lambda^{34} - 130,222,965,528\lambda^{32}$

$+ 332,069,146,453\lambda^{30} - 707,192,500,956\lambda^{28}$

$+ 1,257,989,920,284\lambda^{26} - 1,866,287,443,412\lambda^{24}$

$+ 2,301,545,596,335\lambda^{22} - 2,347,222,219,224\lambda^{20}$

$+ 1,965,105,361,102\lambda^{18} - 1,337,106,330,756\lambda^{16}$

$+ 729,597,602,706\lambda^{14} - 313,604,239,964\lambda^{12}$

$+ 103,654,073,940\lambda^{10} - 25,479,629,340\lambda^{8}$

$+ 4,438,832,481\lambda^{6} - 508,728,588\lambda^{4}$

$+ 33,696,516\lambda^{2} - 960,400.$ (7)

The present author [51] recently extended the computer program for characteristic polynomials to complex hermetian matrices. As a result of this extension it has been possible to evaluate the characteristic polynomials of several organic polymers and periodic networks. Table 2 shows the characteristic polynomial of a one-dimensional square lattice containing 61 unit cells as a function of the crystal momentum n.

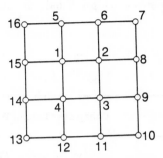

Figure 1. A two tier square lattice graph containing 16 vertices. For the characteristic polynomial of this graph, see expression (6).

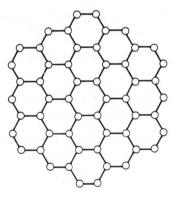

Figure 2. A honeycomb lattice graph containing 54 vertices. For the characteristic polynomial of this graph, see expression (7).

Table 2.
Characteristic polynomial of a one-dimensional square lattice containing 61 unit cells; n is the crystal momentum. The polynomials for only $n \geq 0$ are shown since the polynomials for -k and k are the same.

61

n	Characteristic Polynomial
0	$\lambda^4 - 5\lambda^2 - 4\lambda$
1	$\lambda^4 - 5\lambda^2 - 3.9788\lambda$
2	$\lambda^4 - 5\lambda^2 - 3.9154\lambda$
3	$\lambda^4 - 5\lambda^2 - 3.8105\lambda$
4	$\lambda^4 - 5\lambda^2 - 3.6653\lambda$
5	$\lambda^4 - 5\lambda^2 - 3.4811\lambda$
6	$\lambda^4 - 5\lambda^2 - 3.2601\lambda$
7	$\lambda^4 - 5\lambda^2 - 3.0045\lambda$
8	$\lambda^4 - 5\lambda^2 - 2.7171\lambda$
9	$\lambda^4 - 5\lambda^2 - 2.4009\lambda$
10	$\lambda^4 - 5\lambda^2 - 2.0592\lambda$
11	$\lambda^4 - 5\lambda^2 - 1.6957\lambda$
12	$\lambda^4 - 5\lambda^2 - 1.3142\lambda$

13          $\lambda^4 - 5\lambda^2 - 0.9188\lambda$

14          $\lambda^4 - 5\lambda^2 - 0.5136\lambda$

15          $\lambda^4 - 5\lambda^2 - 0.1030\lambda$

16          $\lambda^4 - 5\lambda^2 + 0.3087\lambda$

17          $\lambda^4 - 5\lambda^2 + 0.7171\lambda$

18          $\lambda^4 - 5\lambda^2 + 1.1179\lambda$

19          $\lambda^4 - 5\lambda^2 + 1.5069\lambda$

20          $\lambda^4 - 5\lambda^2 + 1.8799\lambda$

21          $\lambda^4 - 5\lambda^2 + 2.2330\lambda$

22          $\lambda^4 - 5\lambda^2 + 2.5624\lambda$

23          $\lambda^4 - 5\lambda^2 + 2.8646\lambda$

24          $\lambda^4 - 5\lambda^2 + 3.1365\lambda$

25          $\lambda^4 - 5\lambda^2 + 3.3751\lambda$

26          $\lambda^4 - 5\lambda^2 + 3.5779\lambda$

27          $\lambda^4 - 5\lambda^2 + 3.7429\lambda$

28          $\lambda^4 - 5\lambda^2 + 3.8681\lambda$

29          $\lambda^4 - 5\lambda^2 + 3.9524\lambda$

30          $\lambda^4 - 5\lambda^2 + 3.9947\lambda$

MATCHING POLYNOMIALS AND KING POLYNOMIALS

The matching polynomial of a graph is defined as

$$M_G(x) = \sum_{k=0}^{m} (-1)^k P(G,k) \, x^{N-2k}, \qquad (8)$$

where $P(G,k)$ is the number of ways of choosing k disjoint edges from G containing N vertices. The computation of matching polynomial for any graph is an extremely tedious problem for graphs containing large number of vertices. Ramaraj and Balasubramanian [45] have developed a Pascal program which recursively reduces a given graph into trees. This is based on the following recursive relation for the matching polynomial of a graph G.

$$M_G(x) = M_{G-e}(x) - M_{G\ominus e}(x), \qquad (9)$$

where G–e is the graph obtained by deleting an edge e from G and G$\ominus$e is the graph obtained by deleting e and the vertices of e together with the edges connected to the vertices of e. The characteristic polynomial of a tree is the same as the matching polynomial. The recursive relation (6) is used until the graph we start with reduces to trees. Then computer program for characteristic polynomials [47] is used to generate the characteristic polynomials (matching polynomials) of the trees generated by pruning the graph we start with. Using the relation (9) the polynomials are assembled back to generate the matching polynomial of the graph one starts with.

The computer program for matching polynomials was written in Pascal since this language is most suited for recursive programming. We now illustrate with an example. Consider the bathroom–tile lattice in Fig. 3. The matching polynomial of this lattice is given by expression (10)

$$x^{36} - 48x^{34} + 1044x^{32} - 13628x^{30}$$

$$+ 119223s^{28} - 739404x^{26} + 3354422x^{24}$$

$$- 11327084x^{22} + 28704111x^{20}$$

$$- 54656592x^{18} + 77829184x^{16}$$

$$- 81989532x^{14} + 62781122x^{12}$$

$$- 34032680x^{10} + 12564268x^{8} - 2980608x^{6}$$

$$+ 413985x^4 - 28408x^2 + 648 \qquad (10)$$

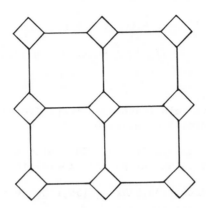

Figure 3. A bathroom-tile lattice of interest in the Ising problem.

A king pattern on a chessboard is simply a way of placing kings on the chessboard so that no two kings take each other. Suppose C_k is the number of ways of placing k non-taking kings on a chess board then the king polynomial K is defined as

$$K = 1 + C_1 x + C_2 x^2 + C_3 x^3 + \ldots + C_n x^n \qquad (11)$$

This polynomial was first defined by Motoyama and Hosoya [57]. The present author and Ramaraj [44] showed that king polynomial is the same as the color polynomial of the associated dual graph which is obtained by joining the centres of the cells. Suppose p_k is the number of ways of coloring k vertices of a graph with one type of color such that no two adjacent vertices carry the color. Then one can define the color polynomial as

$$C = 1 + p_1 x + p_2 x^2 + \ldots + p_n x^n \qquad (12)$$

The color polynomial defined above is different from chromatic polynomials. The recurrence relation for the color polynomial of a graph G is given by

$$C_G = C_{G-v} + C_{G \ominus v} \qquad (13)$$

where G–v is the graph obtained by removing a vertex v and G⊖v is the graph obtained by removing v and all the vertices which are adjacent to v. The given graph is reduced to smaller graphs using the relation (13) until paths of various lengths are obtained. For a path of length n, L_n, the coefficients in the color polynomial are given by a special type of Fibonnaci numbers. An analytical form for the color polynomial C_{ℓ_n} is shown below.

$$C_{\ell_n} = 1 + f(n,1)x + f(n,2)x^2 + \ldots$$
$$+ f(n,k)x^k + \ldots + f(n,m)x^m, \tag{14}$$

$$f(n,k) = \binom{n - k + 1}{k} \tag{15}$$

Using the above relationships a recursive pascal program was developed to generate the color (king) polynomial.

Consider the graph in Fig. 3. The color polynomial of this graph obtained using our pascal program is shown below.

$$1 + 36x + 582x^2 + 35,630x^3 + 159,132x^4 + 513,786x^5$$

$$+ 1,219,984x^6 + 2,148,728x^7 + 2,813,856x^8 + 2,732,684x^9$$

$$+ 1,953,584x^{10} + 1,014,796x^{11} + 375,212x^{12} + 95,634x^{13}$$

$$+ 15,968x^{14} + 1,604x^{15} + 84x^{16} + 2x^{17} \tag{16}$$

The color polynomials thus obtained are useful in exact lattice statistics [44].

Thus with the use of recursive programming techniques and other computational methods, several graph-theoretical polynomials which have several useful applications in chemistry can be obtained. With the advent of these computational techniques, it is expected that polynomials of graphs and lattices of interest in a variety of chemical problems can be obtained. It is then hoped that these polynomials will be exploited in a number of chemical applications.

ACKNOWLEDGEMENT

This research was supported in part by the NIH Biomedical Research Support Fund awarded to Arizona State University.

REFERENCES

[1]   K. Balasubramanian, Chemical Reviews (submitted).

[2]   K. Balasubramanian, J. J. Kaufman, W. S. Koski and A. T. Balaban, J. Comput. Chem. 1, 149 (1980).

[3]   J. J. Kaufman, P. C. Hariharan, W. S. Koski and K. Balasubramanian, in "Molecular basis of cancer, Part A" Allan R. Lyss Press, 1985.

[4]   M. Randić, J. Comput. Chem. 1, 386 (1980).

[5]   M. Randić, SIAM J. Algebraic and Discrete Methods 8, 145 (1985).

[6]   M. Randić, Theor. Chim. Acta 62, 485 (1983).

[7]   M. Randić and H. Hosoya, Theor. Chim. Acta 63, 473 (1983).

[8]   I. Gutman, Theor. Chim. Acta 45, 79 (1977); 50, 287 (1979).

[9]   I. Gutman and N. Trinajstić, Croat. Chem. Acta 47, 507 (1975).

[10] S. S. D'Amato, B. M. Gimarc, and N. Trinajstić, Croat. Chim. Acta, 54, 1 (1981).

[11] A. T. Balaban and F. Harary, J. Chem. Doc. 11, 258 (1971).

[12] J.-M. Yan, Adv. Quantum. Chem. 13, 211 (1981).

[13] H. Hosoya and T. Yamaguchi, Tetrahedron Lett., 4659 (1975).

[14] H. Hosoya and K. Hosoi, J. Chem. Phys., 64, 1065 (1970).

[15] H. Hosoya and N. Ohkami, J. Comput. Chem. 4, 585 (1983).

[16] W. C. Herndon and M. L. Ellzey, Jr., J. Am. Chem. Soc., 96, 6631 (1974).

[17] R. B. King, Theor. Chim. Acta. 44, 223 (1977).

[18] R. B. King, Theor. Chim. Acta. 56, 296 (1980).

[19] S. S. D'Amato, Theor. Chim. Acta, 53, 269 (1980).

[20] R. A. Davidson, Theor. Chim. Acta, 58, 193 (1981).

[21] T. Au-chin and K. Yuan-sun, Sci. Sinica, 1, 49 (1976); 3, 218 (1977).

[22] R. B. Mallion and D. H. Rouvray, Mol. Phys. 36, 125 (1978).

[23] D. H. Rouvray, in Chemical Applications of Graph Theory, A. T. Balaban, Ed., Academic, New York, 1976.

[24] I. Gutman, J. Chem. Soc. Fataday Trans. 2, 76, 1161 (1980).

[25] I. Gutman, Z. Naturforsch., 35a, 453 (1980).

[26] H. Hosoya, Theoret. Chim. Acta. 25, 215 (1972).

[27] K. Balasubramanian, International J. Quantum. Chem. 22, 581 (1982).

[28] K. Balasubramanian and M. Randić, Theor. Chim. Acta 61, 307 (1982).

[29] K. Balasubramanian and M. Randić, International J. Quantum Chem. (in press).

[30] K. Balasubramanian, Studies in Physical and Theoretical Chemistry, Elsevier, New York, 23, 149 (1983).

[31] I. Gutman and N. Trinajstić, Topics Curr. Chem. 42, 49 (1973).

[32] I. Gutman, N. Trinajstić, and T. Živković, Chem. Phys. Lett., 14, 342 (1972).

[33] M. Randić, G. M. Brissey, R. B. Spencer, and C. L. Wilkins, Comput. Chem., 3, 5 (1979).

[34] M. Randić, Match, 5, 135 (1979).

[35] K. Balasubramanian, Studies in Phys. and Theor. Chem., Elsevier, New York, 28, 243 (1983).

[36] K. Balasubramanian, J. Chem. Phys. 73, 3321 (1980).

[37] K. Balasubramanian, Theor. Chim. Acta, 51, 37 (1979).

[38] L. Glass, J. Chem. Phys., 63, 1325 (1975).

[39] R. Amit, C. A. Hall, and T. A. Porsching, J. Comput. Phys., 40, 183 (1981).

[40] I. Gutman and O. E. Polansky, Theor. Chim. Acta, 60, 203
      (1981).

[41] O. E. Polansky and M. Zander, J. Mol. Struct., 84, 361
      (1982).

[42] J. S. Frame, "A simple recursion formula for inverting a
      matrix," presented to American Mathematical Society at
      Boulder, Colorado, September 1, 1949 (as referred in
      reference 41).

[43] P. S. Dwyer, Linear Computations, Wiley, New York, 1951,
      pp. 225-235.

[44] K. Balasubramanian and R. Ramaraj, J. Comput. Chem (in
      press).

[45] R. Ramaraj and K. Balasubramanian, J. Comput. Chem. 6,
      122 (1985).

[46] K. Balasubramanian, Computers and Chem. (in press).

[47] K. Balasubramanian, J. Comput. Chem. 5, 387 (1984).

[48] K. Balasubramanian, Theor. Chim. Acta 65, 49 (1984).

[49] P. Křivka, Ž. Jeričević and N. Trinajstić, Int. J.
      Quantum Chem.: Quantum Chem. Symp., (in press).

[50] I. M. Mladenov, M. D. Kotarov and J. G. Vassileva-
      Popova, International J. Quant. Chem. 18, 339 (1980).

[51] K. Balasubramanian, J. Computational Chem. (submitted).

[52] N. Trinajstić, "Chemical Graph Theory" CRC Press, Boca
      Ratan, Florida, 1983, Vol I and II.

[53] A. T. Balaban (Editor) "Chemical Applications of Graph
      Theory" Academic, N.Y., 1976.

[54] H. Hosoya, Bull. Chem. Soc. Japan 44, 2332 (1971).

[55] B. Mohar and N. Trinajstić, J. Comput. Chem. 3, 28
      (1982).

[56] I. Gutman and H. Hosoya, Theor. Chim. Acta. 48, 279
      (1978).

[57] A. Motoyama and H. Hosoya, J. Math. Phys. 18, 1485
      (1977).

Chapter 5

ALGORITHMS FOR CODING CHEMICAL COMPOUNDS

D. Bonchev[a], O. Mekenyan[a], and A.T. Balaban[b]
[a]Higher School of Chemical Technology, Burgas 8010, Bulgaria
[b]The Polytechnic, Bucharest, Roumania

ABSTRACT

A classification of the coding algorithms is proposed. Two effective coding systems, designed by the anthors are outlined. The HOC system is based on three principles: a hierarchical ordering of vertex-extended connectivities, a unique topological representation of the molecule, and treatment of molecules at three levels of complexity. The algorithm DISTANCE proceeds from three centric criteria and from an iterativecombined specification of the centrik ordering of graph vertices and edges.

INTRODUCTION

The fast computerization of chemistry generated the creation of a multitude of chemical information centres and data banks. The efficient processing of chemical information and, particularly, the information on chemical structures is impossible without an effective coding system. The requirements as to the coding algorithms have recently been outlined (Read, 1978) referring mainly to the code uniqueness and compactness, the fast coding and retrieval, etc. The number of the coding systems developed is rapidly growing raising thus the question of their classification and comparison. An attempt for such a classification is presented here stressing on the classification of the algorithms that are topological in nature. Two original algorithms are also presented and their effectiveness demonstrated.

CLASSIFICATION OF MOLECULAR CODES

The classification of algorithms for coding chemical compounds is a problem of debate. The chemical nomenclature is usually not regarded as a coding system since it does not meet some of the requirements as to coding algorithms, and, first of all, it is not a sufficiently computer-oriented system. We classify molecular codes into three groups: fragment, mixed, and topological codes, two of which coincide with those from the recent classification of Moreau (1980).

Fragment Codes (e. g. the Ring Analysis Index, 1979) contain a list of molecular fragments which are included in the thesaurus of chemical compounds. The connectedness of the fragments is, however, not given.

Mixed Codes list fragments together with some structural connections between them. The Dyson (1949) and Wiswesser (1954) codes are the most widely known representatives of this group.

Topological Codes comprise the whole structural information on the chemical compound, i. e. atoms and their mutual connections are described in the language of graph theory. Codes of this group are considered below in more detail and subjected to an additional classification. Only universal codes applicable to all chemical compounds will be considered.

(i) Algorithms based on topological indices of molecular graphs (Bonchev et al., 1981; Randić, 1984) These algorithms are useful in a preliminary screening of chemical compounds but do not provide unique codes.

(ii) Algorithms based on the connectivity (i. e. the adjacency matrix) of molecular graphs. The FEVA algorithm of Randić (1975) is based on the FirstValue coefficients from the adjacency matrix. Another algorithm proposed by Randić (1974) generates the so-called Smallest Binary Code (SBL) after permutations of rows and columns in the adjacency matrix. A faster version of Largest Binary Code (LBC) has been recently presented by Hendrickson and Toczko (1983).

A large group of algorithms is related to the Morgan's Extended Connectivity Algorithm (1965) which will be discussed later. The improvements or/and modifications of ECA made by Wipke and Dyott (1974), Corneil and Gotlieb (1970), Frieland et al. (1975), Shelly and Munk (1977), Schubert and Ugi (1978), Moreau (1980), Zu and Zang (1982), Herndon (1983), etc. should be mentioned here. The HOC algorithms developed by the authors also belong to this group.

(iii) Algorithms based on the centric properties of molecular graphs. This is another large group of coding algorithms. All of them proceed from the ordering of graph vertices around a central vertex using also connectivity and chromatic properties of graphs. The DENDRAL system (Lederberg, 1966) utilizes the mass centre of the graph while the classical graph centre is used by Jochum and Gasteiger (1977). The central vertex in the DARC-ELCO system of Dubois (1973) is not always chosen on a topological basis, it is rather a characteristic functional focus of the molecule. An extended graph centre definition (Bonchev et al., 1980, 1981a) is applied for obtaining the topological code (Bonchev et al., 1981b, 1983) which forms the major part of the compound's name within the newly proposed universal nomenclature of chemical compounds. The same type of centre is used in our algorithm DISTANCE which will be discussed later. The AVTOGRAF program of Trach and Zefirov (1980) makes use of the so-called binary equidistancy matrix, i. e. the metric properties of the graph are used rather than the centric ones.

(iv) Algorithms based on the clusterization of molecule. Differing from all other topological algorithms these algorithms do not deal with individual atoms only but rather with sóme groups of atoms, called clusters, uniquely selected on a structural basis. Such algorithm have so far been proposed by Read (1978, 1980), as well as by Lozac'h, Goodson and Powell (1979).

(v) Other algorithms. The original work of Golender et al. (1981) is based on the so-called vertex potentials (first, second, etc.) introduced by analogy between graphs and electrical networks.

THE HOC ALGORITHMS BASED ON THE HIERARCHICALLY ORDERED EXTENDED CONNECTIVITIES OF GRAPHS

As mentioned in the previous section, a large number of coding algorithms exploit the idea of extended connectivity, i. e. they account not only for the atoms directly connected with the atom under consideration but also for its second, third, etc. neighbouring atoms. The different approaches vary in the specific realization of this idea. The original algorithm of Morgan (1965) counts first the connectivity of each atom as the total number of the neighbouring nonhydrogen atoms. Then, the sum of the connectivities of the first neighbours, called extended connectivity, EC, is calculated for each atom. The procedure continues in successive stages until the number of different EC values is the same in two iterative steps (Fig. 1).

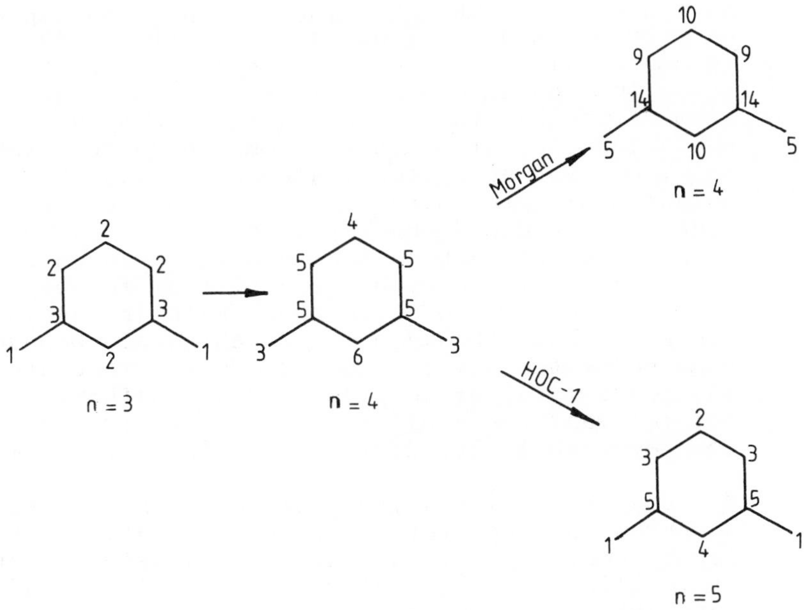

Figure 1. An example illustrating the difference between the Morgan and HOC-1 algorithms

The pitfalls of the Morgan algorithm are mainly in the EC oscillation during the iterations (e. g. the pair of vertices having EC = 10 in the last iteration has different values both in the preceding step (4,6) and in the following step (18,28). Topologically non-equivalent atoms also could acquire the same EC values. These disadvantages have been overcome in some algorithms proposed later. Yet, the

search for a better coding algorithm is one of the
most attractive trends in contemporary chemical graph
theory.

We present here the outline of a coding system based
on two ideas:

(i) treatment of simple compounds by means of a simple algorithm while more complex compounds are treated by
means of a more sophisticated algorithm

(ii) use of *hierarchically ordered* extended connectivities

(iii) use of a Unigue Topological Representation of
molecule (UTR). The first idea is an extension of some previous studies (Corneil and Gotlieb, 1979; Golender et al., 1981, etc.) in which a more powerful
(and time-consuming) algorithm is used *after* the
simpler one has failed in finding the topological
equivalence of atoms in the compound unter examination. Our coding system (Balaban et al., 1985) makes
an automatic *preliminary* selection of out of three
algorithms - HOC 1, HOC - 2, and HOC - 2 A, proceeding from the mathematical proff (Mekenyan et al.,
1985) for sufficiency of each of them for rigorously
specified classes of structures with increasing complexity. Thus, the majority of chemical compounds is
handled by the simple and fast HOC-1 procedure while
the third level of complexity deals with some cases
not relevant to chemistry but frequently used as
counter-examples to different coding algorithms.

A hierarchical iterative procedure can be used as
a convenient tool for avoiding ascillations of extended connectivity, as shown previously by Corneil
and Gorlieb (1970), Shelly and Munc (1977) and Hern-
dan (1983). We present here a different, simple hierarchical approach, called HOC-1. In each iteration
this procedure orders the non-hydrogen atoms into
classes of topological equivalence which receive the
ranks 1, 2, ..., k. The HOC - ranks of the atoms neighbouring a certain atom are ordered in the next iteration in an increasing sequence and then summed up.
The EC value thus obtained and its summands are used
as discrimination criteria which may divide further
the atoms from a certain equivalence class into two
of more such classes. In a few iterations the procedure terminates by finding the orbits of the automor-
phismgroup of the molecular graph, as well as by
ordering them according to their HOC - ranks. As seen
from Fig. 1, in contrast to the Morgan algorithm,

our HOC-1 algorithm detects the topological equivalent atoms and makes it sufficiently fast.

For asymetrical molecules HOC-1 terminates with the
canonical nlmbering of all atoms (Fig.2). In case
of symmetrical molecules this is a job of a special
algorithm HOC-3.

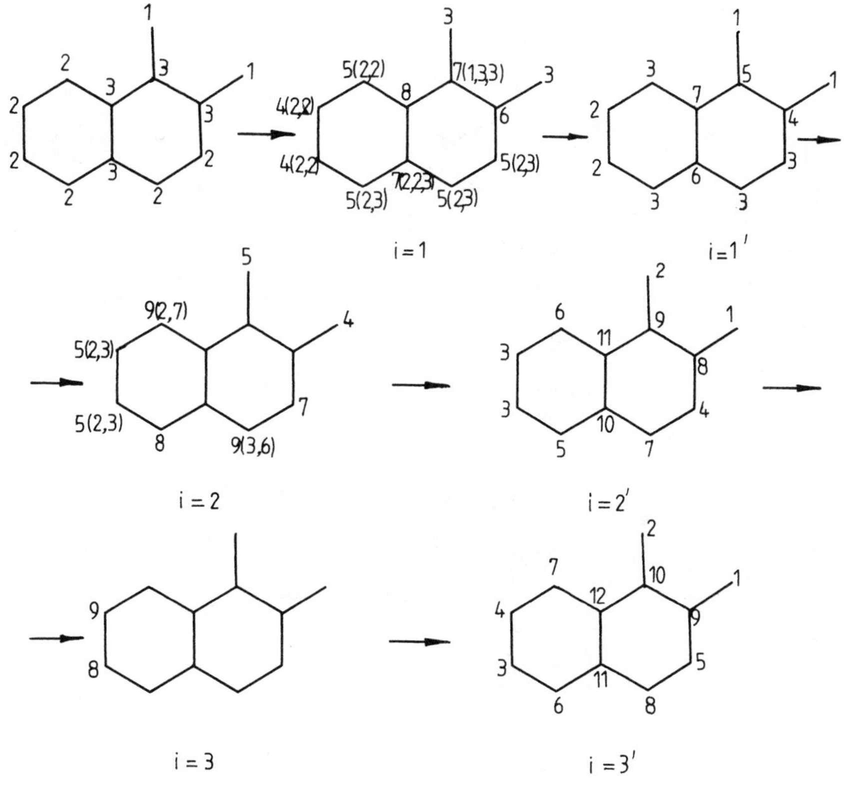

Figure 2. An example illustrating the HOC-1 procedure

The simplest structure which requires the more sophisticated *HOC-2 algorithm* for finding the topological equivalence of atoms is shown in Fig. 3. HOC-1
fails in solving this job puttimg all four vertices
of degree two in the same equivalence class. HOC-2
makes use of supplementary input information on the
cycles to which all graph vertices belong. This information is provided by the program "RING" which

will be described elsewhere (Karabunarliev et al.). This information is introduced in the initial itera- tion as an additional discrimination criterion for the graph vertices with the same connectivity. Then the basic algorithm HOC-1 suffices to specify com- pletely the topological equivalence of atoms in the molecule under examination. For the example conside- red in Fig. 3 one thus arrives at two orbits of two vertices each, instead of the initial clas of four vertices.

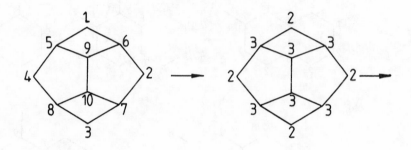

Initial numbering

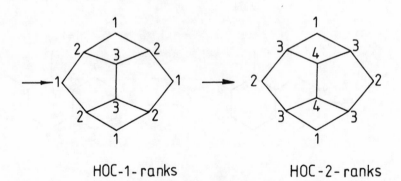

HOC-1-ranks   HOC-2-ranks

$$1,3 \in C_4; \quad 2\ 4 \in C_5; \quad 5,6,7,8 \in C_4,C_5 \quad \text{and} \quad 9,10 \in C_4\,C_5\,C_5$$

Figure 3. An example illustrating the HOC-2 algorithm

In the very rare cases of third-level complexity graphs of the HOC-2A algorithm makes use of some supplementary input data. These are the ordered sets

of the degrees of all vertices that belong to a certain cycle, which are also provided by the program
"RING".

A special subroutine called *"Graph classifier"* is
introduced in the computer's options. Thus, the
HOC-1 algorithm is applied to molecules whose graphs
have a cyclomatic number of lest than four (It was
found that structures of level B of complexity have
at least four peri-condensed cycles of different
sizes). Graphs that are composed by weakly connected

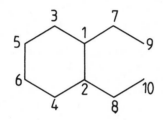

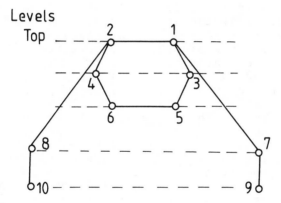

$$TC \quad 8,7;2,1;4.5,3;2,1;1,$$
$$8,7,2,1,4.5,3,2,1,1$$
$$CTC \quad 87214.53211$$

Figure 4. An example illustrating the Unique Topological Representation (UTR)
of a molecule, as well as its Topological Code (TC) and Compressed Toplogical
Code (CTC)

components, each one having less than four cycles are
also classified into the lowest of structural complexity. Many more chemical compounds belong to this

level but their automatic recognition needs time comparable to that for the direct handling of the structures of B complexity level by the HOC-2 algorithm. Almost all regular graphs also belong to this level.

The third basic idea of our approach is the concept of a Unique Topological Representation (UTR) of a molecule. This is a pictorial representation of the molecule in which atoms are located at the levels of topological equivalence (the orbits of the graph's automorphism group)arranged according to their increasing HOC-ranks. At each level, topologically equivalent vertices are arranged from right to left according to their increasing numbering, as specified by the HOC-3 procesure (Fig. 4).

Further to discerning symmetry (Bonchev et al., 1985) the UTR of molecules may be regarded as a very convient basis of molecular codes (Balaban et al., 1985a). As seen from the example in Fig. 4 the *topological code (TC)*, describes in increasing numerical order the neighbours of the vertices from each orbit (topological equivalence level) starting from the left- most vertex at the lowest level. The neighbours to vertices of different orbits are separated in the code by semicolumns while those to vertices of the same orbit are separated by commas. When the adjacent vertex is located on the same orbit this type of connection is denoted by a full stop. Connections to neighbouring vertices at higher levels or at the same level are used only. The advantages of the topological code proposed here are obvious. Its total number of symbols (including the punctiation marks) is close to that of the Morgan's From and Ring Closure list (Morgan, 1965) but the topological code is more informative containing the complete information on the orbits of the graph's automorphism group. If this supplementary information is not given, as all other molecular topological codes do, then the *compressed topological code (CTC)* can be obtained by omitting all puctuation marks except that for separating the vertices adjacent to the same vertex. CTC seems to be one of the shortest molecular codes.

The proposed topological code can be used as a basis for the complete coding of chemical compounds by supplementing information on the kind of atoms and bonds, as well as on stereochemistry, isotopic composition, etc. (Balaban et al., 1985a).

The HOC-system of algorithms incorparates also the HOCCANON algorithm oriented towards fast canonical

numbering of atoms in molecules HOCCANON discrimina-
tes the atoms and bonds still in the initial stage
(and before introducing extended connectivities) by
means of the chemical and isotopic nature of atoms,
their charges, the bond multiplicity, etc., as well
as by means of the size of the rings to which the
vertex belongs (for B complexity level only) and the
vertex degree sequences of the rings (for C complexity level only). Having thus the maximal preliminary
distinction of atoms HOCCANON continues with the basic HOC-1 treatment arriving very fast to classes of
topochemical equivalence of atoms. This equivalence
may be regarded as a result of the combination of the
point group symmetry of the molecule with the local
symmetries of molecular fragments (mainly terminal
groups).

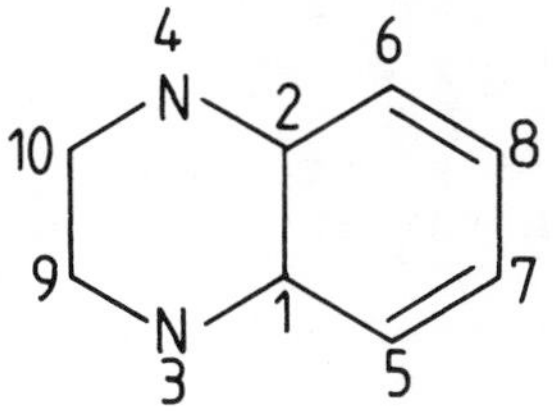

$$\text{TCC} \quad 4/N.9,3/N;6^27,5^2;2,1;2,1;1,$$

Figure 5. An example illustrating the HOCCANON vertex numbering and the
Topo-Chemical Code (TCC)

An example of the HOCCANON canonical numbering of
atoms is presented in Fig. 5, together with the respective *topo-chemical code (TCC)*. The latter resul-
ted from the Unigue Topo-Chemical Representation
(UTCR) of the molecule which is constructed similarly to UTR. The chemical symbols of heteroatoms are
presented there after a slash while the multiplicity
of bonds (higher than one) is specified as exponent
after the vertex number denoting one endpoint of the
bond.

THE ALGORITHM DISTANCE BASED ON THE CENTRIC PROPERTIES OF GRAPH VERTICES AND EDGES

This algorithm based on the gtaph distances and
more spesifically it makes use of the cehtric properties of molecular graphs.

The procedure consists of two parts. The first one
orders centrically the graph vertices and edges into

equivalence classes on the basis of the recently developed (Bonchev et al.,1980, 1981a) generalized graph centre concept. Proceeding from the adjacency matrix of the graph the three centric characteristics are calculated for each vertex: its radius (the maximal distance), r, its ditance number, d_r, which is the sum of distances to the remaining graph vertices, and the distance code, d_c, which includes the frequency numbers of the different for the vertex frequency numbers of the different distances for the vertex tric criteria coincides with the first discriminating criterion, the number of the occupied neighbouring spheres (NOON), in the Jochum and Gasteiger algorirtm (1978). All graph vertices are thus divided into equivalence classes which are ranked according to their centric properties starting with rank 1 for the central vertices. The same procedure is applied then to all graph edges.

An iterative two-step procedure of centric re-ordering of the graph vertices and edges followes.(1). The ranks of the incident edges are taken into account by summing and forming an increasing sequence for each vertex. Some of the vertices classified as equivalent in the first part of the procedure may thus appear as nonequivalent. This increases the number of equivalence classes which receive new ranks. (2). The new ranks of the incident vertices are analogously taken into account for each graph edge. The equivalence classes of graph edges thus may on their turn be partitioned into smaller classes and be re-ranked. Steps (1) and (2) are repeated until two successive iterations result in the same centric ordering of the graph vertices and edges.

The procedure is illustrated in Fig. 6 where the three centric criteria are presented for all graph vertices and edges. It is shown that two iterations suffice for finding the ultimate vertex and edge centric ordering into equivalence classes which coincide with the orbits of the graph's automorphism group. The last assertion is not proven in the general case but it is supported by the lack of a counter-example among the inspected several thousand graphs, including regular ones, as well as all the counter-examples to the known algorithmes (Bonchev et al., in preparation).

Having obtained the levels of topological equivalence of atoms (UTR) one can proceed finding the topological code of the compound which is analogous to (but not identical with) the topological code described in

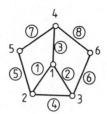

Figure 6. An example illustrating the iterative centric re-ordering of the graph vertices and edges according to the DISTANCE algorithm

the previous section.

CONCLUDING REMARKS

The HOC and DISTANCE algorithms outlined in this paper have a number of advantages. They are free of nontopological conventions   and they can be easily implemented both manually and by computer program (Ralev et al., 1985). Being considerably faster than many of the existing algorithms also provide concise molecular codes. Perhaps the most important feature of the HOC and DISTANCE codes is in preserving the symmetry of molecular graphs, as well as in preserving the ordering of vertices according to the connectivity or centric properties of graphs, respectively. This ordeting reproduces surprisingly well some experimental data such as ^{1}H-NMR chemical shifts in polycyclic benzenoid aromatic hydrocarbons (Bonchev et al., 1981a; Mekenyan et al., 1983, 1985a). Thus, HOC and DISTANCE appear so far as the single procedures, out of the known unique and universal coding algorithms, that have some physico-chemical significance. In addition, the conservation of the topological similarity of molecules of a different chemical composition could help in the search for structure-activity correlations (Mekenyan et al., 1985b), in classification and nomenclature problems, etc.

REFERENCES

Balaban, A. T., Mekenyan O., and Bonchev, D., 1985, 1985a, J. Comput. Chem., in press.

Bonchev, D., Balaban, A. T., and Mekenyan, O., 1980. J. Chem. Inf. Comput. Sci. 20, 106.

Bonchev, D., Mekenyan, O., and Trinajstić, N,, 1981. J. Comput. Chem. 2, 127.

Bonchev, D., Balaban, A. T., and Randić, M., 1981a. Int. J. Quantum Chem. 19, 61.

Bonchev, D. and Balaban, A. T., 1981b. J. Chem. Inf. Comput. Sci. 21, 223.

Bonchev, D., 1983. Pure Appl. Chem. 55, 221.

Bonchev, D., Mekenyan, O., and Balaban, A. T., 1985. Math. Chem. (MATCH), in press.

Bonchev, D., Karabunarliev, St., Mekenyan, O., in preparation.

Corneil, D. G. and Gotlieb, C. C., 1970. J. Assoc. Comp. Mach. 17, 51.

Dubois, J.-E., 1973. "Computer Representation and Manipulation of Chemical Information", Chapter 10, Wiley, New York.

Dyson, G. M., 1949. "A New Notation and Enumeration System for Organic Compounds", Longmans, London (2nd edn.).

Frieland, R. G., Funk, S. A., and O'Korn, L. J., 1975. J. Chem. Inf. Comput. Sci. 19, 2.

Golender, V. E., Drboglav, V. V., and Rosenblit A.B., 1981. J. Chem. Inf. Comput. Sci. 21, 196.

Hendrickson, J. B. and Toczko, A. G., 1983, J. Chem. Inf. Comput. Sci. 23, 171.

Herndon, W. C. and Leonard, J. E., 1983, Inorg. Chem. 22, 554.

Jochum. C. and Gasteiger, J., 1977. J. Chem. Inf. Comput. Sci. 1, 113.

Karabunarliev, S., in preparation.

Lederberg, J., 1966. Instrumentation Research Laboratory, Technical Report No 1140, Stanford University, Palo Alto California.

Lozac'h, N., Goodson, A. L., and Powell, 1979. Angew. Chem. Internat. Ed. *18*, 887.

Mekenyan, O., Bonchev, D., and Balaban, A. T., 1983. Chem. Phys. Lett. *109*, 85.

Mekenyan, O., Bonchev, D., and Balaban, A. T., 1985. J. Comput. Chem., in press.

Mekenyan, O., Balaban, A. T., and Bonchev, D., 1985a. J.Magn. Res. *63*, 1.

Mekenyan, O., Peitchev, D., Bonchev, D., Trinajstić, N., and Bangov, I., 1985b. Drug Research, in press.

Moreau, G., 1980. Nouv. J. Chim., *4*, 17.

Morgan, H. L., 1965. J. Chem. Doc., *5*, 107.

Ralev, N., Karabunarliev, S., Mekenyan, O., Bonchev, D., and Balaban, A. T., 1985. J. Comput. Chem., in press.

Randić, M., 1974. J. Chem. Phys. *60*, 3920.

Randić, M., 1975. J. Chem. Inf. Comput. Sci. *15*, 105.

Randić, M., 1984. J. Chem. Inf. Comput. Sci. *24*, 164.

Read, R. C. and Milner, R.S., 1978. Research Report CORR 78-42, University of Waterloo, Ontario, Canada.

Read, R. C., 1980. Research Report CORR 80-7, University, of Waterloo, Ontario, Canada.

Ring Analisis Index, 1979. Index of Parent Compounds I, Parent Compound Handbook, Chemical Abstract Service, Colombus, Ohio.

Shelley, C. A. and Munk, M. E., 1977. J. Chem. Inf. Comput. Sci. *17*, 110.

Schubert, E. and Ugi, I., 1978. J. Amer. Chem. Soc. *100*, 37.

Trach, S. S., and Zefirov, N. S., 1980. Fifth All-Union Conference on Computer Applications to Molecular Spectroscopy and Chemical Research, Novosibirsk, p. 5.

Wipke, W. T. and Dyott, T. M., 1974. J. Amer. Chem. Soc. *96*, 4825, 4834.

Wiswesser, W. J., 1954."A Line-Formula Chemical Notation", T. Y. Growell, New York.

Zhu S. Y., and Zhang, J. P., 1982. J. Chem. Inf. Comput. Sci. *22*, 34.

Chapter 6

APPLICATION OF THE MONTE CARLO METHOD FOR STUDYING THE HYDRATION OF MOLECULES: BASE STACKING

V.I. Danilov
Institute of Molecular Biology and Genetics,
the Ukrainian SSR Academy of Sciences, Kiev, USSR

ABSTRACT

A Monte Carlo hydration simulation of uracil, thymine and their associates has been performed. The results obtained enable one to elucidate the nature of nucleotide base stacking.

INTRODUCTION

The factors providing the stability of nucleic acid secondary structure in water are already displayed in the interactions between separate nucleotide bases in different associates. Therefore, the investigation of the nature of base associate stability in water is very important to understand the stability conditions of different conformational states and the mechanisms of intramolecular structural transitions in nucleic acids.

With the help of the thermodynamic and spectro-

scopic study of interactions of the bases, their
derivatives, nucleosides and nucleotides it is shown that stacked associates exclusively are formed
in aqueous solutions (see, for example, /1/). It is
impossible as yet to detect in-plane hydrogen-bon-
ded (H-bonded) base pairs in water.

At the same time from theoretical investigations /2,3/ one may conclude that H-bonded base pairs
in vacuum are energetically more preferable than
stacked associates for the majority of dimers of
the same composition. The analysis of the whole variety of the experimental and theoretical data available has allowed us to make as assumption /3,4/
about an important role of water in the formation
of stacked dimers. Sinanoğlu and Abdulnur arrived
at the similar conclusion /5,6,7/ on the basis of
the rough estimates of different solvent contributions to the stability of nucleotide base associates performed by means of solvophobic force theory.

Despite numerous experiments, the energetic advantage of the base association reaction in water,
in particular, the factors stabilizing base stacking,
and the role of water as the solvent for this have
been studied insufficiently. In addition, it remains not realized enough why stacked dimers are more
preferable than H-bonded base pairs; the role of
hydrophobic groups in base stacking being not made
clear enough either. Elucidation of the nature of
stacking interactions may promote considerably the
understanding of the nature of hydrophobic interactions playing so important role in the organization
of biological structures.

To elucidate the mechanism of nucleotide base
stack formation in water, it is necessary to study
water-water, water-base and base-base interactions

on molecular level /8/. Similar investigations on this problem have been already begun. It is even obvious now, however, that using quantum mechanical methods one is hardly able to investigate directly the systems containing more than 150-200 atoms. In such a situation to refuse the obtaining of inter-molecular interactions for real system in numerical form is most likely the only way out. The analyti-cal form of interaction potentials for the diffe-rent pairs of atoms should be then found from the results of quantum mechanical or semiempirical stu-dy of different configurations of the system simu-lating molecular complex.

The analytical potentials available allow us to use statistical thermodynamics, for which the Monte Carlo method is the most effective when averaging temperatures in real system. For recent years this method is widely used in chemistry and biology for calculating the average values of different proper-ties.

In this connection we have begun a Monte Carlo computer hydration simulation of nucleic acid bases and their dimer associates in the cluster of N=200 water molecules /9-12/. The enthropy has not been considered for simulation. This is justified for our aim, since it has been shown experimentally that the association of the bases and their deriva-tives in water is determined by enthalpy /1/.

In the process of the Monte Carlo computations one of the bases in each stacked dimer was moved randomly according to the Metropolis algorithm. As the starting configurations for stacked uracil (Ura) and thymine (Thy) dimers the most preferable for vacuum antiparallel configuration and the anti-parallel one with the further rotation of a base on

180° around the glycosidic bond were chosen. Let us call the stack obtained from the first configuration as A-stack and the one obtained from the second configuration as O-stack. A more detailed description of the computation method is given in /13/.

MONTE CARLO HYDRATION SIMULATION OF BASE ASSOCIATES

Some results of the study of association reaction obtained from the Monte Carlo hydration simulation data of Ura, Thy and their stacked dimers are presented in this work.

The calculation results for the changes of the average magnitudes of the water-water interaction energy ΔU_{ww}, the water-base interaction energy ΔU_{wb}, the base-base interaction energy ΔU_{bb} and the energy of the system ΔU are given in Table 1 (here and hereafter all the values are given in kcal/mol of dimer).

Table 1.
Energetic characteristics of the base stacking reaction in water

Transition	ΔU_{ww}	ΔU_{wb}	ΔU_{bb}	ΔU
Ura+Ura $\rightleftharpoons$ stacked dimer	-52	9.5	-4.2	-47
Thy+Thy $\rightleftharpoons$ stacked dimer	-22	22.3	-5.1	-5

Their analysis shows that the change in the water-water interaction associated with the structural rearrangement of water around monomers during their association is the main factor promoting stacked dimer stabilization. The stacked associate is considerably less stabilized by the base-base interaction. At the same time the stack is destabilized

substantially by the water-base interaction. This
confirms the assumption /3-7/ about the decisive
role of the water-water interaction for base stacking.

Unfortunately, the results of the experimental
study of stacking association enthalpy of Ura and
Thy in water are absent. The data available for the
different methylated derivatives of these bases
/14/ show that the methylation of Ura derivatives
in the C5 position leads to the decrease in the absolute magnitude of association enthalpy. The values ΔU we have computed for Ura and Thy agree
with these data.

Given in Table 2 are the changes in the poten-
tial energy $\Delta U^{s,p}$ and the water-water $\Delta U_{ww}^{s,p}$, water-base $\Delta U_{wb}^{s,p}$ and base-base $\Delta U_{bb}^{s,p}$ interaction
energies upon the transition from the H-bonded di-
mer to the stacked one calculated from data /9,10/.

Table 2.
Energetic characteristics of the transition from H-bonded base pair to
stacked dimer

Transition	N	$\Delta U^{s,p}$	$\Delta U_{ww}^{s,p}$	$\Delta U_{wb}^{s,p}$	$\Delta U_{bb}^{s,p}$
Base pair Ura·Ura →	200	−16	−14	−7.1	5.1
stacked dimer Ura/Ura	39	−21.7	−17.2	−9.6	5.1
	82	−18.7	−22.1	−1.7	5.1
Base pair Thy·Thy →	200	−25	−24	−5.4	4.3
stacked dimer Thy/Thy	39	−9.0	−5.9	−7.4	4.3
	82	−10.1	−9.0	−5.4	4.3

It follows from Table 2 that in water stacked
Ura and Thy associates are energetically more preferable than H-bonded dimers. This preference is
mainly due to value $\Delta U_{ww}^{s,p}$ and is caused by the ene-

rgetically more favourable structure of water aro-
und the stack. The water-base interaction also stabilizes the stack as compared to the base pair; the
stack being at the same time destabilized by the
base-base interaction.

A considerably greater energetic preference of
stacked dimers obtained by us in comparison with H-
bonded ones hampers extremely the detection of the
latter in water. Note that it is true irrespective
of whether the base pair formation from monomers is
favourable or not.

For a more detailed understanding of the nature
of the preference of stacked dimer as compared to
the H-bonded one, we have calculated the energetic
properties of the transition from the base pair to
the stack for the nearest 39 and 82 molecules of
water cluster (Table 2) using data /9,10/. The Table
shows that for the subsystem including 39 water molecules there is almost the same energetic advantage of the stacked Ura dimer when compared to the
corresponding base pair as for the whole water cluster. Term $\Delta U_{ww}^{s,p}$ makes the main contribution to
value $\Delta U^{s,p}$. The data for the subsystems of Ura di-
mer+82 water molecules leads to analogous conclusions. This testifies to the fact that the preference
in the formation of the stack of Ura molecules is
due to the nearest water molecules /9/.

It is appropriate to note that for the subsystems of Ura dimer+39 water molecules the water-base
interaction makes a tangible contribution to a greater stability of the stack as compared to the base
pair. The performed analysis of the radial distribution of the water-base interaction energy shows that
there is a layer of water molecules only 1.5 Å thick
(5.3-6.8 Å from the sphere centre) around the sta-

cked Ura dimer, whose interaction with the bases makes the contribution to U_{wb} that amounts to -46 kcal/mol and makes up 47%. The existence of this layer around the stack and its absence around the base pair is primarily due to the different character of the distribution of hydrophilic groups for stacked and H-bonded Ura dimers.

The analysis of the data for the analogous sub-systems of Thy dimer+water (Table 2) shows that a greater energetic preference of the stack as compared to the base pair is also displayed when considering the nearest 39 and 82 water molecules around dimers. However, for the subsystems considered magnitude $\Delta U^{s,p}$ makes but a small part of this value for the system including 200 water molecules. A greater energetic preference of the Thy stack as compared to the base pair is most likely due to the layers of water molecules more distant from the dimer /10/. Since the magnitudes $\Delta U_{ww}^{s,p}$ for subsystems are small in comparison with value $\Delta U_{ww}^{s,p}$ for the whole cluster (see Table 2), the dominant contribution of the latter to $\Delta U^{s,p}$ is also most likely determined by more distant layers of water molecules. This defines the difference between the hydration of Thy associates and those of Ura considered above.

The differences of the potential energy ΔU^{s}, the water-water ΔU_{ww}^{s}, water-base ΔU_{wb}^{s}, base-base ΔU_{bb}^{s} interaction energies and the differences in the number of water-water H-bonds Δn^{s} for O- and A-stacks in water are given in Table 3.

It is seen from this Table that the energetically most preferable systems among those of Thy dimer+water and Ura dimer+water include different types of stacks.

Table 3.
Energetic and structural characteristics of the hydration of uracil and thymine
stacked dimers

System	N	ΔU^s	ΔU_{ww}^s	ΔU_{wb}^s	ΔU_{bb}^s	Δn^s
Stacked Thy di-	200	−34	−38	5.4	−0.5	6
mer+water	39	−12.1	−21.1	9.7	−0.5	2
	82	−8.2	−15.6	7.6	−0.5	2
Stacked Ura di-	200	−34	−44	7.9	1.1	7
mer+water	39	−9.4	−21.1	10.5	1.1	3
	82	−14.8	−23.8	8.2	1.1	4

The energetic preference of a more stable stacked dimer when compared to a less stable one is almost completely determined by the water-water interaction. Water is more ordered around a more stable dimer. A greater number of water-water H-bonds testifies to this fact (see values Δn^s in Table 3). The base-base interaction does not almost make any contribution to the stability of one type of stacks as compared to another. A less preferable dimer is stabilized by the water-base interaction to a greater extent. Similar conclusions follow from the data of Table 3 for the subsystems including the nearest 39 and 82 molecules of water cluster.

The analysis of the data obtained for stacked Thy dimers in water enables one to make one more important observation. It is seen from Table 3 that the dimer in which the methyl groups of Thy molecules are adhered (O-stack) is more preferable than the one for which these groups are separated from each other (A-stack). This fact is analogous to the known phenomenon that in water two non-polar molecules or groups aim"to adhere".

The "adhering" observed theoretically is most
likely due to the effect of methyl groups and not
to the rotation of one of the base rings around the
glycosidic bond. This is shown by our data that in
water the uracil A-stack is more stable than the O-
stack.

The nature of the methyl group "adhering" that
leads to a more preferable type of the stack of Thy
may be understood qualitatively from the calculated
radial distribution functions of the water-water
interaction energy, a number of water-water H-bonds
and water molecules for the subsystems including 39
and 82 water molecules. The analysis of these functions performed by us for the A- and O-stacks of
Thy shows that due to a small number of water molecules, water structuring around the A-stack for the
layers positioned in a proximity to methyl groups
is low. The transition from the A- to O-stack leads
to the increase in the number of water molecules for
3.3-3.8, 4.8-5.3 and 6.3-6.8 Å layers. This allows
water molecules to form a greater number of water-
water H-bonds in these layers. The decrease in a
number of water-methyl group contacts observed during this transition leads to an additional formation of water-water H-bonds in the 6.3-6.8 Å layer
that raises its structuring. It is evident to lower
tangibly the system energy. Really, the energetic
contribution made by the 6.3-6.8 Å layer to the preference of the O-stack of Thy amounts to -22.5 kcal/
mol and is very close to value ΔU_{ww}^{s} for the system
including 39 water molecules (see Table 3). The additional lowering of the system energy is caused by
the cooperative effect of the adjacent methyl groups
in the O-stack of Thy on the structure of more distant layers of water molecules (Table 3).

The "adhering" of Thy molecule methyl groups
upon the formation of the O-stack detected by us is
a typical manifestation of the hydrophobic effect.
Unlike the classical case, however, the enthalpy
term calculated by us makes a considerable contribution to the stabilization of the O-stack of Thy.
It should be noted that the similar phenomenon is
observed for all the studied cases of nucleotide
base association in water /1,14/. The multiply al-
kylated bases with the volume substituents are an
exception. Their large hydrophobic surface formed
due to alkyl substituents screens greatly a polar
nucleus of the rings that leads to the classical
(enthropy) nature of the base association.

The hydration simulation of stacked dimers performed by us has shown that due to the possible
change of their geometry in water the configurations of stacks differ strongly from the most preferable configurations computated for vacuum (see /13/).
This leads to a tangible decrease of the absolute
value U_{bb} during the transition from vacuum to water. So, for the A-stack of Ura the most preferable
configuration of the bases in vacuum has the value
U_{bb} amounting to -6.31 kcal/mol. During the transition to water the magnitude U_{bb} becomes equal to
-4.15 kcal/mol. The same difference is also observed for the O-stack of Thy during the transition
from vacuum to water (-7.33 and -5.13 kcal/mol,
respectively).

The comparison of our data on the hydration of
the A-stack of Ura with the variable and fixed (vacuum) geometry of the dimer shows that values U_{ww}
and U_{wb} are changed, as well as value U_{bb}. Therefore, the conclusions of the recent paper /15/ devoted to the study of the base association reaction in

water and based on the fixed geometry of the stacks
should be considered with great caution. In addition, the conclusion of authors /15/ that there is no
enthalpy stabilization of stacked dimers by water
obtained from their data is erroneous.

ACKNOWLEDGEMENTS

The author thanks gratefully Dr. N. V. Zheltovsky
and I. S. Tolokh for useful discussions, and G. M.
Ostrovskaya for technical assistance.

REFRENCES

/1/ Ts'o, P. O. P. (1974) In: Ts'o P. O. P. (ed.)
    Basic principles in nucleic acid chemistry.
    London and New York, Academic Press, vol. 1,
    p. 453-584.

/2/ Pullman, A., and Pullman, B. (1968) Adv. Quan-
    tum Chem. 4, 267-325.

/3/ Danilov, V. I., Zheltovsky, N. V., and Kudrit-
    skaya, Z. G. (1974) Stud. Biophys. 43, No. 3,
    201-216.

/4/ Danilov, V. I. (1975) Mol. Biol. Repts 2, No. 3,
    263-266.

/5/ Sinanoğlu, O., and Abdulnur, S. (1964) Photo-
    chem. and Photobiol. 3, No. 4, 333-342.

/6/ Sinanoğlu, O., and Abdulnur, S. (1965) Fed.
    Proc. 24, No. 2, 12-23.

/7/ Sinanoğlu, O. (1968) In: Pullman, B. (ed.) Mole-
    cular associations in biology. New York, Acade-
    mic Press, p. 427-445.

/8/ Sapper, H., and Lohmann, W. (1978) Biophys.

Struct. and Mech. $\underline{4}$, 327-335.

/9/ Danilov, V. I., Tolokh, I. S., Poltev, V. I., and Malenkov, G. G. (1984) FEBS Lett. $\underline{167}$, No. 2, 245-248.

/10/ Danilov, V. I., Tolokh, I. S., and Poltev, V. I. (1984) FEBS Lett. $\underline{171}$, No. 2, 325-328.

/11/ Danilov, V. I. and Tolokh, I. S. (1984) FEBS Lett. $\underline{173}$, No. 2, 347-350.

/12/ Danilov, V. I. (1984) Dok. Akad. Nauk SSSR (Moskva) $\underline{278}$, No. 4, 994-996.

/13/ Danilov, V. I., and Tolokh, I. S. (1985) Biopolimery i Kletka (Kiev) $\underline{1}$, No. 2, 59-69.

/14/ Pleciewicz, E., Stepień, E., Bolewska, K., and Wierzchowski, K. L. (1976) Biophys. Chem. $\underline{4}$, 131-141.

/15/ Pohorille, A., Pratt, L. R. , Burt, S. K., and McElroy, R. D. (1984) J. Biomol. Struct. Dyn., $\underline{1}$, No. 5, 1257-1280.

Chapter 7

STERIC EFFECTS ON RATES AND EQUILIBRIA

DeLos F. DeTar
Department of Chemistry, Florida State University,
Tallahassee, Florida 32306 USA

ABSTRACT

New approaches to the theoretical prediction of relative rate constants based on estimations of relative enthalpies of formation of reactants and of models of transition states provide calculated rate constants that agree with experimental values, in some cases within a factor of two. The underlying principles are described. Formal steric enthalpy (FSE), a new formal definition of steric properties has proved useful for calculating the relative enthalpies and in investigating the origins of the steric effects. Evaluation of steric effects on rates (and on equilibria) has wide applicability. It is a powerful tool that may be expected to serve as a stimulus for new studies of steric effects while providing the means toward a better understanding and a more effective use of these effects.

INTRODUCTION

For reactions controlled primarily by steric effects computations using molecular mechanics can reproduce relative rate constants rather well. Representative references are Bingham and Schleyer 1971, DeTar and Tenpas 1976b, DeTar et al 1978, DeTar and Luthra 1980, DeTar 1981a, 1981b, DeTar and Delahunty

1983, Müller and Perlberger 1976, Perlberger and
Müller 1977, Müller et al 1982a, 1982b, Farcasiu
1978, Beckhaus et al 1978, 1980, Bernlöhr et al 1984,
Ruechardt and Weiner 1979, Ruechardt and Beckhaus
1980, Schneider and Thomas 1979, 1980, Schneider et
al 1983, Müller 1985.

In recent more detailed studies of esterification
of $R_1R_2CHCOOH$ and hydrolysis of the esters, we have
been able to reproduce relative rate constants to
within a factor of 1.7 over a range of 5 powers of
ten for some 85 data values, DeTar et al 1985d. R_i is
H, Me, Et, i-Pr, or t-Bu. These results together with
those published show that chemists now have available
a powerful tool for gaining new insights about steric
effects on rates and equilibria.

The purpose of this discussion is to present the
underlying principles and to indicate some techniques
for applying them to more general problems of evaluating steric effects. The illustrations will be
based primarily on esterification.

PRINCIPLES: RELATIVE RATE CONSTANTS AND TRANSITION STATE THEORY

The estimates of relative rate constants are based
on the thermodynamic state approach of transition
state theory, and they require calculation of relative free energies of formation of reactants and of
models of transition states, DeTar and Tenpas 1976a,
1976b, DeTar et al 1978, DeTar and Luthra 1980, DeTar
et al 1985d.

Several problems have to be solved in order to apply this approach. First off it must be possible to
calculate relative free energies of activation with
high accuracy, preferably to better than 0.4 kcal/-
mole. Attaining this level of accuracy in estimating
enthalpy of activation imposes severe demands on the
computations of relative enthalpies of formation and
on the force fields.

We must devise adequate models for the transition
states. We must use appropriate procedures for applying calculations to reactions in solution, not
just the gas phase. We must treat entropy effects.
And we must develop methods for interpreting the
results.

Turning first to the entropy problem, there are
two approaches: calculate the entropy (DeTar and
Luthra 1980), or ignore it on the grounds that the
entropy of activation is effectively constant
throughout the series of reactions. Both approaches
can work (DeTar and Tenpas 1976b, DeTar et al 1985d).

We cannot usually calculate the free energy of activation for a single reaction nor can we calculate the free energy of individual compounds in solution. What we can do is to calculate the difference of the free energy of activation for two reactions. This amounts to the calculation of a ratio of rate constants, eq 1.

$$\log (k_2/k_1) = (\Delta G_1^{\ddagger} - \Delta G_2^{\ddagger})/(2.3RT) \tag{1}$$

Eq 1 makes use of the powerful standard method of double differences, illustrated further in eqs 2 to 9. A generalization of the double difference method is a linear free energy expression, eq 10. It should perhaps be emphasized that eq 10 is not a catch-all application of the linear free energy approach; it is a theoretically correct generalization of the fundamental double difference method.

$$R_1-COOH + MeOH \longrightarrow R_1-C(OH)_2OMe \tag{2}$$
$$r1 \qquad\qquad\qquad\qquad\qquad t1$$

$$R_2-COOH + MeOH \longrightarrow R_2-C(OH)_2OMe \tag{3}$$
$$r2 \qquad\qquad\qquad\qquad\qquad t2$$

$$\Delta H_1^{\ddagger} = \Delta Hf_g(t1) - \Delta Hf_g(r1) + \Delta H_s(t1) - \Delta H_s(r1) \tag{4}$$

$$\Delta H_2^{\ddagger} = \Delta Hf_g(t2) - \Delta Hf_g(r2) + \Delta H_s(t2) - \Delta H_s(r2) \tag{5}$$

$$\Delta\Delta H_{21}^{\ddagger} = \Delta H_2^{\ddagger} - \Delta H_1^{\ddagger} \tag{6}$$

$$\Delta\Delta H_{21}^{\ddagger}(s) = \Delta Hf_g(t2) - \Delta Hf_g(r2) - $$
$$\Delta Hf_g(t1) + \Delta Hf_g(r1) + \Delta\Delta H_s \tag{7}$$

$$\Delta\Delta H_s = \Delta H_s(t2) - \Delta H_s(r2) - \Delta H_s(t1) + \Delta H_s(r1) \tag{8}$$

$$\Delta\Delta H_{21}^{\ddagger} \cong FSE(t2) - FSE(r2) - FSE(t1) + FSE(r1) \tag{9}$$

$$\log (k_i/k_0) = a + b\cdot\Delta FSE_i \tag{10}$$

$$\Delta FSE_i = FSE(ti) - FSE(ri) \tag{11}$$

Eqs 2 and 3 represent formation of the tetrahedral intermediate in esterification; $r1$ and $r2$ are two acids and $t1$ and $t2$ are the respective transition states on the way to the tetrahedral intermediates. $\Delta H_1^{\ddagger}$ and $\Delta H_2^{\ddagger}$ are the enthalpies of activation in solution, eqs 4 and 5, while $\Delta\Delta H_{21}^{\ddagger}$ is the relative difference of enthalpies of activation in solution and will be equal to the free energy difference of eq 1 if the entropies of activation may be considered constant.

The enthalpy of formation of reactant 1 in solution, $\Delta H(r1)$, may be expressed as the sum of the gas phase enthalpy of formation, $\Delta Hf_g(r1)$, plus the enthalpy of solution, $\Delta H_s(r1)$, and likewise for the transition state 1, $\Delta Hf_g(t1)$ and $\Delta H_s(t1)$. The enthalpy of activation in solution may then be expressed as eq 4. Eq 5 is the corresponding expression for the second reaction.

Expansion of the right hand side of eq 6 yields the summation shown in eq 7. The term $\Delta\Delta H_s$, eq 8, is expected to be nearly 0. The argument runs as follows: The enthalpy of solvation about the R_1 group is nearly the same in reactant $r1$ and in transition state $t1$; likewise for solvation of R_2. The enthalpy of solvation about the COOH group is nearly the same for reactant $r1$ and reactant $r2$; likewise for solvation about the $C(OH)_2OMe$ groups. Any steric effects on solvation will to a considerable extent tend to parallel structural steric effects and hence will be confounded with them.

We conclude therefore that the double difference of free energies of activation in solution is nearly equal to the double difference of free energies in the gas phase, and also nearly equal to the double difference of enthalpies of formation.

I will show presently that the gas phase double difference of enthalpies of formation incorporated into eq 7 may be represented by eq 9 in terms of a double difference of formal steric enthalpies. Eq 10 is the linear free energy equation that generalizes eq 9, while eq 11 defines the independent variable. The slope b of eq 10 should be $1/(RT)$, but may be different if the assumptions are not all met. As examples, an imperfect model of the transition state used in calculating the FSE(ti) terms, failure of solvation effects to cancel, incorrect estimates of FSE values are three possible causes.

For intermolecular esterification the entropy tends to cancel out in the double difference. This is fortunate since the entropy of activation is a large negative quantity for most acyl transfer reactions, and it varies with solvent. There is, moreover, an established linear relationship between enthalpies of activation and entropies of activation for some esterifications, Krug et al 1976a, 1976b, 1976c, 1977. The successes of the calculation of relative rate constants owes a great deal to the various cancellations afforded by the double difference method.

For esterification and ester hydrolysis we have used the tetrahedral intermediate itself as the model of the transition state on the way to the intermediate. This is clearly inexact, since at the transition state the three C-O bond lengths should be unequal. To date we have discovered no trends in the error residuals which would suggest the need for a more sophisticated model or which would permit a valid test of one. Uncertainties about models for transition states must, however, be kept in mind. Even cruder models have been used in some of the studies referenced in the Introduction. It appears that the results are rather insensitive to models if they provide a reasonable representation of the steric effects in the transition state.

There are trade-offs in choosing a model for the transition state. If the model is actually a molecule, then the force field may already include all necessary constants for estimating relative enthalpies of formation. If not, then several additional constants will have to be assigned and often there is no independent way of assessing their validity. Experience to date suggests that the double difference method pretty well cancels out the weaknesses of the models and of the assignments.

FORMAL STERIC ENTHALPY AND ENTHALPY OF FORMATION

Turning now to the estimation of relative enthalpies of formation, we assume that steric effects and bonding effects can be treated independently, as implied in eqs 12 to 14. This assumption is formally stated in eq 12; we postulate that the enthalpy of formation of a single conformer may be represented arbitrarily as the sum of formal bond enthalpy and formal steric enthalpy. We postulate further that the FBE term may be represented as a summation of group increments independently of structure. We interpret FBE as the enthalpy of formation of a hypothetical molecule having the prescribed structural units but having no intramolecular "strain".

$$\Delta Hf(g) \quad = \quad FBE \quad + \quad FSE \tag{12}$$

$$\Delta Hf(single\ conformer) \quad = \quad \sum n_i c_i \quad + \quad FSE \tag{13}$$

$$FSE \quad = \quad SE \quad - \quad \sum n_i d_i \tag{14}$$

Eq 12 is a refinement of and an extension of the traditional representation of the enthalpy of formation of alkanes in terms of group increments based on a count of CH_3, CH_2, CH, and C units plus a correction for steric effects, Stull et al 1969, Cox and Pilcher 1970, Benson 1976. However, the enthalpy of formation defined by eq 12 applies to a single conformer (rather than to the existing mixture of conformers) and the FSE values are defined by the formalism described below.

Eqs 12 and 13 are therefore equivalent and for alkanes the FSE is equivalent to "single conformer strain energy", Schleyer 1970, or to "intrinsic strain energy", Burkert and Allinger 1982.

FSE values may, however, be defined independently of whether enthalpies of formation are available or not, and the assumption of additivity of group increments to give FBE may be extended to many types of molecules containing functional groups. FSE values may be defined for transition state models and for molecules that are too unstable to be studied. Thus the FSE value has many advantages as a specialized and precise definition of what is implied less precisely by the idea of "molecular strain" insofar as it arises from intramolecular nonbonded repulsions and attractions.

Molecular mechanics calculations represent the "steric energy" of a molecule in terms of deviations from standard bonds, angles, and torsions plus nonbonded interactions. The "steric energy" is calculated by purely empirical functions whose constants have been chosen so that at the minimum value the geometry, the enthalpy of formation, and perhaps other molecular properties agree with experiment.

The "steric energy" value obtained for a given conformer by molecular mechanics is force field dependent. For some force fields the steric energies may even decrease with increasing substitution; DeTar et al 1985a. The problem lies in the fact that a raw SE value includes the desired FSE quantity of eq 12 plus a variable and force field dependent admixture of residual FBE. For two molecules that are isostructural, as are conformers, the difference of their steric energies is a valid estimate of $\Delta\Delta Hf$ since the residual FBE component is the same for both conformers. For two molecules that are not isostructural the difference of steric energies is force field dependent and hence theoretically meaningless.

For an alkane the SE value may be converted to an enthalpy of formation, which is, of course, force field independent; Allinger et al 1971, Engler et al 1973, Burkert and Allinger 1982. FSE values are also force field independent; they may be considered as corrected steric energies. Differences of FSE values are equal to differences of enthalpies of formation and are thus significant.

The definition of FSE values for a given set of molecules involves three steps: identification of the

molecular groups necessary to define the FBE value
for any member of the set by additivity, selection of
suitable standard conformers, and arbitrary assignment of FSE values to the standards. This formal (arbitrary) assignment defines the FSE value of every
member of the set of compounds and makes it possible
to derive tables of steric properties that are independent of method of calculation or of experimental
estimate. The assignments provide the information
necessary to calculate the c_i values of eq 13 if the
necessary enthalpy data are available, and they provide for the calculation of the correction terms d_i
of eq 14. Examples are given by DeTar et al 1985a
(alkanes), 1985b (alcohols, ethers, olefins), and
1985c(esters).

Inasmuch as FSE values treat only the steric effects it is necessary either to limit the reaction
set under consideration to compounds for which polar
effects are negligible, or else to correct for polar
effects by a linear free energy relationship, DeTar
1980, DeTar et al 1985d.

It is also necessary to avoid changes in bonding
conditions at the reaction center. For example, al-
pha-beta unsaturated acids may have the complication
of steric effects on resonance.

HOW TO TREAT POPULATIONS OF CONFORMERS

Given, then, that there is a suitable set of compounds to work with, a suitable force field, and a
suitable model for the transition state and given
further that we have calculated formal steric enthalpies for several conformers of each reactant and each
transition state, how do we go about calculating the
rate constants.

A fundamental requirement is to locate the global
minima of energy for both the reactant and for the
model of the transition state. For many acids the
conformer of the R-group is the same for both, but
this correspondence is not always found. In the acid
the R-group opposes an sp^2 carbon and in the transition state a developing sp^3 carbon. On the supposition that the acid conformers are in rapid equilibrium it is still correct as a first approximation to
base calculations on the two global minima.

It is of interest to consider in more detail how
the predicted rate constants are modified due to the
presence of populations of conformers of reactants
and of transition states. One way to visualize the
system is to consider that by least motion the conformation of the R-group does not change during the

activation step. If we then picture the system as a mixture of conformers each with its own reaction path, we can describe the process as "adiabatic". In more general thermodynamic terms we can lift this restriction and need not be concerned with details of the dynamics by which the several transition states are reached. We can accordingly look upon the process as involving an entropy correction. If there are several transition states of low energy, then there are multiple paths and the rate will be greater than through a single path based on the global minima. Conversely, if there are multiple low lying reactant states that match up with high energy transition states, then the reaction will be retarded. In either case the comparison is to the rate calculated from the reactant state and the transition state that are the global minima.

The theory for calculating the rate constant from FSE values for populations of conformers has been described in both kinetic terms and in thermodynamic terms (DeTar et al 1985d). The result is eq 15. Each k_i is to be calculated using the FSE value for the global minimum of the reactant conformers and the FSE value for each transition state in turn. D is the Boltzmann denominator, eq 16. In one set of acids the k derived by eq 15 differed from that obtained from the pair of global minima by factors ranging from about 0.5 to 2.

$$k = (1/D)(k_1 + k_2 + k_3 + \ldots) \tag{15}$$

$$D = 1 + \exp[-(\Delta G_2 - \Delta G_1)/RT] + \ldots \tag{16}$$

MECHANISTIC VS OBSERVED RATE CONSTANTS

Another question to be settled is that of defining clearly the relationship between the rate constant that has been calculated and the phenomenological ("experimental") rate constant. This is not always obvious. A theoretically calculated rate constant is a mechanistic rate constant, which may or may not be equal to or proportional to the phenomenological constant. In esterification, as an example, the observed rate constant is usually a pseudo first order rate constant or a derived value based on the Goldschmidt correction, Goldschmidt 1913, Smith 1939. This value is usually converted by proportion to the rate constant in 1 M catalyst acid, and the derived value is the phenomenological constant. The values are somewhat medium dependent, but this is not a serious problem as long as we work with relative rate

constants, which are independent of medium. Changing
the concentration of HCl catalyst from 0.5 M to 0.1 M
causes the phenomenological rate constant to increase
by from 10 to 20%, Smith and Reichardt 1941.

$$R\text{-}COOH + MeOH \underset{k_{-1}}{\overset{k_1}{\rightleftharpoons}} R\text{-}C(OH)_2OMe \xrightarrow{k_2} Ester$$

$$k_{obsd} = k_1 \times \frac{k_2}{k_{-1} + k_2} = k_1 \times f \tag{17}$$

Esterification and ester hydrolysis involve tetrahedral intermediates, and in consequence the phenomenological rate constant is the product of the mechanistic rate constant for forming the intermediate
and a distribution fraction that represents the fraction going on to products, DeTar 1982.

There is evidence that the distribution fraction f
of eq 17 may be presumed to be relatively constant
for R = alkyl; DeTar 1982, DeTar et al 1985d. The
phenomenological rate constant should closely parallel the mechanistic rate constant as long as the
equilibrium constant for esterification stays constant. This will probably hold for esters of n-alco-
hols since there is no steric interaction between the
alkoxyl group and the R-group of the acid.

Target rate constants that may be calculated include those for esterification in methanol (largest
number of available data), for esterification in
ethanol, for acid-catalyzed hydrolysis (a few examples), and extensive data for base-catalyzed hydrolysis of esters, limited, albeit, to a narrow range of
structures. See DeTar et al 1985d for references.

Relative rate constants, taking acetic acid as the
reference, are pretty well constant for all acid-
catalyzed reactions within each series even though
the data have been obtained at different temperatures
and with different catalysts. For hydroxide-catalyzed
hydrolyses of esters the relative rates are also
constant even under differing conditions of solvent
and temperature, but it is necessary to apply a small
correction for the differing inductive effects of
alkyl groups, DeTar 1980a, in order to bring acid-
catalyzed and hydroxide-catalyzed reactions into
coincidence. The distribution fraction, eq 1, is
probably unity for alkaline hydrolysis, DeTar 1982.

Corrected relative rate constants also appear to
be pretty much the same irrespective of whether the

alkoxyl group is methoxyl or other primary alkoxyl group; if there are trends, they are masked by the experimental uncertainties of the data. We find no evidence of steric effect between the R' of a primary R'O group and the $R_1R_2R_3C$ group of any acid we have investigated. Branched R'O groups do, of course, encounter steric interactions.

For reactions that differ in activation energy the relative rates will, of course, depend on what temperature is selected as the reference temperature. For two esterification reactions having a ratio of rate constants of about 1000 at one reference temperature (log k(rel) = 3.0) the change in log k(rel) is about 0.2 for a 20° change in reference temperature, assuming that the rate difference is due entirely to the enthalpy factor.

Most of the data for esterification were obtained prior to 1940 by many groups. The most extensive of these is by H. A. Smith; see Smith and Burn 1944 and earlier papers. The range of structures investigated was rather narrow. Newman (Loening et al 1952) greatly extended the range. The recent work of the Chapman group is a model of experimental excellence, Burden et al 1980 and earlier papers.

Two recent studies of methanolic esterification of highly hindered acids were based on competitive esterifications, Sniegoski 1976 and MacPhee et al 1978. Although gas chromatography was used in the analysis, material balances were not reported and there are problems with the data. Rates for two key reference compounds, triethylacetic acid and t-butyldimethylacetic acid, are divergent by factors of 30 as reported by the several laboratories, Loening et al 1952, Sniegoski 1976, and MacPhee et al 1978. As a consequence the rate constants for esterification of several highly hindered acids must for the present be considered unknown.

CALCULATION OF RATES OF ESTERIFICATION

To illustrate the method of calculation the above procedures have been applied to rate data for esterification of fifteen substituted acetic acids, R_1R_2CH-COOH, having R_1 and R_2 H, Me, Et, i-Pr, and t-Bu. There are 15 acids in the set, and experimental values have been reported for all. However, data for three of these are uncertain for reasons mentioned above; they are i-Pr-i-PrCH-COOH, i-Pr-t-BuCH-COOH, and t-Bu-t-BuCH-COOH.

The data are summarized in Table 1. The column labelled Delta FSE is the difference in the formal steric enthalpies of the conformer of minimum energy of the tetrahedral intermediate (the transition state model) and of the conformer of minimum energy for the acid.

The column Adiabatic Corr is the adiabatic or entropy correction explained above to correct for populations of conformers.

Literature rate constants have been used in deriving Log k rel avg. Calculated rate constants are based on eq 18. The values calculated for the last three acids may be considered as predicted rate constants.

$$\log k \text{ rel} = -0.72929(.04) - 0.98326(.03) \cdot \Delta FSE + \text{Adiabatic Corr} \tag{18}$$

standard deviation 0.22 R^2 0.9788   28 d.f.

Comparison of the observed and calculated values of log k(rel) show that the refined approach we have described does in fact give a good correlation of the rate constants for the first 12 acids. Experimental problems with the data for the last three acids have been described above.

Making use of formal steric enthalpy it becomes possible to determine the origins of the steric effects. For example, we can compare the effects on the acid of replacing the alpha H of $EtCH_2$-COOMe in turn by Me, Et, i-Pr, and t-Bu. The respective FSE values (for the conformer of minimum energy) are 0.84, 1.01, 1.22, 2.29, 3.66 kcal/mole. We can thus examine in detail the accumulation of steric effects with increasing substitution.

The corresponding values for the model of the transition state are 0.26, 1.28, 2.48, 4.15, 8.39. There is a reasonable proportionality up to the Et-t-Bu-CH group. At this last step there is a large jump in the crowding of the tetrahedral structure. FSE values now make it possible to investigate in detail the effects of structure on steric properties.

It may be noted that the reported FSE values of the acids in Table 1 are actually those of the methyl esters. The two are the same except for a constant. Calculations were performed on the esters in order to discover whether there were any interactions between the methoxyl methyl group and the R-group of the acid. There are none for the acids discussed.

Table 1. Formal Steric Enthalpy Values
and Rate Data for R_1R_2COOH Esterification

$R_1{}^a$	R_2	Delta^b FSE	Adia-^{c} batic Corr	-Log k^d rel avg obsd	-Log k^b Calc
H	H	-.76	0.00	0.01	-.02
Me	H	-.62	0.00	0.05	0.12
Et	H	-.58	-.14	0.31	0.31
i-Pr	H	0.32	-.12	0.93	1.17
t-Bu	H	1.38	0.00	1.65	2.09
Me	Me	-.41	0.00	0.50	0.33
Et	Me	0.27	0.02	0.99	0.97
i-Pr	Me	0.67	-.22	1.90	1.61
t-Bu	Me	2.78	0.00	3.23	3.46
Et	Et	1.26	0.06	1.96	1.91
i-Pr	Et	1.86	-.06	3.25	2.62
Et	t-Bu	4.73	0.28	5.14	5.09
i-Pr	i-Pr	3.32	0.18	4.98	3.81
i-Pr	t-Bu	4.84	-.01	6.53	5.50
t-Bu	t-Bu	8.58	0.00	6.97	9.17

$^a R_1R_2CH-COOH$ acid-catalyzed esterification in methanol. ^{b}Value for tetrahedral conformer of minimum energy - value for ester conformer of minimum energy. ^{c}See text. ^{d}relative to acetic acid; average of several values. See DeTar et al 1985d for references.

CONCLUSIONS

The procedures described have wide applicability. They can be applied to the prediction of steric effects in many types of reactions providing that the calculations are made with due consideration of the limitations of the method, particularly in regard to care with molecules involving steric effects on resonance.

There are many opportunities for new experimental studies of steric effects. One example is the investigation of joint steric interactions in the substrate and in the nucleophile in aminolysis reactions, DeTar and Delahunty 1983. Another area of great interest is the treatment of cyclization reactions, DeTar and Luthra 1980, for example. In lactone formation the differing equilibrium constants implies that there is no longer a proportionality between the phenomenological rate constants and the mechanistic rate constants. Many other types of reactions are controlled by steric effects, and may yield interesting insights.

A potentially very important use of the method can be in the design of enzyme-like catalysts. The fact that enzyme-substrate specificity can be reproduced indicates that potential catalysts can be evaluated computationally, DeTar 1981a, 1981b.

Many of these applications can benefit greatly by development of improved force fields. The potential number of constants needed for molecules containing half a dozen or so types of atoms runs into the hundreds. It is going to require a great deal of evaluation to learn effective ways to work with this problem.

BIBLIOGRAPHY

Allinger,N.L.; Tribble,M.T.; Miller,M.A.; Wertz,D.H. *J.Am.Chem.Soc.* 1971, *93*, 1637

Beckhaus,H.-D.; Hellmann,G.; Ruechardt,C. *Chem.Ber.* 1978, *111*, 72

Beckhaus,H.D.; Kratt,G.; Lay,K.; Geiselmann,J.; Ruechardt,C.; Kitschke,B.; Lindner,H.J. *Chem.Ber.* 1980, *113*, 3441

Benson,S.W. Thermochemical Kinetics, Wiley, New York, 1976, 2d Ed.

Bernlöhr,W.; Beckhaus,H.-D.; Schnering,H.-G.; Rüchardt,C. *Chem.Ber.* 1984 *117*, 1013

Bingham,R.C.; Schleyer,P.v.R. *J.Am.Chem.Soc.* 1971, *93*, 3189

Burden,A.G.; Chapman,N.B.; Shorter,J.; Toynes,K.J.; Wilson,L.M. *J.Chem.Soc.Perkin II* 1980, 1212

Burkert,U.; Allinger,N.L. ACS Monograph No. 177 Am. Chem.Soc. Washington,1982

Cox,J.D.; Pilcher,G. *Thermochemistry of Organic and Organometallic Compounds*, Academic Press, London, 1970

DeTar,D.F.; Tenpas,C.J. *J.Am.Chem.Soc.* 1976a, *98*, 4567

DeTar,D.F.; Tenpas,C.J. *J.Am.Chem.Soc.* 1976b, *96*, 7903

DeTar,D.F.; McMullen,D.F.; Luthra,N.P. *J.Am.Chem.Soc.* 1978, *100*, 2484

DeTar,D.F. *J.Org.Chem.* 1980a, *45*, 5166

DeTar,D.F.; Luthra,N.P. *J.Am.Chem.Soc.* 1980, *102*, 4505

DeTar,D.F. *Biochemistry* 1981, *20*, 1730

DeTar,D.F. *J.Am.Chem.Soc.* 1981, *103*, 107

DeTar,D.F. *J.Am.Chem.Soc.* 1982, *104*, 7205

DeTar,D.F.; Delahunty,C. *J.Am.Chem.Soc.* 1983, *105*, 2734

DeTar,D.F.; Binzet,S.; Darba,P. *J.Org.Chem.* 1985a, *50*
DeTar,D.F.; Binzet,S.; Darba,P. *J.Org.Chem.* 1985b, *50*
DeTar,D.F.; Binzet,S.; Darba,P. *J.Org.Chem.* 1985c, *50*
DeTar,D.F.; Binzet,S.; Darba,P. *J.Org.Chem.* 1985d, *50*
Engler,E.M.; Andose,J.D.; Schleyer,P.v.R. *J.Am.Chem. Soc.* 1973, *95*, 8005
Farcasiu,D. *J.Org.Chem.* 1978, *43*, 3878
Goldschmidt,H.; Theisen,A. *Z.Physikal.Chem.* 1913, *81*, 30
Krug,R.R.; Hunter,W.G.; Grieger-Block,R.A. *Nature* 1976a, *261*, 566
Krug,R.R.; Hunter,W.G.; Grieger-Block,R.A. *J.Phys. Chem.* 1976b, *80*, 2335
Krug,R.R.; Hunter,W.G.; Grieger-Block,R.A. *J.Phys. Chem.* 1976c, *80*, 2341
Krug,R.R.; Hunter,W.G.; Grieger-Block,R.A. *Chemometrics: Theory and Applications*, ACS Symposium Series 52, ACS 1977
Loening,K.L.; Garrett,A.B.; Newman,M.S. *J.Am.Chem. Soc.* 1952, *74*, 3929
MacPhee,J.A.; Panaye,A.; Dubois,J.E. *Tetrahedron* 1978, *34*, 3553
Müller,P.; Perlberger,J.C. *J.Am.Chem.Soc.* 1976, *98*, 8407
Müller,P.; Blanc,J.; Lenoir,D. *Helv.Chim.Acta* 1982a, *6*, 1212
Müller,P.; Blanc,J.; Perlberger,J.C. *Helv.Chim.Acta* 1982b, *65*, 1418
Müller, P. 1985, personal communication; recalculation of bridgehead solvolyses.
Perlberger,J.C.; Müller,P. *J.Am.Chem.Soc.* 1977, *99*, 6316
Rüchardt,C.; Beckhaus,H.-D. *Angew.Chem.Int.Ed.Engl.* 1980, *19*, 429
Ruechardt,C.; Weiner,S. *Tetrahedron Lett.* 1979, 1311
Schleyer,P.v.R.; Williams,J.E.; Blanchard,K.R. *J.Am. Chem.Soc.* 1970, *92*, 237
Schneider,H.J.; Schmidt,G.; Thomas,F. *J.Am.Chem.Soc.* 1983, *105*, 3556
Schneider,H.J.; Thomas,F. *J.Am.Chem.Soc.* 1979, *101*, 1424
Schneider,H.J.; Thomas,F. *J.Am.Chem.Soc.* 1980, *102*, 1424
Sniegoski,P.J. *J.Org.Chem.* 1976, *41*, 2058
Smith,H.A. *J.Am.Chem.Soc.* 1939, *61*, 254
Smith,H.A.; Reichardt,C.H. *J.Am.Chem.Soc.* 1941, *63*, 605
Smith,H.A,; Burn,J. *J.Am.Chem.Soc.* 1944, *66*, 1494
Stull,D.R.; Westrum,Jr.,E.F.; Sinke,G.C. *The Chemical Thermodynamics of Organic Compounds*, Wiley, New York, 1969

Chapter 8

MOLECULAR TOPOLOGY, ELECTRON CHARGE DISTRIBUTIONS, AND MOLECULAR PROPERTIES

Benjamin M. Gimarc and Jane J. Ott
Department of Chemistry, University of South Carolina,
Columbia, S.C.29208, USA

INTRODUCTION

The problem with chemistry is that it has too many examples. Over 100 elements combine with themselves and each other to form an essentially limitless array of compounds, each with its own properties. This diversity is both the fascination and frustration of chemistry. A detailed knowledge of the properties of one compound tells us nothing *a priori* about the properties of others. Thus, at the most fundamental level chemistry is a qualitative science through which we hope to understand trends in properties through broad classes of compounds.

One way to approach a qualitative synthesis of chemical information is by applying qualitative arguments, such as those involving symmetry properties and atomic orbital overlaps, within the framework of molecular orbital theory [1,2]. Another approach starts with molecular topology [3]. The connectivity of atoms within a molecule, or more succinctly if less correctly, molecular topology, is a structural feature of obvious and fundamental importance to chemistry. The pattern of charge densities in a molecule is determined by molecular topology and the number of electrons available. Nature prefers to place atoms of greater electronegativity in those positions where the topology of the structure tends to pile up extra charge. Since such heteroatomic systems are preferentially stabilized by charge distributions established by molecular connectivity, this effect has been called the *rule of topological charge stabilization* [4].

The example of the linear triatomic azide anion N_3^- serves to demonstrate the rule. The ion has 16 valence electrons with two identical but mutually perpendicular pi electron systems each containing 4 electrons. Each pi MO set is composed of three identical 2p AOs. For N_3^- the calculated pi electron charge densities are greater at the terminal nitrogens (1.5) than at the central atom(1.0). This pattern of charge densities can be easily understood form the nodal properties of the occupied pi orbitals. In Hückel theory charge densities are given by the squares of AO coefficients summed over occupied MO's. In the lowest energy or bondng MO $\Psi_1(1)$ the 2p AO on the central nitrogen has a larger coefficient than do the terminal AO's. In the higher occupied or nonbonding MO Ψ_2

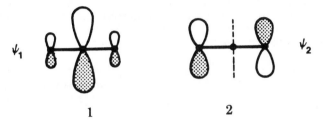

Ψ_1 Ψ_2

1   2

(2), a nodal plane perpendicular to the molecular axis eliminates any contribution from the central atom 2p AO and pushes electron density to the terminal atoms. The net result is a greater electron density at the terminals than on the central atom. Although N_3^- is known it is usually reactive, in some cases explosive. Much more stable are isoelectronic, isostructural heteroatomic species such as CO_2, NO_2^+, and BO_2^- in which constituent atoms occupy positions that match their relative electronegativities with the distributions of charge determined in the homoatomic system N_3^-, for which only topology and electron filling level establish the pattern of electron densities. We refer to such a homoatomic system as the *uniform reference frame* from which we can make rationalizations and predictions concerning atomic positions and relative stabilities of related heteroatomic species.

The validity of the rule of topological charge stabilization transcends simple Hückel theory and indeed even the molecular orbital approximation as the following perturbation argument shows. Consider the uniform reference frame as the unperturbed system, with Hamiltonian $H°$, wavefunction $\Psi°$, and total energy $E^{(o)}$ related by the Schrödinger equation:
$$H°\Psi° = E^{(o)}\Psi°.$$
Now introduce one or more heteroatoms as a perturbation, holding the molecular structure and the number of electrons fixed. The perturbing Hamiltonian H' is a sum of changes in Coulombic nuclear-electron attraction terms due to changes in nuclear charge ΔZ_α that result from substitution of a heteroatom at position α. For the perturbed system described by $H = H° + H'$, the total energy

can be calculated as a sum of higher order corrections, $E = E^{(o)} + E^{(1)} + ---$. The first-order correction is calculated from the unperturbed wavefunction:

$$E^{(1)} = <\Psi^\circ| H'| \Psi^\circ>$$

Since the operator H' involves only simple multiplication, the Ψ° factors can be combined within the integral to give the unperturbed electron density $\rho^\circ = (\Psi^\circ)^*\Psi^\circ$ and $E^{(1)} = <\rho^\circ H'>$. Therefore, to achieve maximum stability or energy lowering through the correction $E^{(1)}$, the heteroatoms of largest ΔZ should match those positions in the molecule where ρ° is already greatest in the unperturbed or reference frame. For qualititave generalizations involving atoms in different rows of the periodic table it is convenient to consider valence electrons only and to replace ΔZ_a by changes in effective nuclear charge $\Delta \zeta a$, or even more simply, to use relative electronegativity as a rough measure of $\Delta \zeta a$.

To demonstrate the general applicability of the rule of topological charge stabilization and to show insights that can be derived from it, the following sections present many examples from both inorganic and organic chemistry.

PLANAR CONJUGATED SYSTEMS

Pentalene (3) has 8 pi electrons. Differences between largest and smallest Hückel pi charge densities are large. Attempts to prepare pentalene itself have failed although 1,3,5-triterbutyl pentalene has been synthesized. The inorganic analogs 4 and 5 have been made. In these examples the electronegativities of the constituent atoms match the charge denisity distributions of 3.

3   4   5

The pentalene dianion has a system of 10 pi electrons. Differences among Hückel pi charge densities (6) are smaller than in pentalene itself (3), probably increasing the stability of the dianion relative to that of pentalene. The unsubstituted dianion has been prepared. The extra pair of electrons in the dianion have gone to positions 1, 2, 4, and 6 which now have the largest charge densities. Consider the series of isomeric thienothiophenes (7-10) which are isoelectronic with the pentalene dianion. All four isomers

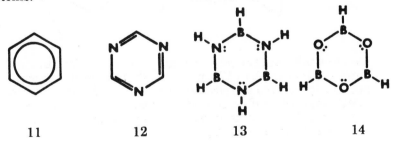

6

7   8   9   10

have now been prepared although 10 is known only as the tetraphenyl derivative. Comparing these structures with the charge distributions in 6 suggests that 7 and 8 should be of comparable stability, 9 less so, and 10 the least stable of all. These conclusions are in excellent agreement with the known reactivities of this series and with the results of several sets of semiempirical resonance energy calculations [6].

The rule of topological charge stabilization does not imply that heteroatoms cannot be introduced into stable systems for which the uniform reference frame has uniform charge densities. Countless examples of such systems exist. For example, all pi charge densities in benzene (11) are unity. The isoelectronic, heteroatomic species 12, 13, and 14 are all known. Topological charge stabilization suggests a decrease in stability through this series, following the trend of increasing localization of charge on more electronegative atoms.

11   12   13   14

THREE-DIMENSIONAL SYSTEMS

In P_4S_3 (15) the four phosphorus atoms are at the four corners of a distorted tetrahedron with an apical phosphorus linked by bridging sulfurs to a basal triangle of phosphorus atoms. As_4S_3 has an analogous structure. The known anion P_7^{3-} serves as the uniform reference frame for P_4S_3 and As_4S_3. Structure 16 displays the Mulliken net atomic populations (or more simply, atomic charges) for P_7^{3-} calculated from extended Hückel wavefunctions [7]. The extended Hückel method is known to yield exaggerated charges [8]

15

but experience shows that they are adequate for the purposes here which require only a qualitative pattern of charge distributions.

The uniform reference frames for three-dimensional systems are often hypothetical and have very large total charge Q. Individual atomic charges q_r sum over all atoms r to the total charge Q. Since we are interested only in charge differences we have introduced *normalized charges* q_r' which sum to zero:

$$q_r' = q_r - Q/N,$$

where N is the number of atoms in the structure. Structure **17** shows the normalized charges of P_7^{3-} from **16**. Since the reference frame in **17** is no longer composed of real atoms we have suppressed

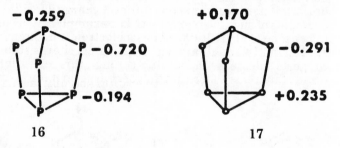

16   **17**

atomic symbols in the corresponding structural formulas. The normalized charges in **17** emphasize that the bridging positions are negative compared to apical and basal sites. Hence, the more electronegative sulfurs should occupy the bridges with less electronegative P or As atoms in the apex and basal triangle as observed in P_4S_3 and As_4S_3.

Topological charge stabilization offers a beautiful explanation for the driving force to equilibrium in the system P_4S_3/As_4S_3. When reactants are mixed in the stoichiometric ratio (1:3) the equilibrium lies predominately to the side of the PS_3As_3 product [9]:

$$P_4S_3 + 3\,As_4S_3 \leftrightarrows 4\,PS_3As_3$$

The structure of the product PS_3As_3 presumably has a phosphorus atom at the apex, three bridging sulfurs, and a basal triangle of arsenic atoms, exactly those position in which electronegativities of constituent atoms match the relative charges in the uniform reference frame [10].

In P_4S_4 (18) there are four equivalent sulfur atoms located at the corners of a square plane and bonded to pairs of phosphorus atoms above and below the plane [18]. Normalized charges calculated for an appropriate uniform reference frame (19) can rationalize the different positions taken by atoms of Groups V and VI. The more electronegative atoms prefer the plane positions with

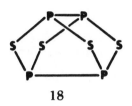

18

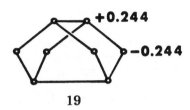

19

the less electronegative atoms occupying sites above and below the plane, an arrangement that is realized in the known examples of this series [11]. In S_4N_4 and Se_4N_4 the Group V element (N) occupies the negative square positions and in P_4S_4, As_4S_4, and As_4Se_4 the Group VI elements (S or Se) are on the plane.

The P_4S_5 cage (20) is related to that of P_4S_4 (9) with an additional sulfur atom bridging one pair of phosphorus atoms. Normalized charges for the uniform reference frame appear in 21. Again, the more electronegative sulfurs occupy the negative sites

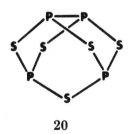

20

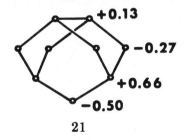

21

while the less electronegative phosphorus atoms take the positive positions. Similarly, in the related structures P_4Se_5 and As_4S_5, the more electronegative Group VI atoms (S or Se) are in the negative sites and the less electronegative Group V atoms (P or As) take up the positive locations. But positions of Group V and VI elements are reversed in $S_4N_5^-$ because N is more electronegative than S.

Insertion of still another sulfur between the bonded pair of phosphorus atoms in 20 leads to the structure which one anticipates for P_4S_6 (22), this particular molecule being as yet unknown. The highly symmetrical structure is that of adamantane. The charges of the uniform reference frame (**23**) show four positive tetrahedral sites and six negative bridging positions that cover the edges of the

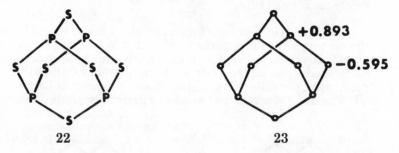

22   23

tetrahedron. The more electronegative element should be located in the bridges with the less electronegative element at the tetrahedral sites, an arrangement that is well known in P_4O_6, As_4O_6, As_4S_6, and Sb_4O_6, as well as in hydrogen and methyl substituted examples such as $P_4(NMe)_6$ and $(HSi)_4S_6$.

The adamantane examples have 56 valence electrons. These structures lead to an empirically based topological rule that says less electronegative atoms go to the sites where they form more bonds. This may be true for most 56 electron systems but just the reverse is observed in some 44 electron species that also have the adamantane structure. In $(HC)_4(BR)_6$ (R = Me, Cl or Br) the carbons are at the tetrahedral sites with borons in the bridges, an arrangement that agrees exactly with the polarity of normalized charges for the uniform reference frame with 44 electrons (**24**).

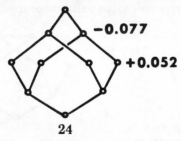

24

The rule of topological charge stabilization can be used to predict the relative stabilities of the positional isomers of the well known *closo*-carboranes, $C_2B_{n-2}H_n$, where $5 \le n \le 12$. Reference frames are required that simulate the structures of *closo*-boron hydrides, $B_nH_n^{2-}$, $5 \le n \le 12$, the homoatomic analogs of the carboranes. To emphasize charge differences and simplify the calculations we used unsubstituted carbon systems C_n^{2-} rather than $B_nH_n^{2-}$.

The $C_2B_3H_5$ system is a trigonal bipyramid. Calculated charges for the uniform reference frame (**25**) are negative at the apical positions and positive at the equatorial sites.

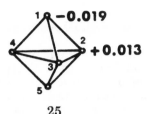

25

Since carbons prefer to occupy the negative sites, the three possible isomeric carboranes should follow the decreasing order of stability: 1,5-$C_2B_3H_5$ > 1,2- > 2,3. This order agrees with the energy order established by *ab initio* MO calculations. More important, it agrees with what is known experimentally about $C_2B_3H_5$ isomers. The 1,5-isomer is the only known unsubstituted isomer. The 1,2-isomer exists only as the methyl substituted form and the 2,3-isomer has not been reported in any form.

The carborane $C_2B_4H_6$ has octahedral geometry. The six vertices of a regular octahedron are equivalent so the charges on the atoms of the C_6^{2-} uniform reference frame must be identical. The rule of topological charge stabilization cannot distinguish between the two possible carborane isomers. But suppose a single electronegative heteroatom is introduced at one of the equivalent vertices of C_6^{2-} to make C_5N^-. In this perturbed system (26) charges are no longer the same everywhere and the rule of topological charge stabilization can be used to predict the prefered locations for the introduction of the second electronegative heteroatom. The perturbing heteroatom at position 1 (indicated by ● in 26) produces another large negative charge at position 6. Therefore 1,6-$C_2B_4H_6$ should be more stable than the 1,2-isomer, again in agreement with *ab initio* calculations and the experimental report that the 1,2-isomer quantitatively rearranges to the 1,6-isomer on heating at 250°.

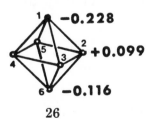

26

We have used similar procedures to predict the order of stabilities for the carborane analogs of $B_7H_7^{2-}$ through $B_{12}H_{12}^{2-}$ and in every case our predictions agree with available experimental data [12]. It is interesting that our predictions based on crude charge densities calculated from extended Hückel wavefunctions, are often in better agreement with experiment on relative isomer stabilities than are total energies calculated from the same extended Hückel wavefunctions.

ONE-DIMENSIONAL EXAMPLES

The pseudo-one-dimensional character of chain-type structures should make their electronic properties easier to interpret using qualitative MO theory. But chains have lower symmetry than most of the molecules mentioned here previously. For example, a five-atom chain has three different kinds of sites compared to only two for the ten-atom adamantane cage. In this section we review a number of five-atom chains with several classes containing different numbers of valence electrons.

The molecules OCCCO and OBOBO have 24 valence electrons. C_3O_2 is almost linear; it is actually a semirigid bender. The isoelectronic species B_2O_3 and NCSCN are V-shaped. Extended Hückel charge densities for the 24-electron five-atom uniform reference frame in linear (27) and V-shaped (28) conformations both show negative charges at the chain ends, directing electronegative atoms to those terminal sites. Notice that the change from linear to

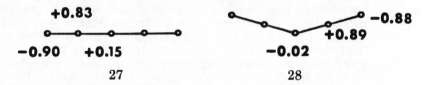

27   28

V-shape increases the electron density at the central atom, providing increased stability for more electronegative atoms at that position such as the central O and S in B_2O_3 and $S(CN)_2$.

N_2O_3 (28 valence electrons) has two isomeric forms, a chain of planar W-shaped conformation (29) and a planar branched structure (30). The branched isomer is the more stable of the two. From the normalized charges for the corresponding uniform reference frames (31, 32) one can see that the pattern of charge densities in the

29   30

31   32

branched form is more compatible with the location of all three oxygen atoms at the most negative sites and suggests topological charge stabilization as the source of extra stability for the prefered branched isomer of N_2O_3.

The ion $ClSNSCl^+$ (33), with 30 valence electrons, is planar and U-shaped. The normalized charges for the uniform reference frame (34) do not follow the order of electronegativities of atoms along the chain, the only significant deviation from the rule in this series.

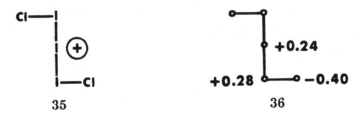

The ions I_5^+, $I_3Cl_2^+$, and $I_3Br_2^+$ (34 valence electrons) are planar chains bent in a zig-zag or Z-shape. In the heteroatomic examples, such as $ClIIICl^+$ (35) the atoms are connected such that the two more electronegative atoms are at the terminals and the three less electronegative atoms occupy the intermediate and central sites. No examples such as $I_2Cl_3^+$ are known. These observations are in accord with the distribution of normalized charges for the uniform reference frame (36).

The 36 electron chains, such as I_5^-, $I_2Cl_3^-$, $I_2Br_3^-$, and $Xe_2F_3^+$, are linear or V-shaped. In $ClICllCl^-$ (37), for example, the three more electronegative atoms are located at the terminal and central positions with the less electronegative atoms at the intermediate sites. Examples such as $IClIClI^-$ are unknown. The observed arrangement of atoms matches perfectly with the pattern of charges in the uniform reference frame (38).

CONCLUSIONS

Atomic charge in a molecule is an old and useful concept. Even the notion that charges might be determined by connectivity goes back at least 35 years [13]. We have used the rule of topological charge stabilization to rationalize the structures and predict the relative stabilities of a wide variety of compounds. For each structural class these conclusions were based on charge density distributions calculated by simple Hückel or extended Hückel theory for a single, homoatomic reference structure. The rule can be justified by first-order perturbation theory. Charge density patterns are often easy to understand with the aid of qualitative MO theory. The rule is easy to apply and could be used to guide synthetic efforts and to point out problems that merit further study by both theory and experiment. Topological charge stabilization can serve as a unifying principle for the organization of chemical information.

REFERENCES

1.   B.M. Gimarc, *Accounts Chem. Res.* **7**, 57 (1974).

2.   B.M. Gimarc, "Molecular Structure and Bonding: The Qualitative Molecular Orbital Approach," Academic Press, N.Y., 1979.

3.   A. Graovac, I. Gutman, and N. Trinajstic, "Topological Approach to the Chemistry of Conjugated Molecules," Springer-Verlag, Berlin, 1977; N. Trinajstic, "Chemical Graph Theory," CRC Press, Boca Raton, Fla., 1983.

4.   B.M. Gimarc, *J. Am. Chem. Soc.* **105**, 1979 (1983).

5.   B.M. Gimarc and J.J. Ott, submitted for publication.

6.   I. Gutman, M. Milun, and N. Trinajstic, *J. Am. Chem. Soc.* **99**, 1692 (1977); B.A. Hess, Jr., L.J. Schaad, and C.W. Holyoke, Jr., *Tetrahedron* **28**, 3657 (1972); **31**, 295 (1975); M.J.S. Dewar and N. Trinajstic, *J. Am. Chem. Soc.* **92**, 1453 (1970).

7.   R. Hoffmann, *J. Chem. Phys* **39**, 1397 (1963).

8.   A. Pullman, *Intern. J. Quantum Chem.* **2**, 1875 (1968).

9.   R. Blachnick and U Wickel, *Angew. Chem. Int. Ed. Engl.* **22**, 317 (1983).

10.   B.M. Gimarc and P.J. Joseph, *Angew. Chem. Int. Ed. Engl.* **23**, 506 (1984).

11.   J.K. Burdett, *Accounts Chem. Res.* **15**, 34 (1982).

12.   J.J. Ott and B.M. Gimarc, submitted for publication.

13.   H.C. Longuet-Higgins, C.W. Rector and J.R. Platt, *J. Chem. Phys.* **18**, 1174 (1950).

Chapter 9

SYSTEMATIC SYNTHESIS DESIGN BY COMPUTER

J.B. Hendrickson and A.G. Toczko
Department of Chemistry, Brandeis University

The primary problem confronting the development of any system for synthesis design is the vast size of the necessary search space, a huge "synthesis tree" composed of the possible intermediate structures and their interconversion reactions. Hence the criteria applied to select the best pathways must be not only clear but very stringent. Establishing adequate viable criteria, to be applied impartially by the computer rather than left to the user, is therefore the central and hardest task. As a basis for this selection we have chosen economy: the shortest, most efficient paths from the cheapest available starting materials. Since a synthesis is a sequence of reactions starting with small starting material molecules and leading to a large target structure, the only obligatory reactions are those that link those starting material molecules together, and so the shortest synthesis then will employ only those reactions. Such shortest paths then become our search goal.

To see the importance of this conclusion we must understand an organic structure as the sum of its skeleton and its functional groups. The skeleton is its σ-bonded framework of carbon atoms, and the functional groups its attached heteroatoms and carbon-carbon Π-bonds. This dichotomy is also seen in reactions: construction reactions are those that build the skeleton, i.e., that create C-C σ-bonds; refunctionalization reactions alter the functional groups without changing the skeleton. Our goal then is to find sequences of construction reactions only, from starting materials to target. This is a very stringent criterion as indicated by the fact that the average synthesis has twice as many refunctionalization steps as constructions and constructs 1/4-1/3 of the skeletal bonds.

Approaching this huge tree search problem with the intent to assess all its many possible combinations, we need a clear, linear, digital expression of these organic molecules to define and simplify the structures and reactions in the search space, as well as to manipulate them rapidly. If we grant a canonically numbered list of the skeletal carbons and their connectivity, then the functional groups could be expressed by a simple number at each carbon and the functionality of the molecule as an ordered list of such numbers. The net functional change in any reaction may then be annotated simply as the arithmetic change in such a functionality list from substrate to product, or vice-versa. Conversely, the substrate functionality lists can be generated from the product lists (or vice-versa) by adding a "generator" list characteristic of a particular reaction. This assumes that the connectivity changes attendant on construction reactions are separately recorded. The necessary digital description of structure envisioned here should be fast and simple: fast for rapid computer manipulation; and simple implying abstracted or generalized from normal functional group description to coalesce trivial distinctions and so reduce the number of items to examine in the search space.

An overview of the procedure derived has two phases. In the first only the skeleton of the target is dissected in order to find the most efficient modes of assembly of that skeleton from the largest skeletons of available starting materials. This designates a set of bonds that must be constructed (the bondset). In the second phase the functional groups necessary to initiate constructions of those designated bonds are laid onto the skeletons so as to proceed, by a sequence of construction reactions only, from real starting materials to the target. Now we are dealing not only with the correct skeletons but also with the correct functional groups in the right positions on the skeletons as well. Both the best modes of skeletal assembly (the bondsets), and the demands on functionality to create seqential construction reactions only, together constitute a very stringent basis for selection of the optimal synthesis pathways.

In the first place we seek to define which skeletal bonds should best be constructed, and in what order. Thus we are looking for the best ordered bondsets, and the problem is not trivial.[1] To construct λ bonds from a target of b bonds, there are $b!/(b-\lambda)!$ ordered bondsets which may be followed. For a C_{21} steroid, therefore, there are 3×10^{10} ways to construct

1/3 of its 24 bonds. The most efficient synthesis, however, is a convergent one and the number of these is much smaller. In order to find the fully convergent syntheses we must first dissect the target skeleton into two pieces and then each piece into two again, etc. If we were to stop this process at two such levels of dissection there would be four starting material skeletons. If we also apply the constraint that the only acceptable sets are those for which all four skeletons are found in the catalog of starting materials, then we have developed a very demanding criterion which will result in relatively similar starting skeleton sizes and reject many other sets. For a C_{20} target the average starting skeleton will have five carbons, and the variety of available starting materials falls off sharply above C_5. In the case of larger target it may therefore be necessary to look to a third level of dissection.

Comparison of the skeletal pieces from target dissection with skeletons of the starting materials in the catalog now becomes a primary need for our program. To this we sought a procedure which can create an unambiguous canonical numbering of any skeleton for its identification.[2] The skeleton is a graph and so may be fully characterized by its adjacency matrix (an nxn connectivity matrix), but there are n! different matrices equally representing any skeleton and these differ in the numbering of the skeletal atoms. Therefore, we require a clear definition of a single, unique numbering such that any two molecules may be so numbered and then compared for identity.

The adjacency matrix may be strung out into a binary string of $n(n-1)/2$ bits and we may treat this string as a binary number. The particular numbering of the skeletal atoms (and its corresponding matrix) which affords the maximum numerical value of this binary string is then a unique numbering and so may be used for our comparisons of skeleta. Owing to symmetry there may be several equivalent such maximal numberings for a given skeleton (cyclohexane has twelve), but they all have the same maximal binary string and so comparison of the strings still identifies like skeleta. A procedure was developed for row-wise generation of the maximal matrix and its corresponding skeletal numbering. The catalog of starting material skeletons was then set up as a listing of their corresponding maximal binary strings in numerical order. Such an ordered listing can then be rapidly searched for the presence of a particular skeleton generated by the dissection of the target, and identified by its maximal binary number. Our procedure for making these comparisons is very fast and appears to be error-free, in contrast to some other methods for identifying such structural isomorphism.

By separating the skeleton and functional groups this dissection procedure first derives all skeletal bondsets corresponding to fully convergent orders of construction of the skeleton from four smaller pieces found in the catalog. The whole synthesis tree is now rendered easier to search since each of these optimal bondsets so derived represents a small subtree of the whole, rooted on real starting materials; and each bondset is examined separately. For each bondset we now have a set of target bonds which must be constructed and the order in which they are to be made. Essentially, the next task is to examine the target functional groups out from either end of the last bond to be constructed; and then to ask what reactions will produce those groups and what substrate functional groups are required to do it. Then each bond in the bondset, in order back from the target, is so examined, to identify for each product all possible reactions and their corresponding substrates. Thus the elements to establish a given reaction are both the position of the bond to be made as well as its neighboring functional groups.

In the second phase of the procedure we examine a particular bondset, i.e., an ordered sequence of skeletal bonds to construct, and we must find all ways to construct each bond without functional group repair reactions between the constructions. The requirement that we assess all possible reactions, coupled with the huge search space, mandates an initial generalization or abstraction of the involved functionality to a broad, simple description which only in the successful cases needs to be refined later.

The system developed for this purpose,[3] outlined in Figure 1, consists of a definition of four generalized kinds of attachment on any carbon atom: H for hydrogen or other electropositive elements; R for σ-bond (skeletal bond) to another carbon; π for a π-bond to carbon; and z for a bond (π- or σ-) to electronegative heteroatom. For any carbon, then, the number of attachments of each kind is respectively, h, σ, π, and z, such that $\underline{h+\sigma+\pi+z = 4}$. The functionality is then $\overline{\pi+z}$ and,

Figure 1. Characterization of Structures

Kind	Number	Oxidation State
H	h = 0-4	x = z - h
R	σ = 0-4 = Skeletal	
π	π = 0-2 ⎫	
Z	z = 0-4 ⎬ = Functional	$\pi + z = 4-\sigma-h$

$\Sigma = 4$

since the skeleton is given, σ is known and h derives by subtraction. The result is that any (connected) carbon functionality can be described by two digits, z (=0-3) and Π(=0-2), requiring only four bits per carbon to designate in the computer. Any structure is then easily described as a $z\Pi$-list of the carbon values ordered by their skeletal numbering. In this way crotonic acid, linearly numbered (IUPAC rules), becomes 30•01•01•00 and its 2,4-dichloro-derivative is 30•11•01•10. With this description structures can be very economically and rapidly manipulated as lists in the computer. The numbering of the carbons for these lists can be either the given numbering as input by the user or that derived by maximization of its adjacency matrix binary list. The fundamental nature of this description by attachment types is substantiated by the observation that the oxidation state (x) at any carbon is given by $x=z-h$, and so the oxidation state change in any reaction is quickly calculated by $\Sigma\Delta x$ for all changing carbons.

Reactions are characterized in this system very clearly and simply. A unit reaction is defined as a unit exchange of attachments on one carbon, and can be expressed as two letters, the first being the kind of attachment bond which is made and the second that which is broken. Thus the reduction of alkyl halide to alkane is an HZ unit reaction, as is reduction of ketone to alcohol. The oxidation state change is found from $\Delta h=+1$, $\Delta z=-1$ and so $\Delta x=\Delta z-\Delta h=-2$. Some reactions must involve more than one carbon at the same time, as in alkene reduction, $H\Pi\cdot H\Pi$ ($\Sigma\Delta x=-1+(-1)=-2$), and of course in all constructions, such as alkyl-lithium addition to ketone, which is RH•RZ, with $\Sigma\Delta x=0$. There are 16 possible unit exchanges which may be written from combinations of the four kinds of attachments. This system makes possible a very clear and simple basis for characterizing and cataloguing all possible organic reactions in terms of their net structural change, i.e., exchange of attachment types at the several involved carbons. Such a system for organizing reactions is analogous to the Beilstein system for organizing structures in that all possible reactions, presently known or unknown, have a defined place in the catalog. This can certainly be a very useful basis for defining, creating and searching a compendium of organic reactions.[4]

For our purposes we focus on the construction reaction types which may forge the skeletal bonds designated by a bondset. Using these definitions we may easily derive all possible constructions of a given C-C σ-bond. We may generalize any construction reaction with up to six involved carbons as shown in Figure 2. The position of the bond formed is already located on the skeleton by the bondset; we need only find the possible functionality changes on each side characteristic of construction reaction families. The carbons are labeled α, β, γ out from the constructing bond on each side. The functionality change on each side (α, β, γ) may be separately considered, as a construction "half-reaction", such that the combination of two half-reactions will constitute a full construction.[5]   All possible half-reactions on up to three carbons on either side may be generated by unit exchanges, as summarized here with the oxidation state changes appended:

Simple: RH($\Delta x=+1$)   RZ($\Delta x=-1$)   at one carbon(α)
Addition:R$\Pi \cdot$ZΠ($\Sigma \Delta x=+1$)   R$\Pi \cdot$HΠ($\Delta x=-1$)   at two carbons(α, β)
Allylic: R$\Pi \cdot \Pi\Pi \cdot$HΠ($\Sigma \Delta x=+1$ R$\Pi \cdot \Pi\Pi \cdot$IZ($\Delta x=-1$)at three carbons(α, β, γ)

If we accept only full constructions with no overall redox change, there are three possible half-reactions of each oxidation state change and so only nine $\pm$ combinations as full constructions. The oxidative half-reactions ($\Delta x=+1$) are characteristic of nucleophiles, reductive ($\Delta x=-1$) of electrophiles.

Figure 2. Generalized Form of Construction Reactions

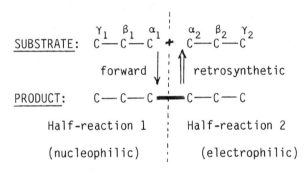

The digital representation of structure as $z\Pi$-list can now be applied to those construction reactions. Each construction demands a minimal characteristic $z\Pi$-list across the six carbons of the substrate and of the product, hence a characteristic change or $\Delta z\Pi$-list, which may be regarded as a generator to generate the product from the substrate, or vice versa. In Figure 3 are shown two examples: the Michael reaction shown from substrate to product and the Claisen rearrangement from product to substrate, both with the $z\Pi$-lists written parallel to the corresponding carbons. These $z\Pi$-lists are now treated as numbers so that a generator, Δ, may be derived such that its addition to the product $z\Pi$-list yields the substrate $z\Pi$-list (or vice-versa), i.e., PROD + Δ = SUB. In both examples the generator, Δ, is this retrosynthetic one which generates substrate by addition to product. The generators are shown as decimal numbers to illustrate the addition although in the computer the generator and $z\Pi$-lists are all binary numbers. Under each example is shown the basic unit reaction as a string of unit exchanges at each changing carbon. It may be observed that higher levels of functionality generate equally well as long as the minimum is present, i.e., addition of the generator creates the correct reaction partner. Indeed the right-hand β-carbon in the Claisen example is shown at higher than the minimum necessary functional group ($z=2\rightarrow3$ instead of $z=1\rightarrow2$).

When we apply these nine constructions systematically to the carbons on each side of designated construction bonds in various targets, we obtain proper substrates in many cases. But we also find that a number of known reactions do not turn up, while on the other hand some of the produced substrates represent constructions that are mechanistically unacceptable. Each kind of divergence from reality can be rectified after a closer examination.[5,6] The known constructions which are not produced are those which incorporate a spontaneous refunctionalization with the construction. Thus a Wittig reaction produces also a Π-bond across the σ-bond first constructed, in our terms a construction (RH·RZ) followed by an elimination ($\Pi Z \cdot \Pi Z$). An organometallic carbanion like a Grignard reagent is first created by a reductive refunctionalization (RCl + Mg $\rightarrow$ RMgCl, or HZ half-reaction) followed <u>in situ</u> by its use in construction (half-reaction RH). We examined the whole spectrum of formal refunctionalization reactions for possible two-step combinations with construction and found three types to be general or useful: prior reduction, elimination after construction, and tautomerism. Using these options we expanded the six fundamental construction half-reactions to include nine more such two-step combinations with concomitant refunctionalization, offering a set of 15 half-reactions which results in 32 possible $\pm$ full constructions, which are isohypsic (no overall oxidation or reduction). In this way all of the "known" construction reactions were produced.

With respect to the mechanistically unacceptable results, one can easily see that application of the nine constructions to the central bond of 1-butene would produce (among others) RH·RZ and generate as substrates an ethyl carbanion (for RH) and vinyl chloride (for RZ), a reaction unlikely to succeed. Whereas our generators are only designed to produce all possible combinations of net structural change, we also perceive (as in the correlation of oxidation state change with nucleophile/-electrophile) that these changes have a mechanistic basis, i.e., that simple bond/electron movements are implied. With this recognition it is possible to apply mechanistic tests or qualifications to the generation of substrates to establish the viability of a generated reaction. Furthermore, with our numerical description of structure in terms of h,σ,π,z, these tests can be rapidly made simply by evaluating these numbers.

Mechanistically, these tests fall into two categories: required activation and disallowed functional groups, the first needed for a given reaction to proceed, the second rejecting it on the grounds of intervention of a different course of reaction (as in β-elimination from a carbanion or incorrect regioselectivity for addition). In order to apply these mechanistic tests we found that the umbrella definition of z as any bond to heteroatom was too broad to convey mechanistic function. Accordingly, we added a subset qualification of z to indicate leaving group (L), electron-withdrawing (E), electron-donating (O), or, for z>1, carbonyl-type withdrawing (W). The tests of these mechanistic qualifications must now be made not only on the carbons which change functionality in the construction but also on the carbons attached to these, since their functions may modify the reaction even though they are not changed. For illustration, both the ketone and nitrile in the Michael reaction (Figure 3) do not change but are required to activate the construction, whereas a leaving group on another β-carbon of the ketone (not shown) would vitiate its success and should be rejected.

Figure 3. Examples of Reaction Generators

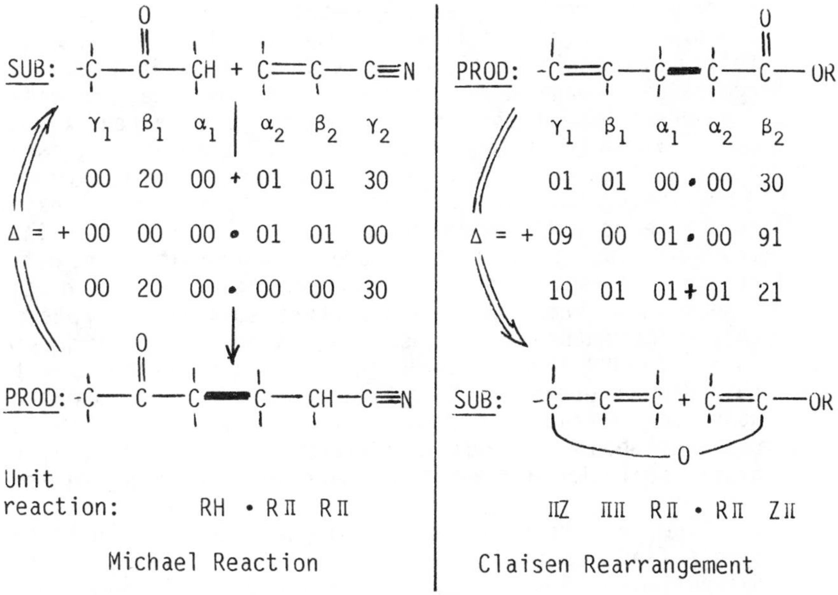

Michael Reaction   Claisen Rearrangement

For each half-reaction substrate or product we can test the viability of each of the 15 half-reactions by applying to a list of $Z\amalg$-LEOW, not only for α, β, γ but also for the relevant attached atoms, a parallel test-list, one for required activation features and one for disallowed functions.[5,6] This is a simple AND operation of these two binary lists and applies all of the required mechanistic tests on all the atoms at once and very rapidly. Furthermore, this introduction of mechanistic test-lists now makes it possible to introduce heteroatoms (N,O,S) into the skeleton itself. They are treated as if they were carbon but the test-lists also include tests of whether α, β, γ and their connected atoms are heteroatom instead of carbon, and so can require or reject reactions on that basis as well. These test-lists consist of an "R-list" (require) and "X-list" (reject) for each of the 15 half-reactions and so constitute a kind for tuning of the quantity and quality of reactions which are generated and which will appear as synthetic steps in the output. These tests are separated as modules which may be easily changed, made more or less demanding, as may be felt necessary. The less demanding they are, the more "new" reactions will appear but more output will have to be scrutinized. Made more demanding, only common, "reliable" sequences are likely to be generated.

We have incorporated the logic developed above into a program named SYNGEN.[6] The program, written in FORTRAN, was developed on a DEC 11/23 mini-computer which utilizes a megabyte of active memory. Presently SYNGEN is being converted to a micro-VAX computer in more efficient form. It generally analyzes a given target structure in under ten minutes and stores the completed results for display by a second program, SYNOUT. In order to illustrate the operation of SYNGEN, we can follow its analysis of a particular target in Figure 4. The economical Torgov-Smith synthesis of estrone proceeds by a sequence of construction reactions only, to a penultimate precursor which is shown at the top of Figure 4. It is labeled "Testrone" since it was commonly used as a test of our procedures, which must at least generate this known synthesis. The structure is entered graphically on a Tektronix terminal with thumb-wheels, as a fast, crude drawing with the heteroatom attachments shown as z-values on their attached carbons. SYNGEN then normalizes the structure so that it appears as in (A). The nature of z (as LEOW, above) is then queried for each. SYNGEN then proceeds independently of the user to seek out all convergent construction routes from no more than four starting materials available in our catalog. This catalog contains about 4000 unique $z\Pi$-list entries numerically ordered by skeletal size and their maximized matrix binary lists.

The procedure followed by the program is illustrated with sample findings in Figure 4. The first phase is the skeletal dissection of the target into two pieces all ways such that each piece is larger than three carbons. This is the first level, and shown down the left side of Figure 4 is one such first-level bondset (B) with the bonds ordered one way. Here the two pieces in each set are compared with skeletons in the catalog: in this set (C) is not found, but (D) is found and so the set is marked for priority. Precursor skeleton (C) is now cut again at second level in all ways which yield found starting skeletons, one such set shown as (E) and (F).

For each of these bondsets the functionalized target (A) is now queried for viable construction reactions, shown down the right side in Figure 4. The two sequential constructions shown which are found at first level are annotated for priority since they are perceived as being capable of proceeding in one laboratory operation, i.e., a true annelation procedure. These produce precursors (G) and (H), with skeletons (C) and (D), and (H) is searched in the catalog as a functionalized variant of found skeleton (D). Since (H) is now found as a true starting material, the priority for this path is maintained. At second level the intermediate (G) is further queried for construction (3) in this bondset and the several resultant starting materials are again looked up in the catalog. Here real starting materials (J) and (K) are also found and so the whole route is stored for SYNOUT display, including annotation of the particular half-reaction pairs which generated these intermediates in the three successive construction reactions. The three actual starting materials found in the catalog are shown below in conventional notation as (H),(J) and (K).

At the bottom of Figure 4 is a summary map of the successive $z\pi$-list changes undergone by the six skeletal carbons (8-14) in the retrosynthetic direction. These changes are the result of adding the generators for these successive constructions, which were found viable by the mechanistic qualification tests. They end in implicit cleavage of the marked bondset bonds and so the three separate starting materials (H,J and K) with their generated functional groups at the six changing carbons.

The SYNOUT program now displays all the successful findings as starting materials, first-level intermediates, and the reactions interconverting them. The mechanical nature of the generation procedure often produces minor variants of many reactions, such as both substrates for allylic reactions or β-halo-ketone displacement as well as conjugate addition. Such "chemical equivalents" are sorted out from their "primary" reactions to be looked at separately. For testrone SYNGEN found the route in Figure 4 and eleven other primary true annelations at first level, from three successful bondsets. All of the output generated in SYNGEN may be examined on the screen in a variety of displays allowing deletion of unwanted starting materials, intermediates or reactions, and the best final elected routes drawn out on a plotter. Not only is the known synthesis of testrone found, but also a number of other routes equally short.

The procedure outlined here is retrosynthetic and based on convergent bondsets to assemble the skeleton, followed by generation of functionality to create sequences of construction reactions only from found starting materials. These represent stringent criteria and the protocol used must find all possible routes which fit these criteria. In the event that none are found, or practical difficulties exist with the reactions generated, another option is available, i.e., a forward search which allows a limited number of refunctionalizing reactions to intervene. In this option the best bondsets are first assembled as at present. Then <u>all</u> starting materials available with

Figure 4. Analytical Steps in Generating a Synthesis

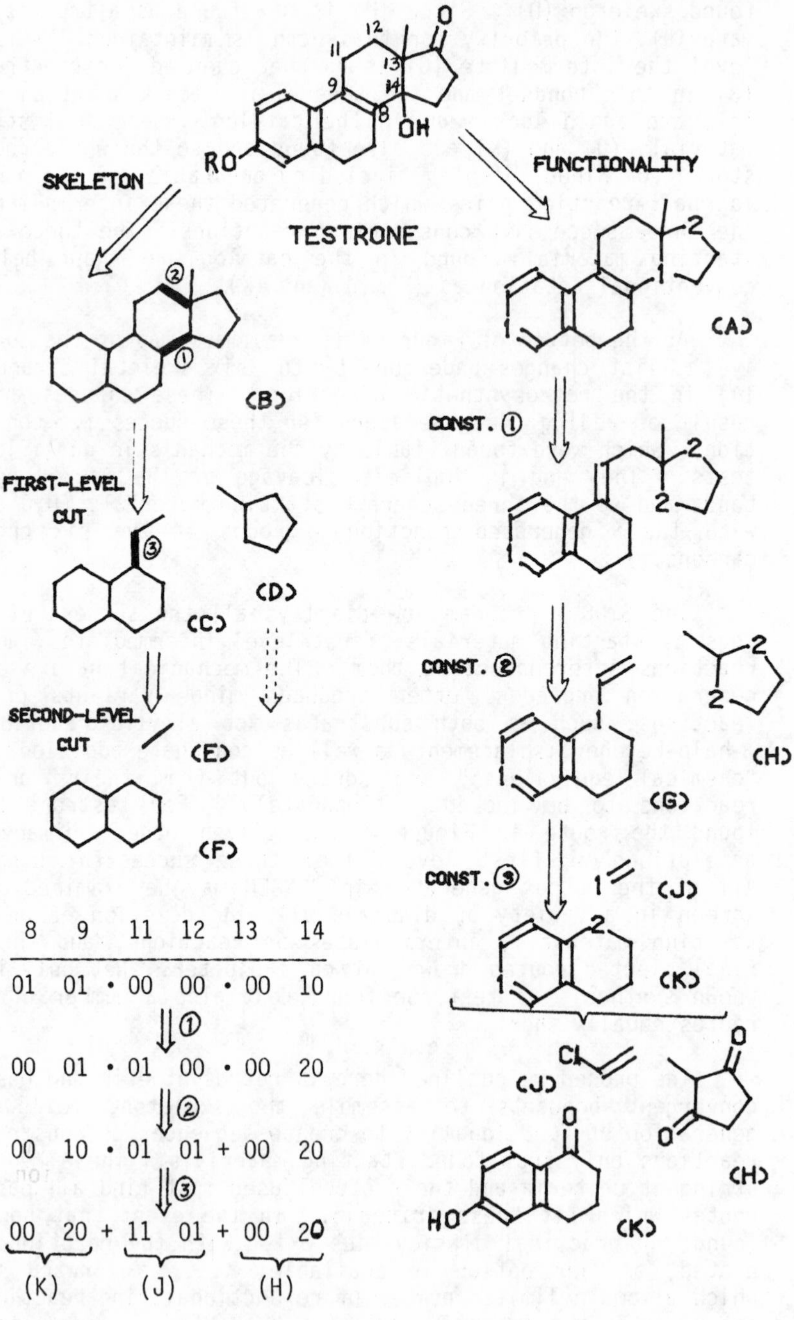

SKELETON

TESTRONE

FUNCTIONALITY

(A)

(B)

CONST. ①

FIRST-LEVEL CUT

(D)

(C)

CONST. ②

(H)

SECOND-LEVEL CUT

(E)

(G)

(F)

CONST. ③

(J)

(K)

8	9	11	12	13	14
01	01 · 00	00 · 00			10
00	01 · 01	00 · 00			20
00	10 · 01	01 + 00			20
00	20 + 11	01 + 00			20

(K)   (J)   (H)

(J)

(H)

(K)

the skeletons so derived are allowed to react together pairwise in all viable constructions of the bonds designated. The same functionality generation, via the 15 half-reactions, is used here but in the forward direction, with the corresponding R-list and X-list qualification tests. This will generate all possible functionalized variants on the skeletons of the bond-set-designated intermediates, and hence all the functionalized variants of the target skeleton which can arise from combinations of actual starting materials. These then must undergo refunctionalization reactions either to repair the intermediates or the final target variant to produce the true target functionality.

In practice this is usually an enormous combinatoric task owing to the variety of functional groups as starting materials on most small skeletons. It may be reduced to a practical range, however, by virtue of the fact that the system for designating structure by h, σ, π, z allows a simple calculation of "chemical distance," i.e., the number of unit reactions which are required to convert one structure to another.[7] This chemical distance, or number of steps, is given by $N = 1/2\Sigma_i (|\Delta h_i| + |\Delta z_i|)$. Using this formula we can compute the chemical distance of the carbons in each starting material from the same carbons in the target and so eliminate many from consideration at the outset. The same consideration can be applied to the intermediates, removing those which are intrinsically too many refunctionalization steps away from the target. This process in the forward direction then produces complementary syntheses which are a few steps longer than the retrosynthetically derived routes based on constructions only. The power of this approach is to assemble all routes through any bondset from all of its starting materials.

Acknowledgment. The authors gratefully acknowledge the computer expertise and enthusiasm of their coworkers David L. Grier, Elaine Braun-Keller and Zmira Bernstein, as well as financial support provided by the National Science Foundation.

REFERENCES

1. Hendrickson, J. B., J. Am. Chem. Soc., 99, 5439 (1977).
2. Hendrickson, J. B. and Toczko, A. G., J. Chem. Inf. Comp. Sci., 23, 171 (1983); 24, 195 (1984).
3. Hendrickson, J. B., J. Am. Chem. Soc., 93, 6847 (1971).
4. Hendrickson, J. B., J. Chem. Inf. Comp. Sci., 3, 129 (1979).
5. Hendrickson, J. B., Braun-Keller, E., Toczko, A. G., Tetrahedron (Suppl. 1), 37, 359 (1981).
6. Hendrickson, J. B., Grier, D. L., Toczko, A. G., J. Am. Chem. Soc., 107, 0000 (1985).
7. Hendrickson, J. B., Braun-Keller, E., J. Chem. Comp., 1, 323 (1980).

Chapter 10

VALENCE BOND STRUCTURE-RESONANCE THEORY FOR BORANES. PYROLYTIC INTER-MEDIATES AND REACTIONS

W.C. Herndon, M.L. Ellzey, Jr., R.L. Armstrong, and I.S. Millett
Department of Chemistry, The University of Texas at El Paso, El Paso, Texas USA, 79968

ABSTRACT

Structure-resonance theory is used to estimate the relative stabilities of potential intermediates in the pyrolysis of diborane to yield the higher boron hydrides, and the results are compared with those obtained from recent molecular orbital calculations. In most cases where alternative isomeric structures are possible, the MO and this empirical VB method lead to qualitative agreement regarding the structure of the most stable isomer. Possible mechanisms of pyrolytic borane reactions are discussed. Some limitations of a VB resonance theory for boranes are delineated.

INTRODUCTION

The structure-resonance description of bonding in the boron hydrides, first proposed by Pauling [1,2], has recently been reconsidered [3,4], and a valence bond structure-resonance theory (VBSRT) for boranes based solely on two-electron two-center bond structures has been shown to provide realistic first-order descriptions of known borane structures. Graph theoretical algorithms and computer programs for counting neutral and ionic two-center bond structures were described. Experimental thermodynamic data were used to parameterize a four-term ΔH(atomization) equation in which one important factor was a VBSRT algorithmic estimate of resonance energy [5-7].

In this paper we use the ΔH(a) scheme to examine several possible mechanistic steps and intermediates that have been adduced to account for the formation of higher boron hydrides in the pyrolytic polymerization of diborane. The reaction was discovered by Stock [8]; among many significant subsequent thermodynamic and/or kinetic studies were those carried out by Gunn [9,10], Bauer [11,12], Schaeffer [13-15], and their respective coworkers. The kinetics of particular important elementary steps have been investigated by Fehlner [16,17] and more recently by Greenwood, et al. [18,19]. General mechanistic schemes have been advanced by Long [20] and Schaeffer [21]. The structures and heats of formation of many of the postulated intermediates have been calculated as part of the extensive theoretical research on boranes carried out by Lipscomb and his coworkers [22-25].

PARAMETERIZATION

We assume that the heat of atomization of a gaseous boron hydride can be assigned to four structure dependent features: terminal BH bonds H(t), bridging BH bonds H(b), coordinating BB atom pairs, and the resonance energy RE [3,4]. The RE's are calculated by counting the neutral two-center bond covalent structures and using the algorithm for resonance energies C×ln(SC), where C is a determined constant and SC is the structure count [3,7]. The $\Delta H°$(f) data for six of the boranes seem to be well established [26,27], and these data are used in a multiple linear regression procedure to determine the numerical values of parameters. The previous work [3,4] made use of an approximate value for the heat of sublimation of boron; a more precise and accurate value, +134.4 kcal/mole, is now available [26]. Additional required data are $\Delta H°$(f) of hydrogen atoms = +52.1 kcal/mole, and $\Delta H°$(f) of carbon atoms (from graphite) = +171.3 kcal/mole.

Results of the analysis and the supporting data are given in Table 1; the derived linear equation is

$$\Delta H(a) = 90.53 \times H(t) + 66.82 \times H(B) \qquad (1)$$
$$+ 13.89 \times BB + 91.84 \times \ln(SC).$$

The multiple correlation coefficient for eq(1) is unity (5 significant figures), and the standard error of a calculated ΔH(a) value is 2.2 kcal. The predictive power of eq.(1) can be corroborated by comparing the calculated ΔH(a) of borane, 271.6 kcal/mole, with

Table 1.
Heats of Atomization (Kcal) and Structural Factors for Boron Hydrides

Compound	H(t)	H(b)	BB	SC	ΔH(a)expt.	ΔH(a)calc.
B_2H_6	4	2	1	2	573.0	573.3
B_4H_{10}	6	4	5	6	1043.1	1044.5
B_5H_9	5	4	8	24	1123.8	1122.9
B_5H_{11}	7	4	7	11	1220.8	1218.4
B_6H_{10}	6	4	10	42	1305.3	1306.5
$B_{10}H_{14}$	10	4	21	678	2062.9	2063.0

the value, 269.1 kcal/mole, derived from kinetic experimental data [17]. The low standard error and the fact that no single compound exhibits an exceptional deviation from its calculated ΔH(a) may indicate that the experimental data are known with good precision. In particular, the present analysis does not lend any support to the recent suggestion [28] that the ΔH°(f) of decaborane(14) should be substantially revised.

Eq(1) is used in the following section to calculate the ΔH(a) of all borane species considered except for those which have formulas that are multiples of BH_3 and which have monocyclic structures with SC=2, e.g. triborane(9) and tetraborane(12) shown in 1.

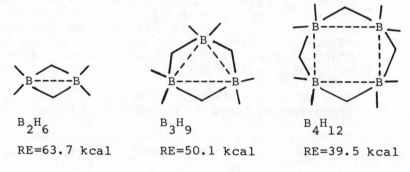

B_2H_6	B_3H_9	B_4H_{12}
RE=63.7 kcal	RE=50.1 kcal	RE=39.5 kcal

1

Previous VBSRT results for benzene and azulene [29], both also with SC=2, and several types of theoretical calculations [30-32] demonstrate that resonance integrals (and RE's) decrease in size as the number of electrons involved in a cyclic permutation increases. This factor requires that RE's of the two larger compounds in 1 be obtained as a calculated fraction of the RE of diborane. The values given were calculated following the method of Coulson and Dixon [33], using bond lengths taken from McKee and Lipscomb [24], and using the potential functions recommended by Housecraft and Wade [34]. Eq(1) can then be used to calculate the $\Delta H(a)$ after replacing the last term with the modified RE's.

Only three compounds of those to be discussed have full covalent BB single bonds, viz one each of isomers of diborane(4), triborane(7) and tetraborane(8). We estimate the energy of such a bond as 51.9 kcal, this estimate also being based on the Housecraft and Wade potential functions and a calculated bond length [24] of 1.689 Å for the triborane(7). The theoretical MO bond index procedure of Laurie and Perkins [28] would yield a BB bond energy of 71.2 kcal. A single thermochemical measurement [10] carried out for decaborane(16), which contains pentaborane(8) moities connected by a full BB single bond, is the only available piece of relevant experimental information. However, its non-inclusion in a recent compilation of thermodynamic data [26] suggests that caution be exercised in using its value for calibration. The average of the two theoretical results, 61.6 kcals, is therefore tentatively taken as the energy value for a single bond of this type.

RESULTS AND DISCUSSION

The structures and calculated $\Delta H(a)$'s of the new species considered in this work are listed in 2, where all energies are given in kcal. The structures for compounds not depicted are given in earlier papers [3,4]. Alternative structures of several of the smaller possible transient species have been investigated by MO techniques using extended basis sets including polarization and correlation correction terms (Lipscomb, et al. [23-25]). These MO relative calculated energies are shown in 2 labelled $\Delta E(MO)$. Some relative MNDO energies from calculations carried out by Dewar and McKee are also available [35,36] and are referred to in context. In 2, where only a single isomer corresponding to a formula is depicted, the structure given is the calculated most stable structure among several viable possibilities.

The relative calculated energies of the isomers of diborane(4) and triborane(7) are found in reasonable

	B_2H_4	B_2H_4	B_3H_9	$3*BH_3$
	SC=1	SC=2	SC=2	SC=1
$\Delta H(a)$	434.2	392.2	835.4	814.8
$\Delta E(MO)$	0.0	25.2	0.0	26.3

	B_3H_7	B_3H_7	B_3H_7
	SC=3	SC=2	SC=2
$\Delta H(a)$	728.9	725.4	715.3
$\Delta E(MO)$	0.0	7.4	4.4

	B_4H_8	B_4H_8	B_4H_8	B_4H_8
	SC=12	SC=7	SC=2	SC=3
$\Delta H(a)$	927.1	901.3	867.5	847.2
$\Delta E(MO)$	5.7	0.0	?	4.2

	B_6H_{12}	B_7H_{11}	B_7H_{13}
	SC=23	SC=156	SC=106
$\Delta H(a)$	1418.4	1484.0	1629.6

	B_8H_{12}	B_8H_{14}
	SC=129	SC=75
$\Delta H(a)$	1646.2	1777.5

2

agreement; both MO and VBSRT theories give nonbridged and bridged most stable structures, respectively. MNDO gives the structure with a single bridged hydrogen as the most stable triborane(7) [36]. The relative VBSRT and MO stabilities of borane and triborane(9) are also in satisfactory agreement. However, the tetraborane(8) results are not in consonance the VBSRT method finds that a tetrabridged structure is 25.8 kcal lower in energy than the MO lowest energy tribridged structure. Furthermore, MNDO predicts a monobridged tetraborane(8) to be 13.0 kcal lower in energy than the tribridged structure. These differences appear irreconcilable at present, and it may be that these results are indicative of deficiencies in VBSRT. Nevertheless, the use of the VBSRT structure for tetraborane(8) seems to give the most sensible calculated endothermicity for the reaction of tetraborane(10) to give hydrogen and tetraborane(8), as will be discussed below. Also, since the VBSRT approach is parameterized with experimental data, we believe it is unlikely that it will give extremely large errors, as would be required if the other calculations are to be accepted without any further examination.

An attractive but very simplistic mechanism for the build-up of higher boranes in the borane pyrolysis involves sequences of addition of borane(3) and loss of hydrogen from intermediates and sometimes isolable species [17,20,21,37]. A thermochemical outline of this mechanism through nonaborane(15) is given in 3. One can see from the diagram that shortcomings of this reaction scheme cannot be ascribed to factors involving the thermodynamic feasibility of the reaction steps. Initial reactions are not prohibitively endothermic, and many of the subsequent steps are calculated to be exothermic. In particular, the known compounds tetraborane(10), pentaborane(11), and pentaborane(9) are formed in exothermic reactions. Only the known hexaborane(10) is predicted to be produced by an endothermic process, and in this case the endothermicity is less than 8 kcal.

The Lipscomb MO [23-25] and the VBSRT calculations are not in good agreement for the important sequence from diborane(6) to triborane(7); MO theory gives the first step endothermic by +4.1 kcal, but the second is found to be exothermic by -5.4 kcal. The available MP3/6-31G total energies for borane(3), triborane(7), and tetraborane(10) [24,38] also allow one to calculate a reaction enthalpy for the subsequent formation of tetraborane(10), which is found to be +176.1 kcal. This result would infer substan-

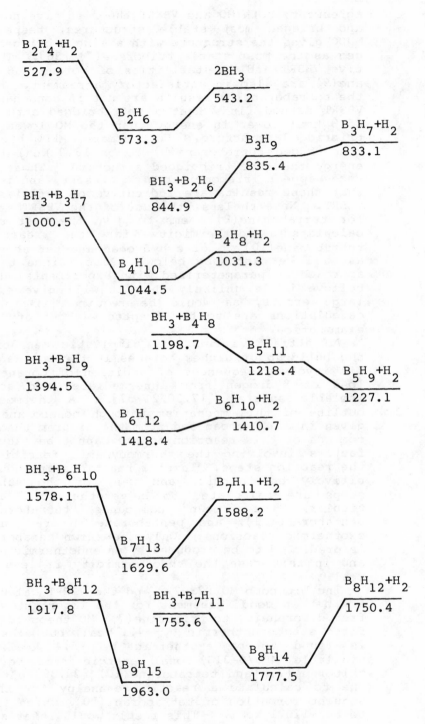

3

tial thermodynamic stability for triborane(7), which does not seem to be reasonable considering its known reactivity [39,40]. All of the VBSRT results in 3, however, can be easily rationalized except perhaps for the fact that boranes with 7 to 9 boron atoms are not usually isolated as products in the normal diborane pyrolysis [20,21,37]. At the least, the calculated VBSRT $\Delta H(a)$'s indicate that no thermodynamic barriers exist that would bar the formation of the higher boranes.

The major deficiency of scheme 3 is the neglect of possible bimolecular reactions that could take place between the labile and stable intermediates. In fact, many reactions of this type have been studied [21,37], and they provide a synthetic source for many of the boranes with more than 6 carbon atoms. The source of decaborane(14) produced in the pyrolysis of diborane is most likely a reaction of this type. In other examples, decaborane is formed in good yield by co-pyrolysis of diborane with pentaborane(9) [41] and of hexaborane(10) with tetraborane(10).

Table 2 contains a list of several bimolecular reactions that could mediate the borane pyrolysis and their calculated VBSRT enthalpies. In each reaction involving either of the labile compounds triborane(7) or tetraborane(8), the intermediate would be presumed to be formed as outlined in 3. Most of the reactions in the table have been previously postulated, and in some cases, kinetic and product studies support the postulates as chemically realistic [8-21,37,39-41]. In addition, an MO calculation for the heat of the reaction of tetraborane(8) with pentaborane(11) gives a value of -30 kcal [25], which is in very good agreement with the VBSRT value.

On the basis of the VBSRT calculations, all the listed reactions seem thermochemically reasonable except for the fragmentation of nonaborane(15) to give octaborane(12) and borane, $\Delta H = +45.2$ kcal. An independent value for the enthalpy of this reaction can be obtained since Housecraft and Wade have calculated both higher borane heats of atomization [34] based on bond lengths in the crystal structures [42,43]. The resulting heat of reaction is +61.9 kcal [44], an even higher value than that from VBSRT. Since one of the best syntheses of octaborane(12) corresponds precisely to this fragmentation carried out at low temperatures [45], the calculations may indicate that a reinvestigation is warranted, and it is possible that the synthesis involves some unidentified reaction or catalytic agent.

Schaeffer tentatively suggested [21,40] that formation of decarborane(14) could involve condensation

Table 2.
Heats of Borane Pyrolytic Reactions

Reaction	Enthalpy (kcal)
(1) $B_2H_6 + BH_3 = B_3H_7 + H_2$	+12.8
(2) $B_3H_7 + B_2H_6 = B_5H_{11} + B_2H_6$	-37.6
(3) $B_3H_7 + B_4H_{10} = B_5H_{11} + B_2H_6$	-40.0
(4) $B_3H_7 + B_5H_{11} = B_6H_{12} + B_2H_6$	-68.1
(5) $B_3H_7 + B_5H_9 = B_6H_{10} + B_2H_6$	-51.7
(6) $B_3H_7 + B_6H_{10} = B_9H_{15} + H_2$	-55.5
(7) $B_4H_{10} = B_4H_8 + H_2$	+13.3
(8) $B_4H_8 + B_2H_6 = B_5H_{11} + BH_3$	+10.4
(9) $B_4H_8 + B_5H_{11} = B_5H_9 + B_4H_{10}$	-21.9
(10) $B_4H_8 + B_6H_{10} = B_9H_{15} + BH_3$	+ 1.0
(11) $B_4H_8 + B_6H_{12} = B_5H_9 + B_5B_{11}$	+ 4.2
(12) $B_9H_{15} = B_8H_{12} + BH_3$	+45.2
(13) $B_4H_8 + B_8H_{12} = B_{10}H_{14} + B_2H_6$	-63.0
(14) $B_9H_{15} + BH_3 = B_{10}H_{14} + 2H_2$	-36.8

of tetraborane(8) with hexaborane(10), and this possibility is certainly supported by the large calculated VBSRT negative heat of reaction, which is -35.8 kcal for a combination of reactions (10) and (14) in table 2. However, a reaction scheme for formation of decaborane later proposed by Gibb, et al. [18], also involving the condensation between tetraborane(8) and hexaborane(10) is not supported. The steps in the Gibb mechanism are (7), (10), (12), and (13) from table 2 taken in turn with an overall ΔH(reaction) = -3.5 kcal. Nevertheless this reaction sequence also includes the large positive enthalpy reaction to produce octaborane(12) discussed above, and for that reason, this particular proposed mechanism would seem unlikely. Again, the possibility that octaborane(12) is produced by an, as yet, unidentified process should be considered.

A combination of the initial reactions depicted in scheme 3 (through pentaborane(9)) with the exothermic or thermoneutral reactions summarized in table 2, accounts for the gross overall results of borane pyrolysis. The steps involved in the reaction or reactions that yield octaborane(12) are a remaining problem, since the VBSRT calculations infer that a thermal unimolecular reaction of nonaborane(15) would be highly endothermic. Any existing activation barrier would further add to the difficulty of such a reaction. After octaborane(12) is formed, subsequent reaction to give decaborane(14) would seem to be insured by the large negative enthalpies associated with decaborane-forming reactions.

REMARKS ON THE USE OF VBSRT

VBSRT is essentially an empirical quantification of the resonance theory approach used so successfully in teaching elementary chemistry. We have shown in this work that VBSRT provides a good first-order description of thermal borane chemistry. It must be noted that the successful use of the theory for boranes was critically dependent upon the availability of experimental thermodynamic data and the results from the high-level MO calculations of Lipscomb and his coworkers[22-25]. An empirical theory of this type must be parameterized, and the accuracy of the parametric data governs the suitability of extending the theory to new chemistry. Of course once verified, the VBSRT procedures can be extended to additional, larger molecules with little expense and effort.

The boranes, with their sigma delocalized structures, constitute a difficult test for the effectiveness of a valence-bond approach. The results outlined in this paper indicate that a successful VBSRT correlation of thermochemical borane data and reactions is possible. The transition states involved in thermal borane reactions [25] will constitute an even more stringent test, but the ability to treat problems of this type by essentially additive methods would allow a more detailed consideration of borane chemistry. Attempts to address the calculations of borane transition state energies using VBSRT are in progress.

ACKNOWLEDGEMENT

Financial support of the Robert A. Welch Foundation of Houston, Texas, is gratefully acknowledged.

REFERENCES

[1]   L. Pauling, "The Nature of the Chemical Bond", Third Ed.; Cornell University Press: Ithaca, N.Y., 1960, pp. 367-379.

[2]   L. Pauling, J. Inorg. Nucl. Chem. 32, 3745 (1970).

[3]   W.C. Herndon and M.L. Ellzey, Jr., Inorg. Nucl. Chem. Letts. 16, 361(1980).

[4]   W.C. Herndon, M.L. Ellzey, Jr., and M.W. Lee, Pure Appl. Chem. 54, 1143 (1982).

[5]   R. Swinbourne-Sheldrake, W.C. Herndon, and I. Gutman, Tetrahedron Letts., 755 (1975).

[6]   S.E. Stein, D.M. Golden, and S.W. Benson, J. Phys. Chem. 81, 314(1977).

[7]   J.-I. Aihara, J. Am. Chem. Soc. 100, 3339(1978). Also see Bull. Chem. Soc. Jap. 52, 2202(1979).

[8]   A. Stock, "Hydrides of Boron and Silicon"; Cornell University Press: Ithaca, N.Y., 1933.

[9]   S.R. Gunn and L.G. Green, J. Phys. Chem. 65, 779(1961); Ibid. 65, 2173(1961); J. Chem. Phys. 36, 1181(1962).

[10] S.R. Gunn and J.K. Kindsvater, J. Phys. Chem. 70, 1114(1966).

[11] S.H. Bauer, J. Am. Chem. Soc. 78, 5775(1956); Ibid. 80, 294(1958).

[12] K.-R. Chien and S.H. Bauer, Inorg. Chem. 16, 867(1977).

[13] J.A. Dupont and R. Schaeffer, J. Inorg. Nucl. Chem. 15, 310(1960); R.E. Enrione and R. Schaeffer, Ibid. 18, 103(1961); G.L. Brenan and R. Schaeffer, Ibid. 20, 205(1961).

[14] J. Rathke and R. Schaeffer, J. Am. Chem. Soc. 95, 3402(1973).

[15] J. Rathke, D.C. Moody, and R. Schaeffer, Inorg. Chem. 13, 3046(1974).

[16] T.P. Fehlner, J. Am. Chem. Soc. 87, 4200(1965).

[17] T.P. Fehlner in "Boron Hydride Chemistry"; E.L. Muetterties, Ed.; Academic Press: New York, 1975, Chpt. 4.

[18] T.C. Gibb, N.N. Greenwood, T.R. Spalding, and D. Taylorson, J. Chem. Soc. Dalton Trans., 1392, 1398(1979).

[19] R. Greatex, N.N. Greenwood, and G.A. Jump, J. Chem. Soc. Dalton Trans., 541(1985).

[20] L.H. Long, J. Inorg. Nucl. Chem. 32, 1097(1970).

[21] R. Schaeffer, XXIVth Internat. Cong. Pure and Appl. Chem. 4, 1(1970).

[22] See the Nobel Prize lecture (1976) by W.N. Lipscomb, Science 196, 1047(1977) for a summary and references 23-25 for recent calculations.

[23] M.L. McKee and W.N. Lipscomb, J. Am. Chem. Soc. 103, 4673(1981).

[24] M.L. McKee and W.N. Lipscomb, Inorg. Chem. 21, 2846(1982).

[25] W.N. Lipscomb, Pure Appl. Chem. 55, 1431(1983).

[26] J. of Physical and Chemical Reference Data 11, Suppl. 2, (1982).

[27] JANAF Thermochemical Tables, National Bureau of Standards (USA), Natl. Stand, Ref. Data Ser. No. 37 (1971).

[28] D. Laurie and P.G. Perkins, Inorgan. Chim. Acta 63, 53 (1982).

[29] W.C. Herndon and M.L. Ellzey, Jr., J. Am. Chem. Soc. 96, 6631(1974).

[30] L. Pauling, J. Chem. Phys. 1, 280(1933).

[31] B.A. Hess, R., and L.J. Schaad, J. Am. Chem. Soc. 93, 305(1971); Ibid. 93, 2413(1971); J. Org. Chem. 37, 4179(1972).

[32] J. Aihara, J. Am. Chem. Soc. 98, 2750(1976).

[33] C.A. Coulson and W.T. Dixon, Tetrahedron 17, 215(1962).

[34] C.E. Housecraft and K. Wade, Inorg. Nucl. Chem. Letts. 15, 339(1979).

[35] M.J.S. Dewar and M.L. McKee, J. Am. Chem. Soc. 99, 5231(1977).

[36] M.J.S. Dewar and M.L. McKee, Inorg. Chem. 17, 1569(1978).

[37] S.G. Shore in "Boron Hydride Chemistry"; E.L. Muetterties, Ed.; Academic Press: New York 1975, Chpt. 3.

[38] M.L. McKee and W.N. Lipscomb, Inorg. Chem. 20, 4453(1981).

[39] R.T. Paine, G. Sodek, and F.E. Stafford, Inorg. Chem. 11, 2593(1972).

[40] J. Rathke and R. Schaeffer, Inorg. Chem. 13, 3008(1974).

[41] M. Hillman , D.J. Mangold, and J.H. Norman, J. Inorg. Nucl. Chem. 24, 1565(1962).

[42] R.E. Enrione, F.P. Boer, and W.N. Lipscomb, J. Am. Chem. Soc. 86 1451(1964).

[43] J.H. Hall, Jr., D.S. Marynick, and W.N. Lipscomb J. Am. Chem. Soc. 96, 770(1974).

[44] See ref. [4] for recalculated values of the heats of atomization.

[45] J. Dobson, P.C. Keller, and R. Schaeffer, Inorg. Chem. 7, 399(1968); J. Dobson and R. Scheaffer, Ibid. 7, 402(1968).

Chapter 11

TOPOLOGICAL INDEX AS A COMMON TOOL FOR QUANTUM CHEMISTRY, STATISTICAL MECHANICS, AND GRAPH THEORY

Haruo Hosoya
Department of Chemistry, Ochanomizu University, Tokyo, Japan

INTRODUCTION

The Z-index was originally proposed in 1971 by the present author under the name of "Topological Index" for characterizing the topological nature of the carbon atom skeleton of saturated hydrocarbons [1]. It was later found to be applicable to many different problems, not only in chemistry, but also in mathematics, informatics, and physics, e.g., coding and identification of graphs, structure-activity relationship, analysis of π-electronic structure of unsaturated hydrocarbon molecules, dimer statistics, etc. As nowadays a number of topological indices have been proposed and good review articles are available [2-6], here various aspects of the Z-index will be described with particular reference to its mathematical properties which relate various concepts in different fields of science, i.e., quantum chemistry, statistical mechanics, and graph theory, with each other. The advantage of the Z-index over other topological indices comes from the fact that it is defined through the counting polynomial $Q(x)$, which is closely related to the characteristic polynomial $P(x)$. Thus the idea of Z-index can be extended to several counting polynomials, such as the matching polynomial $\alpha(x)$, sextet polynomial $B(x)$, distance polynomial $S(x)$, etc., for a wide variety of problems.

NON-ADJACENT NUMBER AND Z-INDEX [1]

In the application of the graph theory to chemistry, a graph may represent a molecular skeleton, crystal lattice, or reaction network. For the graph-theoretical terms adopted in this paper without any definition, consult the standard text books and monographs [6-9]. We will be mainly concerned with connected nondirected simple graphs. Define a non-adjacent number, $p(G,k)$, as the number of ways for choosing k disconnected lines from graph G, with $p(G,0)$ being taken as unity. The Z-counting poynomial $Q_G(x)$ is defined as

$$Q_G(x) = \sum_{k=0}^{m} p(G,k) \, x^k \tag{1}$$

where m is the maximum number of k. For G with an even number of points, N=2m, let us denote $p(G,m)$ as $K(G)$ and call it the perfect mathcing number, or Kekule number irrespective of the fact that G is derived from a conjugated unsaturated hydrocarbon or not. The Z-index is the sum of the $p(G,k)$ numbers,

or $$Z_G = \sum_{k=0}^{m} p(G,k) = Q_G(1). \tag{2}$$

The set of $p(G,k)$ numbers and Z values have been tabulated extensively [10-16]. The polynomials $Q(x)$ for various series of graphs are shown to be transformed into a family of orthogonal polynomials, such as the first and second kinds of Chebyshev, Hermite, Laguerre, and associated Laguerre polynomials [17-19]. As will be shown later the Z values for several series of typical graphs form widely known integer series. For example, the Z values of the path graphs, $\{S_n\}$, and cycle graphs, $\{C_n\}$, respectively, form Fibonacci and Lucas series.

These coincidences are the outcomes of the simple recurrence relations existing among the above-defined quantities. Namely, the inclusion-exclusion principle leads to the following recurrence relation for a given G and an arbitrarily chosen line ℓ:

$$p(G,k) = p(G\text{-}\ell,k) + p(G\ominus\ell,k\text{-}1), \tag{3}$$
$$\ell\text{-exclusive} \quad \ell\text{-inclusive}$$

where G-ℓ and G$\ominus\ell$, respectively, denote the subgraphs of G obtained by deleting ℓ(leaving its terminal points) and all the lines incident to the two points that define ℓ(See Fig.1).

Figure 1. Subgraphs G-ℓ and G$\ominus\ell$ derived from G

It is straightforward to get the recurrence relations for the other qauntities, such as

$$Q_G(x) = Q_{G-\ell}(x) + x\, Q_{G\ominus\ell}(x) \qquad (4)$$

and

$$Z_G = Z_{G-\ell} + Z_{G\ominus\ell}. \qquad (5)$$

Note the factor x in the second term of Eq. (4). Several recurrence relations are also found including the "jumbo" one, which is useful for treating highly symmetrical or highly branched large networks [15,20].

RELATION BETWEEN $Q_G(x)$ and $P_G(x)$

Although there is no one-to-one correspondence between the set of p(G,k)'s (or Z) with graph G, one can easily and roughly differentiate among the isomeric graphs with these quantities. Good examples are shown in Table 1, where the Z values of the nine heptane isomer graphs form a set of integers descending stepwise from 21 to 13, just as in the same order of the boiling point of the corresponding hydrocarbons with only one minor exception [1,21]. Several important issues come out from Table 1, e.g., on i) relation between Q(x) and P(x), the characteristic polynomial, ii) topological dependency of the coefficients of these polynomials, iii) identification and discrimination of graphs, iv) QSAR analysis, etc. These problems will be briefly explained in the following sections.

The p(G,k) numbers appearing in Table 1 are nothing else but the coefficients of the characteristic polynomial. Namely for a tree graph G the following relation is shown to be valid:

$$P_G(x) = \sum_{k=0}^{m} (-1)^k\, p(G,k)\, x^{N-2k}, \quad (G\!:\!\text{tree}) \qquad (6)$$

where

$$P_G(x) = (-1)^N \det(A-xE) \qquad (\text{for all } G) \qquad (7)$$

Table 1.
Relation between the branching and several topological quantities of heptane isomers

Graph	$p(G,k)$ $k=0$	1	2	3	Z_G	$bp(°C)$
	1	6	10	4	21	98.4
	1	6	9	4	20	93.4
	1	6	9	3	19	91.9
	1	6	9	2	18	90.0
	1	6	8	2	17	89.7
	1	6	7	2	16	86.0
	1	6	8	0	15	80.5
	1	6	7	0	14	79.2
	1	6	6	0	13	80.9

with N being the number of points in G, det the determinant, and A the adjacency matrix. Also for a non-tree graph we have the following closed form:

$$P_G(x) = \sum_{k=0}^{m} (-1)^k \, p(G,k) \, x^{N-2k}$$

$$+ \sum_{i}^{R_i \subseteq G} (-2)^{r_i} \left\{ \sum_{k=0}^{m} (-1)^k \, p(G \ominus R_i, k) \, x^{N-n_i-2k} \right\}, \quad (8)$$

where R_i denotes a ring or a set of disjoint rings composed of n_i points [22,23]. In principle this expression is essentially the same as Sach's theorem [6,24,26]. However, Eq. (8) is superior to Sach's theorem in its explicit representation of the effect of the component rings to the counting polynomials $Q(x)$ and $P(x)$ and in its potentiality to yield useful recurrence formulas.

It is to be noted here that the definition of the matching polynomial $\alpha(x)$ [17,27-29] is the same as $P(x)$ in Eq. (6) but for any graph. Thus the conclusions derived from $\alpha(x)$ are automatically related to $Q(x)$ through the relation [30],

$$\alpha_G(x) = x^N Q_G(-1/x^2).$$ (9)

One can quite easily analyze the origin of the mathematical interpretation of the so-called topological resonance energy [25,27,28,30], and also of the topological bond order [31,32] for bonds in the π-electronic network of unsaturated hydrocarbons.

TOPOLOGICAL ANALYSIS OF π-ELECTRONIC STRUCTURES OF UNSATURATED HYDROCARBONS

Consider the π-electronic structure of unsaturated hydrocarbons. The secular determinant of the Hückel molecular orbitals is known to be identical to the characteristic polynomial of the graph G for the carbon atom skeleton. Total π-electronic energy, charge density, bond-order, and other π-electronic properties are obtained from the solutions of $P_G(x)=$ 0. Then Eq. (8) means that the magnitudes of all these quantities can rigorously be explained in terms of the set of the p(G,k) numbers. The topological index Z also reflects indirectly these π-electronic properties. Especially for tree graphs, where the direct one-to-one correspondence between P(x) and Q(x) holds as in Eqs. (1) and)6), one can quite easily estimate the stabilitiy of the π-electronic structures from the following relation with a proper set of parameters and b [30]

$$E = a \log Z + b \quad (tree).$$ (10)

Even for some selected series of cyclic compounds Eq. (10) can be applied. The extra stability or unstability which cannot be accounted for by Z comes from the second summation in Eq. (8), which is nothing else but what is implied by the topological resoance energy. The Z-index can then be modified into $Z' = Z + \Delta Z$ by adding the following terms representing the contributions from the component rings and also from the set of disjoint rings.

$$\Delta Z = 2 \sum_{R(r=1)}^{n=4k+2} Z_{G\ominus R}$$

$$- 2 \sum_{R(r=1)}^{n=4k} Z_{G\ominus R}$$

$$+ 4 \sum_{R(r=2)}^{n=4k} Z_{G\ominus R}$$

$$- 4 \sum_{R(r=2)}^{n=4k+2} Z_{G \ominus R} \quad \frac{\triangle \quad \square \quad \pentagon \quad \triangle}{\triangle \, , \, \hexagon \, , \, \pentagon \, , \, \bigcirc \, ,} \cdots$$

$$+ 8 \sum_{R(r=3)}^{n=4k+2} Z_{G \ominus R} \quad \frac{\triangle\triangle \quad \square\square \quad \pentagon\pentagon}{\square \, , \, \hexagon \, , \, \square \, ,} \cdots$$

$$- \cdots \tag{11}$$

Positive and negative terms, respectively, contribute the stability and instability of the π-electronic system. The first and second summations in Eq. (11) simply express the so-called Hückel's 4n+2 rule. Note the signs of the third and fourth summations showing that a set of two disjoint rings with a total of 4k and 4k+2 carbon atoms, respectively, stabilizes and destabilizes the π-electronic network. On the other hand, the fifth and sixth (not shown but apparent) terms show that a set of three disjoint rings with a total of 4k+2 and 4k carbon atoms, respectively, stabilizes and destabilizes the total system. These findings constitute the "extended Hückel rule". Except for a few cases the modified Z'-index is shown to be obtained by the sum of the absolute values of the coefficients of the even terms of P(x) as [30],

$$Z_G' = \sum_{k=0}^{N/2} |a_{2k}| \tag{12}$$

Encouraged by these findings the topological bond order for bond ℓ in graph G was proposed to be defined as [31]

$$p_\ell^T = Z_{G \ominus \ell}' / Z_G' \tag{13}$$

which was shown to be well correlated with the Coulson's bond order p_ℓ^C. The correlation is greatly improved by adding a small contribution of the Pauling's bond order p_ℓ^P as

$$p_\ell^C = a(p_\ell^T + b \, p_\ell^P). \quad (b=0.14:\text{tree, } 0.16:\text{non-tree}) \tag{14}$$

$$p_\ell^P = K(G \ominus \ell)/K(G). \tag{15}$$

Note here that the lengthy definition of p_ℓ^P originally proposed by Pauling [33] is not only turned into a compact form but also given a well-defined graph-theoretical meaning by using the concept and notation of the subgraph $G \ominus \ell$ introduced in Fig. 1 (Compare Eqs. (13) and (15)).

The empirical relation (14) can almost rigorously be derived [32] by combining the enumeration technique

of the graph theory and the complex integral developed by Coulson and Longuet-Higgins [34]. The esse-
nce of these findings is that if a function $F_{G,\ell}(y)$
is defined for bond ℓ in G as,

$$F_{G,\ell}(y) = \Delta_{r,s}(iy)/P_G(iy), \tag{16}$$

$$\Delta_{r,s}:\text{adjunct of } P_G(x)$$

the three different bond orders can be expressed in
terms of $F_{G,\ell}(y)$ as

$$p_\ell^C = (2/\pi)\int_0^\infty F_{G,\ell}(y)\ dy \quad \text{(alternant hydrocarbon)} \tag{17}$$

$$p_\ell^P = F_{G,\ell}(0) \tag{18}$$

$$p_\ell^T = F_{G,\ell}(1) \tag{19}$$

Topological factors causing the non-uniform π-elec-
tron charge distribution in non-alternant hydrocarbons can also be analyzed in terms of the properly
defined topological charge density [35].

Recently Aono and his coworkers have developed the
theory using the propagator technique for clarifying
the topological analysis of various π-electronic
quantities derived from the Huckel molecular orbitals. Their method leads one to the same conclusions
as those introduced above, since they also calculate
the integral of the functionals of P(x) over the so-
called "Coulson contour" [36-38].

TOPOLOGICAL DEPENDENCY OF Z-INDEX

For the QSAR analysis of the thermodynamic properties
of saturated hydrocarbons, turn to Table 1. Note
that the p(G,2) value is a function of the numbers of
tertiary (Y) and quarternary (X) carbon atoms, as

$$p(G,2) = p(\overline{N},2) - (Y) - 3(X), \tag{20}$$

where $\overline{N}$ denotes the path graph S_N, with N points, or
the carbon atom skeleton of normal hydrocarbon [1].
Similar but more complicated relations can be obtained between other p(G,k) values and the topological structure of a graph. Thus one can show that
among the isomeric tree graphs the unbranched path
graph has the largest Z value, and that the more
branches a graph has the smaller its Z values.

A number of empirical rules on the relation between
the topological structure and the thermodynamic quantities such as the boiling point and entropy can be

explained in terms of the topological dependency of the Z-index, if one admits the relation between the Z-index and such quantity f as [21,39]

$$f \propto \log Z. \qquad (21)$$

Systematic QSAR (quantitative structure-activity relationship) analysis [40] revealed that for saturated acyclic hydrocarbon molecules most of the thermodynamic quantities can be classified into several groups, i.e., Z-dependent, p-dependent, Z,p-dependent, and Z,p-independent types, where p is another topological index, polarity number, proposed by Wiener in as early as 1947 [41].

Although Wiener [41], Platt [42], Cramer [43], and Randić and Wilkins [44] have similarly argued that many physico-chemical properties of chemical substances can be approximated with a set of two selected parameters, their chemical interpretation is not clear. Our "two-parameter" classification of the thermodynamic properties of acyclic saturated hydrocarbons with Z and p suggests that most of these properties are determined by the combinations of two different topological factors, i.e., dynamical factor, Z, which accounts for the properties depending on the number of rotational degree of freedom, and static factor, p, which reflects the bulkiness of branches. One of the reasons for Z and p being a good pair of indices in this respect is ascribed to their "orthogonal" characters with a small correlation coefficient.

SEARCH FOR NON-REDUNDANT TOPOLOGICAL INDEX

Although Table 1 shows a good discriminating power of Z for heptanes, for tree graphs with eight points we get a pair of isospectral graphs [6,7,10,11,26,45,46]

$$Q_G(x) = 1 + 7x + 9x^2$$
$$P_G(x) = x^8 - 7x^6 + 9x^4.$$

For eighteen isomers of octane, the Z-index ranges from 17 to 34 with a little disorder and redundancy [1]. This trend smoothly increases with the number of points. However, the Z-index is shown to be useful for rough sorting of graphs, especially with complicated structure [47].

Several different approaches have been chosen for getting as little redundant topological indices as possible [48]. Razinger et al. has shown that for alkane series Balaban's averaged distance sum connectivity [49] has the best structural selectivity among

the currently used topological indices [50]. They,
however, conclude that an index that would be highly
selective and could at the same time be successful in
correlating many different properties is not yet
discovered.

DISTANCE POLYNOMIAL AND HIGHER GRAPHS

Now let us take another algebraic starting point for
the analysis of the graph. A distance matrix D is
defined as a matrix whose ij element is assigned the
number or the shortest path between points i and j.
One can apply the definition of the characteristic
polynomial of Eq. (7) to D instead of A yielding the
distance polynomial S(x) [51-53] as

$$S_G(x) = (-1)^N \det(D - xE) = \sum_{k=0}^{N} b(k)x^{N-k} \qquad (22)$$

This polynomial is conjectured to have unique characterization ability for graphs. Although we could not
yet encountered any "isospectral" graph pair among
the set of more than a thousand graphs studied, the
above conjecture is still open. One of the most
interesting features of the distance polynomical S(x)
is that the last term of S(x), not necessarily the
determinant of D, depends only on the number of
points and the ring skeleton of the graph. Examples
are shown in Table 2.

Table 2. Examples of distance polynomial

Graph	$S_G(x)$
	$x^5 - 50x^3 - 140x^2 - 120x - 32$
	$x^5 - 38x^3 - 116x^2 - 112x - 32$
	$x^5 - 28x^3 - 88x^2 - 96x - 32$
	$x^5 - 35x^3 - 88x^2 - 74x - 20$
	$x^5 - 30x^3 - 82x^2 - 72x - 20$
	$x^5 - 25x^3 - 70x^2 - 66x - 20$
	$x^6 - 44x^4 - 162x^3 - 201x^2 - 80x$
	$x^6 - 49x^4 - 176x^3 - 209x^2 - 80x$
	$x^6 - 49x^4 - 180x^3 - 220x^2 - 80x$

It is known that the half sum of the off-diagonal elements $D(ij)$ of D is the Wiener's path number w [1]. The half sum of the squares of $D(ij)$ is found to be equal to $-b(2)$, while $-b(N-1)$ is expressed in terms of the numbers of secondary, tertiary, quarternary, $\cdots$ carbon atoms [51].

The distance matrix carries all the informations on the shortest distance between any pair of points. By picking out the term 1's in D one gets the adjacency or 1-neighbor relation of points yielding the original graph, or first order graph, $G_1 = G$. Then by picking out the term 2's in D one gets the set of pairs of points whose shortest distance is two. This relation gives the graph of second order G_2. Similarly we get the graph of third order G_3, the number of whose edges is Wiener's polarity number, p. The set of the Z-indices $\{z^1, z^2, z^3, \ldots\}$ obtained from $\{G_1, G_2, G_3, \ldots\}$ may well represent the topological features of the original graph. For example the second and third order grphs, G^2, G^3, of the last entry of Table 1 are as follows

G^1: ┴┴ G^2: ▷ G^3: ◇ G^4: none

$\{z^1, z^2, z^3\} = \{13, 40, 13\}$

In Fig.2 are shown the set of patterns $(k-z^k)$ for the nine isomers of heptane (Table 1), which is very similar in shape with the original graph. However, we are still in a position to check the mathematical properties of these new topological indices.

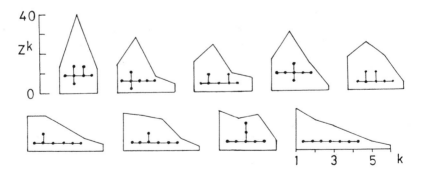

Figure 2. Higher graphs and higher Z-indices of heptane isomers. See Table 1.

RELATIONS OF Q(x) WITH PARTITION FUNCTIONS

Enumeration of the number of ways for placing indis-

tinguishable dumbbells on various periodic lattice
space like polyominoes (rectangular lattices) has a
key role in solving various statistical problems,
such as adsorption of diatomic molecules on a crys-
talline surface, magnetic properties of antiferro-
magnetic metals, stability of ionic crystals [54-56].
For complete covering on special lattices such as
rectangular lattices and tori closed forms and/or
recursion formulas have been obtained [57-59]. However, for incomplete covering problems very limited
number of cases have been solved [15,56,60,61]. Our
Z-counting polynomial is mathematically equivalent
to the partition function for these statistical
problems, if one substitutes x in Eq. (1) with exp
$(-\varepsilon/kT)$, where k is the Boltzmann constant, T the
absolute temperature, ε appropriate energy corresponding to the p(G,k) selection for each model.
By use of the operator technique proposed by us [62]
the recursion formulas for various series of periodic graphs such as mxn (m=1-4) rectangular and 3-
dimensional 2x2xn lattices have been obtained [15].
Extension to larger lattices and further elaboration of this method are still under way.

From quite a similar standpoint to the above study
one can define the rotational polynomial for counting the number of the rotational isomers of alkanes
and also for obtaining their absolute entropy systematically [63]. Sextet polynomial and king polynomial have also been proposed by us, respectively
for counting the numbers of the perfect matching on
the hexagonal and rectangular lattices [64-69]. By
use of these counting polynomials one can get not
only the systematic view of the problem but also many
important mathematical features involved. Application of these enumeration techniques to a number of
challenging problems is still open to us.

ACKNOWLEDGEMENT

The author expresses his sincere thanks to all the
collaborators of this project, especially to my
students for their painstaking tasks. Thanks are also
due to Professor Y. Fukuda and Assistant K. Takano
for their technical assistance.

REFERENCES

[1] H. Hosoya, Bull. Chem. Soc. Jpn., 44, 2332 (1971).

[2] L. B. Kier and L. H. Hall, Molecular Connectivity in Chemistry and Drug Research, Academic Press, New York, 1976.

[3] D. H. Rouvray and A. T. Balaban, in Application of Graph Theory, R. J. Wilson and L. W. Beineke, ed., Academic Press, London, 1979, p. 177.

[4] A. T. Balaban, A. Chiriac, I. Motoc, and Z. Simon, Lecture Notes in Chemistry, No. 15, Springer-Verlag, Berlin, 1980, p. 22; A. T. Balaban, Theor. Chim. Acta, 53, 355 (1979).

[5] A. T. Balaban, I Motoc, D. Bonchev, and O. Mekenyan, Topics in Current Chemistry, No. 114, Springer-Verlag, 1983, p. 21.

[6] N. Trinajstić, Chemical Graph Theory, CRC Press, Boca Raton, Florida , 1983, Vol. II, p. 105.

[7] F. Harary, Graph Theory, Addison-Wesley Publ. Co., Reading, Mass. (1969).

[8] A. T. Balaban, ed., Chemical Application of Graph Theory, Academic Press, London (1976).

[9] A. Graovac, I. Gutman, and N. Trinajstić, Topological Approach to the Chemistry of Conjugated Molecules (Lecture Notes in Chemistry, No. 4), Springer-Verlag, Berlin (1976).

[10] K. Mizutani, K. Kawasaki, and H. Hosoya, Natl. Sci. Rept. Ochanomizu Univ., 22, 39 (1971).

[11] K. Kawasaki, K. Mizutani, and H. Hosoya, Natl. Sci. Rept. Ochanomizu Univ., 22, 181 (1971).

[12] T. Yamaguchi, M. Suzuki, and H. Hosoya, Natl. Sci. Rept. Ochanomizu Univ., 26, 39 (1975).

[13] I. Gutman, S. Petrović, and B. Mohar, Coll. Sci. Papers Fac. Sci. Kragujevac, 3, 43 (1982); 4, 189 (1983).

[14] H. Hosoya and N. Ohkami, J. Comput. Chem., 4, 585 (1983).

[15] H. Hosoya and A. Motoyama, J. Math. Phys., 26, 157 (1985).

[16] H. Hosoya, Math. Comput. Appl., to appear.

[17] O. J. Heilmann and E. H. Lieb, Comm. Math. Phys., 25, 190 (1972).

[18] I. Gutman, Math. Chem., 6, 75 (1979).

[19] H. Hosoya. Natl. Sci. Rept. Ochanomizu Univ., 32, 127 (1981).

[20] H. Hosoya, Fibonacci Quaterly, 11, 255 (1973).

[21] H. Hosoya, K. Kawasaki, and K. Mizutani, Bull. Chem. Soc. Jpn., 45, 3415 (1972).

[22] H. Hosoya, Theor. Chim. Acta, 25, 215 (1972).

[23] H. Hosoya and K. Hosoi, J. Chem. Phys., 64, 1065 (1976).

[24] H. Sachs, Publ. Math. (Debrecen), 9, 270 (1964).

[25] A. Graovac, I. Gutman, and N. Trinajstić, Theor. Chim. Acta, 26, 67 (1972).
[26] D. Cvetković, M. Doob, and H. Sachs, Spctra of graphs, Academic Press, Berlin (1979).
[27] I. Gutman, M. Milun, and N. Trinajstić, math. Chem., 1, 171 (1975); J. Am. Chem. Soc., 99, 1692 (1977).
[28] J. Aihara, J. Am. Chem. Soc., 38, 2750 (1976).
[29] E. J. Farrell, J. Comb. Theor., B27, 75 (1979).
[30] H. Hosoya, K. Hosoi, and I. Gutman, Theor. Chim. Acta, 38, 37 (1975).
[31] H. Hosoya and M. Murakami, Bull. Chem. Soc. Jpn., 48, 3512 (1975).
[32] H. Hosoya and K. Hosi, J. Chem. Phys., 64, 1065 (1976).
[33] L. Pauling, Nature of the Chemical Bond, Cornell Univ. Press, Ithaca, N. Y. (1960).
[34] C. A. Coulson and H. C. Longuet-Higgins, Proc. R. Soc. London, A191, 39 (1947).
[35] I. Gutman, T. Yamaguchi, and H. Hosoya, Bull. Chem. Soc. Jpn., 49, 1811 (1976).
[36] S. Aono, T. Ohmae, and K. Nishikawa, Bull. Chem. Soc. Jpn., 53, 3418 (1980).
[37] S. Aono and K. Nishikawa, Bull. Chem. Soc. Jpn., 54, 1645 (1981).
[38] K. Nishikawa, M. Yamamoto, and S. Aono, J. Chem. Phys., 78, 5031 (1983).
[39] H. Narumi and H. Hosoya, Bull. Chem. Soc. Jpn., 53, 1228 (1980).
[40] H. Narumi and H. Hosoya, Bull. Chem. Soc. Jpn., 58, 1778 (1985).
[41] H. Wiener, J. Am. Chem. Soc., 69, 17, 2636 (1947); J. Phys. Chem., 52, 425, 1082 (1948).
[42] J. R. Platt, J. Chem. Phs., 15, 419 (1947); J. Phys. Chem., 56, 328 (1952).
[43] R. Cramer, III, J. Am. Chem. Soc., 102, 1837, 1849 (1980).
[44] M. Randić and C. L. Wilkins, J. Phs. Chem., 83, 1525 (1979).
[45] L. Collatz and U. Sinogowitz, Abh. Math. Semin. Univ. Hamburg, 21, 63 (1967).
[46] I. Gutman and Trinajstić, Topics Curr. Chem., 42, 49 (1973).
[47] H. Hosoya, J. Chem. Doc., 12, 181 (1972).
[48] L. Spialter, J. Am. Chem. Soc., 85, 2012 (1963); J. Chem. Doc., 4, 261 81964).
[49] A. T. Balaban, Chem. Phys. Lett., 89, 399 (1982).
[50] M. Razinger, J. R. Chretien, and J. E. Dubois, J. Chem. Inf. Comput. Sci., 25, 23 (1985).
[51] H. Hosoya, M. Murakami, and Gotoh, natl. Sci. Rept. Ochanomizu Univ., 24, 27 (1973).
[52] R. L. Graham and L. Lovasz, private communica-

tion.

[53] R. L. Graham, A. J. Hoffmann, and H. Hosoya, J. Graph Theory, 1, 85 (1977).

[54] P. W. Kasteleyn, in Graph Theory and Theoretical Physics, F. Harary, ed., Academic Press, London (1967), p. 43.

[55] H. N. V. Temperley, Graph Theory and Applications, Ellis Horwood, Chichester, England (1981).

[56] R. B. McQuistan and S. J. Lichtman, J. Math. Phys., 11, 3095 (1970).

[57] M. Gordon and W. T. Davison, J. Chem. Phys., 20, 428 (1952).

[58] H. N. V. Temperley and M. E. Fisher, Phil. Mag., 6, 1061 (1961).

[59] P. W. Kasteleyn, Physica, 27, 1209 (1961).

[60] J. H. Hock and R. B. McQuistan, J. Math. Phys., 24, 1859 (1983).

[61] A. J.Phares, J. Math. Phys., 25, 1756 (1984).

[62] H. Hosoya and N. Ohkami, J. Comput. Chem., 4, 585 (1983).

[63] H. Hosoya and C. Ichida, unpublished.

[64] H. Hosoya and T. Yamaguchi, Tetrahedron Lett., 1975, 4659.

[65] N. Ohkami, A. Motoyama, T. Yamaguchi, H. Hosoya, and I. Gutman, Tetrahedron, 37, 1113 (1981).

[66] N. Ohkami and H. Hosoya, Theor. Chim. Acta, 64, 153 (1983).

[67] N. Ohkami and H. Hosoya, Natl. Sci. Rept. Ochanomizu Univ., 35, 71 (1984).

[68] A. Motoyama and H. Hosoya, J. Math. Phys., 18, 1485 (1977).

[69] A. Motoyama and H. Hosoya, Natl. Sci. Rept. Ochanomizu Univ., 27, 107 (1976).

Chapter 12

VALENCE – A MEASURE OF USED COVALENT BONDING CAPACITY OF ATOMS IN MOLECULES

Karl Jug, Theoretische Chemie, Universität Hannover

ABSTRACT

A definition of atomic valence in molecules is presented as a measure of used covalent binding capacity of atoms in molecules. The proposed method is based on an analysis of the density matrix. It is general enough to be applicable in semiempirical or ab initio calculations on SCF and CI level. Calculations of selected examples demonstrate its use for structure and reactions of molecules. In particular it is possible to measure the radical, diradical and zwitterionic character of molecular states. More generally the ionicity, i.e. the extent of ionic character of molecular wave functions can be determined by this method. For acid base reactions a correlation between energies and valence numbers can be established. Woodward-Hoffmann allowed and forbidden reactions can be distinguished by valence number changes during the reactions. Also nonconcerted and photochemical reactions can be analyzed with this method.

INTRODUCTION

The concept of valence dates back to the early days of quantum chemistry. It appeals to a conceptual understanding of bonding in molecules. This line of thought was primarily pursued by Coulson and had its highlight in his famous book by the same name /1/. With the advent of computers and the possibility of increasingly accurate calculations on single molecules, concepts

like valence gradually lost importance, since they were cast in
a framework which was no longer adequate for present day needs.
In order to revive the idea of valence it was necessary to
develop a formalism which could be applied to self-consistent-
field (SCF) and configuration interaction (CI) wave functions
built from linear combinations of atomic orbital (LCAO) basis
sets. It seemed natural to establish a new theoretical framework of valence on the following premises /2/: 1. It must be
invariant under coordinate transformation. 2. It should be a
measure of the actual covalent bonding in molecules. 3. It
should reflect saturation of bonding. 4. It should be related
to the covalent reactivity of atoms in molecules. 5. It should
be derived from the density matrix.
Such a formalism was introduced by the definition of valence
numbers as the sum of squares of bond order elements from pairs
of atoms of density matrices over orthogonalized atomic orbitals /2/ and applied to a variety of compounds containing first-
row atoms /3/. This idea originated from a footnote in a paper
by Wiberg /4/ who advocated a bond index as a measure of covalent bonding. Independently, Perkins and coworkers /5/ and
Semyonov /6/ had recognized the potential of Wiberg's formula
and combined it with atomic valence considerations. Both groups
applied the resulting formulas on a low level of computation,
but failed to develop the formalism in an advanced form from
first principles. After we had already investigated various properties of valence numbers in our initial papers, a more elegant and comprehensive way to obtain valence numbers for general wavefunctions on the LCAO SCF CI level was introduced /7/.
We shall present this general theory of valence in the following section. It will then be demonstrated how structure and
reactivity of molecules can be related to the valence concept.

THEORY

We now analyze the valence properties of a CI wave function
given in the form

$$\Psi = \sum_I A_I \Psi_I \tag{1}$$

The density operator P of wave function (1) is a projection
operator of the form

$$P = |\Psi\rangle\langle\Psi| \tag{2}$$

The configurations are constructed from molecular orbitals
(MO's) ψ_i by single or double substitution of an SCF wave
function Ψ_0. The MO's are in turn expanded in orthogonalized
atomic orbitals λ /7/

$$\psi_i = \sum_\rho c_i \lambda_\rho \tag{3}$$

The density operator P can then be reformulated conveniently in MO form /8/. It is now essential to realize that covalent bonding between two atoms A and B is related to the portion P^{AB} of the density operator linking these two atoms

$$P^{AB} = \frac{1}{2}(|\psi^A\rangle\langle\psi^B| + |\psi^B\rangle\langle\psi^A|) \tag{4}$$

P^{AB} is not a projection operator. After reduction of the MO's to diatomic portions

$$\psi_i^{AB} = \psi_i^A + \psi_i^B , \tag{5}$$

We can define the covalent bonding between the two atoms as an expectation value V_{AB} of operator P^{AB} over occupied MO's ψ_i^{AB}

$$V_{AB} = \sum_i^{occ} n_i \langle\psi_i^{AB}|P^{AB}|\psi_i^{AB}\rangle \tag{6}$$

Here n_i is the generally fractional occupation number of MO ψ_i. Evaluation of (6) in LCAO form leads to

$$V^{AB} = \sum_{\mu,\nu} (P_{\mu\nu}^{AB})^2 \tag{7}$$

with

$$P_{\mu\nu}^{AB} = \sum_i^{occ} n_i c_{i\mu}^A c_{i\nu}^B$$

This proves that the heuristic form of sum of squares of density matrix elements related to the pair of atoms A and B is applicable even on the CI level. Alternatively one can write V^{AB} as the sum of squares of bond order eigenvalues p_μ /2,7/

$$P^{AB} h_\mu^{AB} = p_\mu^{AB} h_\mu^{AB} \tag{8}$$

This relates bond order and valence. Whereas bond order is linear in p_μ, valence is quadratic.
The valence of an atom A is the sum over all contributions from the other atoms

$$V_A = \sum_B V_{AB} \tag{9}$$

The bond number of the total covalent bonding in the molecule is then given as

$$M = \frac{1}{2} \sum_A V_A \qquad (10)$$

We shall show in the following that this number can be used to measure the ionicity of the molecular wave function.

HYPERVALENCE AND SUBVALENCE

The chemist's intuition assigns a normal valence to each atom. This number represents the normal number of single bonds that this atom is able to form. In the first row the normal valence is 1 for Li and F, 2 for Be and O, 3 for B and N, 4 for C. In a heteropolar molecule deviations from these standard values of covalency will occur. We call an atom hypervalent if its actual valence exceeds the normal valence and subvalent if its actual value falls below the normal value. The sign and magnitude of ΔV_A is a measure of this property

$$\Delta V_A = V_A^{actual} - V_A^{normal} \qquad (11)$$

Subvalence with negative ΔV_A will occur for radical and polar molecules. Hypervalence with a positive ΔV_A can occur if lone pair of empty orbitals participate in the bonding.
The following examples in Table 1 may illustrate the different situations. Calculations for all molecules were performed on the SCF level with SINDO1 /9,10/. Valence numbers for all atoms were determined along the lines of the previous section. It is quite expected that we find valence number 4 for carbon in CH_4, 1 for fluorine in F_2 and lithium in Li_2. Also NH_3 and H_2O have normal values. But even B in B_2F_4 and Be in BeF_2 have normal valence numbers. The polarization of the σ orbitals towards F is balanced on the π level by backbonding from the F atoms. Quite different is the situation in HNO_3 and HNC. Here the nitrogen atom does not form three bonds plus one lone pair, but the lone pair is also involved in the bonding. The lithium compounds with unusual coordination numbers have been extensively investigated by Schleyer /11/. The bonding in CLi_6 can be explained if we assume that LiLi bonding is essential in these compounds due to the long range of the diffuse Li orbitals and the involvement of p orbitals. The valence number of 1.50 for Li classifies Li as hypervalent. Participation of the p orbitals in the bonding is the reason for the increase above 1. In this sense also CLi_4 has hypervalent lithium.

In CO_2 the polarization of the CO bond decreases the valence number of C below normal. In $^3CH_2$ we find the diradical character of carbon characterized by the reduction of the valence number. O_2 is a triplet in its ground state. The valence number 1.50 of oxygen is due to this open shell character and means that oxygen in O_2 is not saturated like in H_2O. Be_2 is very weakly bound, so the valence number of beryllium is close to 0.

Table 1. Normal, hypervalent and subvalent atoms in molecules

Molecule	Atom	Valence	Type
CH_4	C	4.00	normal
C_6H_6	C	3.98	
Pyridine	N	3.02	
N_2	N	3.00	
NH_3	N	2.97	
B_2F_4	B	3.02	
H_2O	O	1.96	
BeF_2	Be	1.95	
F_2	F	1.00	
Li_2	Li	1.00	
CH_4	H	1.00	
HNO_3	N	3.75	hypervalent
HNC	N	3.54	
Pyrrole	N	3.39	
BH_3NH_3	B	3.43	
B_2H_6	B	3.41	
O_3	O^a	2.80	
H_3O^+	O	2.53	
CBe_2	Be	2.36	
CLi_6	Li	1.50	
CLi_4	Li	1.40	
CO_2	C	3.74	subvalent
$^3CH_2$	C	1.91	
CN	N	2.78	
NO	N	2.13	
BF_3	B	2.80	
BF	B	1.27	
O_2	O	1.50	
F_3NO	O	1.23	
Be_2	Be	0.03	
LiF	Li	0.81	

a central atom

DIRADICALS AND ZWITTERIONS

Salem and Rowland /12/ and Döhnert and Koutecky /13/ have given global criteria for diradicals and zwitterions. These are either singlet-triplet degeneracy for diradicals and a pair of accompanying zwitterionic states /12/ or the occurence of occupation number 1 for two natural orbitals. With the information in the previous section it is now possible to define diradicals and zwitterions with a local criterion /14/. The two lone electrons of a diradical do not contribute to covalent bonding. In consequence the sum of atomic valence numbers should be reduced by approximately two. Alternatively, the bond number M^{actual} of equ. (9) should be reduced by approximately 1 compared with M^{normal}. We take again $^3CH_2$ as an example : $V_C^{normal} = 4$, $V_H^{normal} = 1$, $M^{normal} = 3$, $V_C^{actual} = 1.91$, $V_H^{actual} = 0.96$, $M^{actual} = 1.92$. If the valence reduction is distributed over many centers, the following general formula derived from (11) must be used.

$$\Delta V = \sum_A \Delta V_A \tag{12}$$

Such a case is the quadratic triplet of cyclobutadiene where $\Delta V = 2.16$. But also the lowest triplet of benzene with a quinoidal D_{2h} structure must be classified as a diradical because the valence of the two atoms in para position is 3.27 and the valence of the other four atoms is 3.80. This amounts to $\Delta V = 2.26$. Since no singlet-triplet degeneracy is present, Salem's criterion cannot arrive at this answer. But even Koutecky's natural occupation number does not give a clue in this case. We have investigated a series of monosubstituted benzenes with substituents CH_3, NH_2, OH, F and NO_2 /15/. The valence numbers of the carbons at the substitution sites are 3.26, 3.20, 3.07, 3.05 and 3.05. So the diradical character is gradually more pronounced in this sequence. We explain the nonplanarity of the last four compounds by the presence of a radical center at the substituent group.

In the case of zwitterions a donor atom should transfer an electron to an acceptor atom. The covalent bonding should be modified in such a way that the donor atom is subvalent and the acceptor atom hypervalent. A typical case is ammonia oxide H_3NO, the isomer of hydroxylamine H_2NOH. The migration of the hydrogen atom causes a substantial change of valence numbers. Whereas the latter molecule has valence numbers close to normal, the former has a hypervalent N with $V_N = 3.67$ and a subvalent O with $V_O = 0.85$. Also the dipole moment of H_3NO is 4.80 Debye compared to 0.56 of H_2NOH. Dipole moments should be substantial in zwitterions if they are not vanishing for symmetry reasons.

IONICITY

Valence numbers can also be used to compare the relative polarization of atoms in molecules for different geometries. If the ionic character of a bond increases, the bond number M of covalent bonding decreases. In Table 2 we compare the changes ΔM in bond number, Δn in occupation number and ΔE in energy for different geometries of some hydrogen and lithium compounds.

Table 2. Changes of bond number, occupation number and energy (kcal/mol) in strong and weak overlap binders dependent on geometry

Molecule	Geometry	ΔM	Δn_{1S}	Δn_{2S}	Δn_{2P}	ΔE
H_2O	linear bent	0.248	0.346	0.239	-0.585	- 72.5
NH_3	planar pyramidal	0.061	0.136	0.277	-0.413	- 13.5
CH_4	planar tetrahedral	0.586	0.802	0.065	-0.867	-145.2
Li_2O	linear bent	0.020	-	0.109	-0.109	- 12.1
Li_3N	planar pyramidal	0.000	-	0.137	-0.137	- 10.6
CLi_4	planar tetrahedral	-0.044	-	0.064	-0.064	- 31.9

Epiotis has classified the hydrogen compounds as strong overlap binders and the lithium compounds as weak overlap binders /16, 17/. The first group is characterized by stabilization through deexcitation from 2p to 2s and 1s. H_2O, NH_3 and CH_4 have less bond ionicity in the more stable form. In the lithium compounds the trend is just the opposite. The more stable form has the higher bond ionicity. But even in this case deexcitation from 2p to 2s takes place. Because the lithium compounds are weak overlap binders, the effect of changes in valence, occupation and energy is much smaller than in the hydrogen compounds.

ACID-BASE REACTIONS

The reactivity of molecules can be studied with valence numbers in particular cases /18/. One such case are acid-base reactions. We define the absolute deviation of actual valence numbers from normal valence numbers

$$|\Delta V| = \sum_A |V_A^{actual} - V_A^{normal}| \tag{13}$$

It is now possible to compare the absolute deviations for reactands and products

$$\Delta\Delta V = |\Delta V_{product}| - |\Delta V_{reactand}| \tag{14}$$

These data are compared with energy changes ΔE in Table 3 for some simple acid-base reactions.

It is apparent from these numbers that small changes in valence deviations are usually accompanied by small changes in energy. Stabilization leads to more normal values of valence numbers. For investigations in solution, a more comprehensive set of neutral molecules and ions has to be considered.

Table 3. Valence and binding energy changes (kcal/mol) in acid-base reactions

Reaction	$\Delta\Delta V$	ΔE^a
$H_3O^+ + OH^- \rightarrow 2\ H_2O$	-1.94	- 27.3
$NH_3 + H_3O^+ \rightarrow NH_4^+ + H_2O$	0.03	- 36.1
$NH_4^+ + OH^- \rightarrow NH_3 + H_2O$	-1.89	-191.2
$F^- + H_3O^+ \rightarrow HF + H_2O$	-1.83	-202.0
$HF + OH^- \rightarrow F^- + H_2O$	-0.03	- 25.3
$N_3^- + H_3O^+ \rightarrow HN_3 + H_2O$	-1.27	-191.5
$HN_3 + OH^- \rightarrow N_3^- + H_2O$	-0.58	- 35.8
$HCO_2^- + H_3O^+ \rightarrow HCOOH + H_2O$	-1.44	-163.5
$HCOOH + OH^- \rightarrow HCO_2^- + H_2O$	-0.42	- 63.8

^{a}adjusted for negative ions

CONCERTED AND NONCONCERTED REACTIONS

Woodward and Hoffmann /19/ classified various types of reactions, e.g. electrocyclic and cycloaddition reaction, as allowed or forbidden according to orbital topology rules. In this section we wish to show that allowed reactions proceed with small changes in bond numbers M, whereas forbidden reactions undergo substantial reductions in valence numbers. The latter are an indication of the breaking of bonds without simultaneous formation of new bonds. So forbidden reactions which show bond breaking without bond formation are nonconcerted and involve diradicals. A typical electrocyclic reaction is the cyclobutene → butadiene rearrangement. Whereas the allowed pathway shows a valence number increase of 0.01 from reactand to transition state and another 0.01 from transition state to product, the forbidden transition state of C_1 symmetry shows a reduction of 0.98 for M compared to the reactand. It is clearly diradicaloid in the sense that we have defined in a previous section. We have also studied the cyclopropyl cation → allyl cation /20/. In this case the difference in valence number changes is much less pronounced.
Much more pronounced is the effect of valence number change in fragmentation reaction of unsubstituted and substituted cyclobutane as well as in the retro Diels-Alder reaction of cyclohexene. In all these cases reaction pathways involve transition states and intermediates accompanied with orbital crossing.

Such stationary points show reduction of the bond number in the
order of 1 and consequently involve diradicals.
Although we had found that the Diels-Alder reaction involved
intermediates in shallow wells, it was not clear whether these
had any bearing on the selectivity of the reaction. A more pronounced case of stepwise reaction is the fragmentation of cyclobutane to two ethylene. From the analysis of the valence numbers
it becomes clear that a large portion of the reaction pathway of
the unsubstituted and substituted reaction is involving diradicals, starting from the twisting of one CC bond and up to the
breaking of the second CC bond. Contrary to expectation from
experiments in solution /22/, even the transition states of the
donor-acceptor complex with an OCH_3 group on one carbon and a
CN group on the adjacent carbon did not show zwitterionic behaviour.
Roth has studied rearrangement reactions which involve the
2,3-dimethylene-1,4-cyclohexadiyl as an intermediate /23/. With
the present method it is possible to assign each stationary
state its degree of diradical character. From the twelve transition states and three intermediates of a previous study /24/
nine transition states and two intermediates have pronounced
diradical character.
The method is not limited to ground state reactions. The photochemical conversion of cyclopentanone /25/ involves three diradical intermediates on the triplet surface which are not degenerate with the ground state. Further reaction leads to cyclobutane or two ethylene plus CO via singlet-triplet degenerate
diradicals or to 4-pentenal via a diradical transition state.
Salem /12/ would not predict some of these and Koutecky /13/
could not identify the diradical centers with his method.

CONCLUDING REMARKS

The analysis of SCF and CI wavefunctions in terms of atomic
valence numbers is a helpful procedure to understand structure
and reactivity in molecules. Valence can be reintroduced as a
generalization of early ideas of Coulson. Different from Coulson
we emphasize that it is not the bond orders which are additive
for the determination of actual covalent bonding of an atom in
a molecule, but the bond valences. This avoids valence numbers
much larger than 4 for carbon e.g. in trimethylenemethane or
neopentane. Since the bond valences are quadratic in diatomic
density matrix elements, whereas the bond orders are linear,
one orbital can supply not more than one valence for binding to
all atoms in a molecule. Lone electrons lead to a decrease and
partial occupation of empty atomic shells to an increase in
atomic valence numbers. This opens the way to trace radicals,
diradicals and zwitterions. The polarity of bonding, i.e. ionicity of the wave function, can also be measured by valence numbers for different geometries isomers or reactivity of molecules.
If covalency or ionicity are related to stability of molecules,
this index can be used to classify groups of molecules. Finally

it is possible to use this index also in thermochemical and photochemical reactions to follow concerted and nonconcerted pathways.

REFERENCES

/ 1/ Coulson, C.A., Valence, Oxford University Press, London 1952
/ 2/ Gopinathan, M.S., Jug, K., Theor.Chim.Acta 63, 497 (1983)
/ 3/ Gopinathan, M.S., Jug, K., Theor.Chim.Acta 63, 511 (1983)
/ 4/ Wiberg, K.B., Tetrahedron 24, 1083 (1968)
/ 5/ Armstrong, D.R., Perkins, P.G., Stewart, J.J., Chem.Soc. 883, 2273 (1973)
/ 6/ Semyonov, S.G., Theory of Valence in Progress, p. 150ff., V. Kuznetsov, Ed., Mir, Moscow 1980
/ 7/ Jug, K., J. Comp. Chem. 5, 555 (1984)
/ 8/ Davidson, E.R., Reduced Density Matrices in Quantum Chemistry, p. 31ff., Academic Press, New York 1976
/ 9/ Nanda, D.N., Jug, K., Theor.Chim.Acta 57, 95 (1980)
/10/ Jug, K., Nanda, D.N., Theor.Chim.Acta 57, 107, 131 (1980)
/11/ Schleyer, P.von R., New Horizons in Quantum Chemistry, p. 95ff., Ed. P.O. Löwdin, B. Pullman, Reidel: Dordrecht 1983
/12/ Salem, L., Rowland, C., Angew.Chem.,Int.Ed.Engl.11, 92 (1972)
/13/ Döhnert, D., Koutecky, J., J.Am.Chem.Soc. 102, 1789 (1980)
/14/ Jug, K., Tetrahedron Letters 26, 1437 (1985)
/15/ Malar, E.J.P., Jug, K., J.Phys.Chem. 88, 3508 (1984)
/16/ Epiotis, N.D., Unified Valence Bond Theory of Molecular Structure, Lecture Notes in Chemistry, Vol. 29, Springer, Berlin, Heidelberg, New York 1982
/17/ Epiotis, N.D., Unified Valence Bond Theory of Molecular Structure, Applications, Lecture Notes in Chemistry, Vol. 34, Springer, Berlin, Heidelberg, New York 1983
/18/ Jug, K., Buss, S., J. Comp. Chem., in press
/19/ Woodward, R.B., Hoffmann, R., The Conservation of Orbital Symmetry, Verlag Chemie, Weinheim 1971
/20/ Jug, K., Gopinathan, M.S., Theor.Chim.Acta, in press
/21/ Jug, K., Müller, P.L., Theor.Chim.Acta 59, 365 (1981)
/22/ Huisgen, R., Acc. Chem. Res. 10, 117, 199 (1977)
/23/ e.g.Roth, W.R. Biermann, M., Erker, G., Jelich, K., Gerhartz, W., Görner, H., Chem.Ber. 113, 586 (1980)
/24/ Jug, K., Iffert, R., Theor.Chim.Acta 62, 183 (1982)
/25/ Müller-Remmers, P.L., Mishra, P.C., Jug, K., J.Am.Chem. Soc. 106, 2538 (1984)

Chapter 13

QUANTUM CHEMICAL STUDIES AND PHYSICO-CHEMICAL STUDIES OF 2-PAM AND DEPROTONATED 2-PAM (SYN AND ANTI)

Joyce J. Kaufman, P.C. Hariharan, W.S. Koski and Nora M. Semo
Department of Chemistry, The John Hopkins University, Baltimore, Maryland 21218

ABSTRACT

Ab-initio MODPOT/VRDDO/MERGE calculations were carried out for different conformations of syn and anti 2-PAM (2-pyridiniumaldoxime methiodide) and on the deprotonated 2-PAM's. Ab-initio all-electron calculations were also carried out on syn 2-PAM and deprotonated syn 2-PAM. Comparisons of our ab-initio MODPOT/VRDDO and ab-initio all-electron calculations confirmed again that there is excellent agreement of orbital energies and gross atomic populations betweeen the two methods when the same valence shell atomic basis set is used and the same inner shell basis set is used for the ab-initio effective core model potentials. There is also excellent agreement between the three-dimensional electrostatic molecular potential contour (EMPC) maps calculated from the ab-initio MODPOT/VRDDO/MERGE and the ab-initio all-electron calculations.

Computed conformational profiles indicated that there were multiple maxima and minima, thus emphasizing that merely optimizing geometry for such pyridinealdoximes and pyridiniumaldoximes by derivative methods from an initial starting guess would not be sufficient, especially for the great majority of cases where there is no experimental crystal structure data.

Three-dimensional electrostatic molecular potential contour (EMPC) maps were generated from the quantum chemical wave functions. The deprotonated species of oxime reactivators had been suggested by pharmacologists to be the probable

physiologically effective species for reactivation of acetylcholinesterase (AChE) inhibited by organophosphorus compounds,
although the pyridiniumaldoxime reactivators have only proven
effective in the peripheral nervous system.

The 2-PAM's themselves are charged quaternary compounds
and thus would not be expected to partition through the lipid
blood-brain barrier and hence not to be effective reactivators
of inhibited AChE in the central nervous system. The
deprotonated 2-PAM is an overall neutral compound. However,
our three-dimensional EMPC maps of deprotonated 2-PAM's indicate that these species will be strongly dipolar ions and thus
would not be expected to partition from aqueous to lipid phase.

These three-dimensional EMPC maps around the deprotonated
2-PAM's can also help to delineate stereoelectronic requisites
for effective reactivating action and in addition they can help
to determine the stereoelectronic complementary requisites of
the AChE active sites - which have not yet been determined
crystallographically.

Since the reactivation of phosphorylated acetylcholinesterase is strongly dependent on the acidity of the oxime
reactivators, we carried out accurate measurements for the pK_a
and the lipophilicity of 2-PAM because of its importance in
transport across membranes and partition through the blood-
brain barrier into the central nervous system.

Our EMPC map findings on the deprotonated 2-PAM's are
confirmed by the experimental measurements of the lipophilicity
and pk_a which show that there is very little tendency for
either the 2-PAM's or the deprotonated 2-PAM's to partition
from aqueous to lipid phase.

INTRODUCTION

Organophosphorus compounds inhibit acetylcholinesterase
(AChE). If the aging of the phosphorylated (inhibited) AChE is
not too rapid, then it is possible to reactivate the phosphorylated AChE with compounds such as 2-PAM (2-pyridiniumaldoxime
methiodide.)

SYN 2-PAM

or related compounds. These
compounds were designed
originally in the 1950's by
Wilson to be complements of the
AChE active site based on available knowledge at that time [1-
3].

The question of the relation of calculated quantum chemical indices to the conformations and pharmacological activity
of the PAM's has been of interest to us for the past two
decades. Beginning in 1964, we carried out the first all-
valence-electron three-dimensional quantum chemical

calculations on any drug or biological molecule for conformational analysis of the PAM's, both protonated and deprotonated, and the charges on the atoms and total overlap populations [4-6].

From that time until we first presented in the spring of 1984 the results of our ab-initio MODPOT/VRDDO/MERGE calculations on syn and anti 2-PA and syn and anti 2-PAM and deprotonated 2-PAM and the electrostatic molecular potential contour (EMPC) maps calculated from these wave functions, [7a] apparently there had not been any other quantum chemical investigations of the PAM's. There had been experimental research on the PAM's and development of related pyridine oxime reactivators of phosphorylated AChE over these past twenty years [8,9].

During this past year we have continued our ab-initio MODPOT/VRDDO/MERGE investigations on a variety of other pyridiniumaldoxime and methyl imidazoliumaldoxime reactivators of inhibited AChE. We also carried out ab-initio all-electron calculations on syn 2-PAM and deprotonated syn 2-PAM.

2-PAM and other pyridiniumaldoxime reactivators of inhibited AChE are effective only in the peripheral nervous system, not in the central nervous system (CNS) [10]. While 2-PAM is a charged quaternary species and thus would not be expected to partition into the CNS through the lipid blood-brain barrier, deprotonated 2-PAM is a net neutral species. To investigate why deprotonated 2-PAM also does not partition into the CNS, we carried out experimental measurements of the pK_a and lipophilicity (as a function of pH), for both of the 2-PAM species.

METHODS

Theoretical
    1. AB-INITIO SCF

The ab-initio MODPOT/VRDDO/MERGE calculations were carried out with our MOLASYS computer program [11] which incorporates as options (which can be used or not used) several features desirable for ab-initio calculations on large molecules or those containing heavy atoms: ab-initio effective core model potentials (MODPOT) which permit calculations of valence electrons only explicitly, yet accurately; a charge conserving integral prescreening evaluation to determine whether or not to calculate integrals explicitly, (which we named VRDDO - variable retention of diatomic differential overlap) especially useful for spatially extended molecules; and an efficient MERGE technique which permits reuse of common skeletal integrals [12].

The ab-initio all-electron calculations were also carried out with the MOLASYS program. The same valence shell atomic basis set was used for both calculations and the same all-electron inner shell basis set was used for the ab-initio effective model core potentials.

ELECTROSTATIC MOLECULAR POTENTIAL CONTOUR MAPS

For molecular interactions involving molecules with net charges or permanent dipoles, useful information may be obtained by examination of the electrostatic potential, [13] arising from one of the partners, and by simple calculations, involving a potential and a simplified description of the charge distribution of the other molecule involved in the interaction [14]. This concept is also very useful for generating electrostatic molecular potential contour (EMPC) maps around molecules which react with receptor sites or enzyme active sites where nothing at all is known about the molecular structure of the site. Even by calculating the EMPC map from the electrostatic potential and a test unit positive charge, the salient stereoelectronic features of a molecule in three-dimensional space around the molecule are revealed. Also, the EMPC maps for various molecules generated in this way can be compared to one another.

The electrostatic contribution is dominant in molecules such as endogenous biomolecules, drugs, toxicants, etc. where there are numerous heteroatoms and hence strong charge redistribution. Even when polarization is appreciable, the electrostatic term is by far the larger contribution when molecule A has significant charge redistribution [15].

We had long ago demonstrated that the EMPC maps generated from ab-initio MODPOT, ab-initio VRDDO and ab-initio MODPOT/VRDDO wave functions matched the EMPC maps generated from the ab-initio all-electron calculations (inner shell and valence electrons) with the same atomic valence basis set (and where the MODPOT input was matched to the same inner shell atomic basis set) [16]. Thus, in this present study we have reconfirmed this by generating the EMPC maps around deprotonated 2-PAM from our ab-initio MODPOT/VRDDO calculations and from ab-intio all-electron calculations, using the same atomic basis set as described in section 1.

The isopotential EMPC maps are generated in three dimensions around the entire molecule. These isopotential EMPC maps can be displayed from any angle, since we can rotate the molecule along with its three-dimensional isopotential EMPC maps.

EXPERIMENTAL

The lipophilicity of a drug (its ability to partition from water to lipid) influences the ability of that drug to penetrate lipid barriers such as the lipophilic CNS blood-brain barrier.

The pK_a values and partition coefficients as a function of pH for 2-pyridiniumaldoxime methochloride were determined using a microelectrometric titration technique, described previously

[17]. Since the partition coefficients of 2-PAM were so small at all pH values, the absolute partitioning of 2-PAM was measured spectrophotometrically.

RESULTS AND DISCUSSION

Theoretical

When we initiated our recent studies on 2-PAM, there was only an old experimental crystal structure of syn 2-PAM reported [18] which did not place the hydrogen atoms. There were, however, more recent experimental crystal structures reported for syn and anti 4-PA (4-pyridinealdoxime) [19,20]. For an initial geometry for quantum chemical calculations we ansatzed the structure of the syn-aldoxime group onto the pyridine ring structure in the 2 position to give 2-PA and then ansatzed a CH_3 group onto the pyridine N, using the experimen-

tal crystal structure of a methylpyridinium cation as a guide [21].

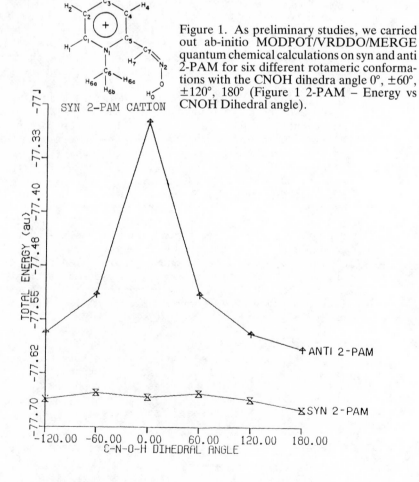

Figure 1. As preliminary studies, we carried out ab-initio MODPOT/VRDDO/MERGE quantum chemical calculations on syn and anti 2-PAM for six different rotameric conformations with the CNOH dihedra angle 0°, ±60°, ±120°, 180° (Figure 1 2-PAM – Energy vs CNOH Dihedral angle).

For the lowest energy conformation of syn 2-PAM to date, we have generated electrostatic molecular potential contour (EMPC) maps in three dimensions around syn 2-PAM. From these potentials we then generated isopotential contour EMPC maps around the syn 2-PAM. A copy of those three-dimensional, isopotential EMPC maps around syn 2-PAM is included (Figure 2 Syn 2-PAM Isopotential Contour EMPC Map Ab-initio MODPOT/VRDDO/MERGE). Since 2-PAM is a positively charged quaternary species, all the EMPC isopotential contours are positive.

Figure 2. Syn 2-PAM

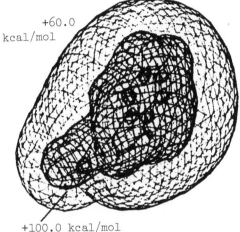

+60.0 kcal/mol

+100.0 kcal/mol

We also carried out ab-initio MODPOT/VRDDO/MERGE calculations for anti 2-PAM for six rotameric conformations with the CNOH dihedral angle $0°$, $\pm 60°$, $\pm 120°$, $180°$ and and for deprotonated anti 2-PAM and generated the EMPC maps [7], (Figure 3 Deprotonated Anti 2-PAM Isopotential Contour Map Ab-initio MODPOT/VRDDO/MERGE).

Figure 3. Anti 2-PAM Deprotonated

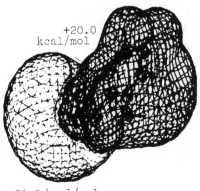

+20.0 kcal/mol

-20.0 kcal/mol

It is obvious from Figure 3 that even though deprotonated 2-PAM is overall neutral, the isopotential EMPC maps show a distinct positive and negative region. The negative region of the EMPC map is located around the ring carbon attached to the ring nitrogen and around the entire oxime side chain. The positive EMPC region embraces the rest of the molecule. Thus, the EMPC maps around the deprotonated anti 2-PAM indicate that this species should behave experimentally as a dipolar ion - rather than as a neutral molecule. This predicted behavior is consistent with our experimental lipophilicity measurements.

Recently an experimental crystal structure of 2-PAM was reported [22]. We have carried out the ab-initio MODPOT/VRDDO/MERGE and ab-initio all electron calculations for 2-PAM at its crystal structure coordinates.

We also carried out calculations for deprotonated syn 2-PAM starting from the crystal structure coordinates and varying the bond lengths and bond angles in the deprotonated aldoxime side chain using our MERGE technique.

We have more recently generated the EMPC maps for deprotonated syn 2-PAM both from the ab-initio MODPOT/VRDDO/MERGE calculations (Figure 4 Deprotonated Syn 2-PAM Isopotential EMPC Contour Map Ab-initio MODPOT/VRDDO/MERGE) and from the corresponding Ab-initio all-electron calculations (Figure 5 Deprotonated Syn 2-PAM Isopotential Contour EMPC Map Ab-initio all electron).

Figure 4. Syn 2-PAM Deprotonated   Figure 5. Syn 2-PAM Deprotonated
                                                            all-electron

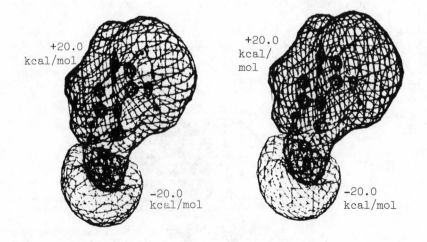

Again, as with anti 2-PAM, while deprotonated syn 2-PAM is overall neutral, the isopotential EMPC map shows the same distinct positive and negative regions as does the EMPC map around deprotonated anti 2-PAM.

Moreover, the EMPC maps around deprotonated syn 2-PAM from the ab-initio all-electron calculations and from the ab-initio MODPOT/VRDDO/MERGE calculations are so close that they are indistinguishable. This agreement between the MODPOT/VRDDO/MERGE calculations and the corresponding ab-initio all-electron calculations could also be seen from comparisons of the gross atomic populations and the valence orbital energies.

The gross atomic populations confirm our very early work [5,6] that the positive charge in 2-PAM is not localized on the nitrogen. Also comparison of the charges on syn 2-PAM and deprotonated syn 2-PAM indicate there is considerable redistribution of charge in both species.

There is also excellent agreement in the orbital energies for syn 2-PAM or for deprotonated syn 2-PAM calculated by the ab-initio MODPOT/VRDDO/MERGE and by the ab-initio all-electron technique.

However, the ab-initio all-electron calculation took about four times longer than the corresponding ab-initio MODPOT/VRDDO calculation with THR 1 $= 10^{-4}$ (the size of the VRDDO pseudo-overlap for prescreening integrals; 10^{-4} is sufficient to allow all integrals 10^{-8}) and THR 2 $= 10^{-6}$ (the size of the 2-e integrals) retained. This is a higher threshold than we normally use since we showed long ago that THR 1 $= 10^{-2}$ and THR 2 $= 10^{-4}$ were sufficient to reproduce gross atomic populations to ~0.02 e, valence orbital energies to ~0.002 a.u. and potential energy curves or isomer total energies to ~0.001-0.0001 a.u. (compared to not using the VRDDO prescreening threshold). The ab-initio all electron calculation on syn 2-PAM took about five times longer than the corresponding ab-initio MODPOT/VRDDO/MERGE with THR 1 $= 10^{-2}$ and THR 2 $= 10^{-4}$.

EXPERIMENTAL
Microelectrometric Technique

The ionization constants and the octanol/water partition coefficients were determined for 2-pyridiniumaldoxime methochloride (2-PAM) at 20°C, 30°C, and 37°C.

Below is the summary of the results. The pK_a values represent the average of at least 4 determinations.

TEMP	CONDITIONS	pK_a
20°C	Aqueous	7.904 ± 0.01
	n-octanol/water	7.895 ± 0.01
30°C	Aqueous	7.785 ± 0.01
	n-octanol/water	7.773 ± 0.01
37°C	Aqueous	7.675 ± 0.01
	n-octanol/water	7.674 ± 0.01

Since the difference in the pK_a's determined in aqueous and n-octanol/water solution is within experimental error, we conclude from our microelectrometric determinations that to within experimental error there is virtually no partitioning of any of the species of 2-PAM from water to oil at any pH in the range from 20°C to 37°C.

A brief survey of the literature revealed the following pK_a values assigned for 2-PAM.

Ginsburg and Wilson [23] determined a pK_a of 8.0 at 25°C by measuring the pH of a half neutralized solution; Mason [24] reports a value of 8.00 ± 0.01 at 20°C measured by potentiometric titration; Hagedorn et al. [25] report a value of 7.68 ± 0.03 by potentiometric titration (no temperature stated).

Since some authors indicate that the reactivation efficiency of AChE is related to the concentration of the oxime anion at physiological pH, we calculated the fraction of the anion in the range 6.80 - 7.70 from the relationship:

$$\alpha = 1 - \frac{antilog\ (pK_a - pH)}{antilog\ (pK_a - pH) + 1} \qquad (\text{for 2-PAM at 37°C})$$

pH	% oxime anion
6.80	11.8
7.10	21.0
7.20	25.1
7.35	32.1
7.40	34.7
7.45	37.3
7.50	40.1
7.60	45.7
7.675	50.0
7.70	51.4

This might prove especially useful in the case of dioximes, where the true proportion of the oxime anion to molecules at physiological pH can be calculated knowing the two ionization constants.

UV Spectophotometric Technique

2-PAM was one of the oximes for which our electrometric measurements indicated a very small distribution coefficient between octanol and water.

We began by first investigating the UV spectra of 2-PAM in HCl, NaOH and buffer at pH = 7.4, in order to establish the molar extinction, at λ = 293 nm and from there a calibration curve for different concentrations.

$$\varepsilon = 1.04 \times 10^4 \text{ at } \lambda = 293 \text{ nm}$$

which means that the lower limit of the detectable amount of 2-PAM in octanol is 2×10^{-6} g/l.

We determined the solubility of 2-PAM in octanol, which is 2×10^5 g/l (1.2×10^{-4} M).

The following characteristic constants were determined for 2-PAM from the optical measurements:

SOLUTION	pH	% OXIME ANION	λ_{max}(nm)	ε
0.1N HCl	1	0%	293	1.219×10^4
0.1N NaOH	13	100%	336	1.810×10^4
Phosphate Buffer	7.4	34.7%	295	1.064×10^4

The distribution coefficient at 37°C was found to be 3.6×10^{-3}.

ACKNOWLEDGEMENT

The theoretical and experimental research on the pyridiniumaldoxime reactivators of AChE was supported by the Army Medical Research and Development Command under Contract # DAMD 17-83-C-3219.

REFERENCES

1. I.B. Wilson and S. Ginsburg, Biochem. Biophys. Acta 18, 168 (1953).

2. I.B. Wilson and S. Ginsburg, Arch. Biochem. and Biophys. 54, 569 (1955).

3. S. Ginsburg and I.B. Wilson, J. Am. Chem. Soc. 79, 481 (1957).

4. Joyce J Kaufman, Principal Investigator, "A Quantum Mechanical Evaluation of the Mechanism of Action of 2-PAM Chloride and Analogs," Contract DA18-035-AMC-745(A), Final Report, RIAS, Baltimore, Maryland, February 1967.

5. W. Giordano, J.R. Hamann, J.J. Harkins and Joyce J. Kaufman. "Quantum Mechanically Derived Electronic Distributions in the Conformers of 2-PAM." In PHYSICO-CHEMICAL ASPECTS OF DRUG ACTION, Ed. A. Ariens, on the proceedings of the IIIrd International Pharmacological Congress, Sao Paulo, Brazil, July 1966, Pergamon Press, New York, NY, Vol. 7, pp.327-354, 1968.

6. W. Giordano, J.R. Hamann, J.J. Harkins and Joyce J. Kaufman, Mol. Pharmacol. 3, 307 (1967).

7a. Joyce J. Kaufman, "Quantum Chemical and Physiochemical Studies of Oxime Reactivators of Inhibited Acetylcholinesterase" An invited plenary lecture presented at the International Sanibel Symposium on Quantum Biology and Quantum Pharmacology, Palm Coast, Florida, March 1984.

7b. Joyce J. Kaufman, "Quantum Chemical and Physicochemical Studies of Old and New Oxime Reactivators of Inhibited Acetylcholinesterase", An invited plenary lecture presented at the International Sanibel Symposium on Quantum Biology and Quantum Pharmacology, Marineland, Florida, March 1985.

8. I. Hagedorn, J. Stark, K. Schoene and H. Schenkel, Arzneim.-Forsch. 28, 2055 (1978).

9. K. Schoene, "Pyridinium Salts as Organophosphate Antagonists," IN MONOGRAPHS IN NEUTRAL SCIENCES, NEUROLOGY OF CHOLINERGIC AND ADRENERGIC TRANSMITTERS," Vol. 7, S. Karger, Basel, Switzerland, 1980, pp.85-98.

10. C.A. Broomfield, B.E. Hackley, F.E. Hahn, D.E. Lenz and D.M. Maxwell, "Evaluation of H-Series Oximes," Proceedings of a Symposium held September 19, 1981. Biomedical Laboratory Technical Report, USA BML-59-81-001, U.S. Army Medical Research and Development Command, Aberdeen Proving Ground, Maryland, April 1981.

11a. H.E. Popkie, "MOLASYS: A Computer Program for Molecular Orbital Calculations on Large Systems," The Johns Hopkins University, 1974.

11b. H.E. Popkie, "MOLASYS-MERGE," The Johns Hopkins University, 1978.

12. Joyce J. Kaufman, H.E. Popkie and P.C. Hariharan, "New Optimal Strategies for Ab-Initio Quantum Chemical Calculations on Large Drugs, Carcinogens, Teratogens and Biomolecules," In COMPUTER ASSISTED DRUG DESIGN, Eds. E.C. Olsen and R.E. Cristoffersen, ACS Symposium Series 112, Am. Chem. Soc., Washington, D.C., 1979, pp.415-435.

13. E. Scrocco and J. Tomasi, Top. Curr. Chem. 21, 97-127 (1973) and references therein.

14. C. Petrongolo, Gazz, Chim. Ital. 108 445-478 (1978) and references therein.

15. W.A. Sokalski, S. Roszak, P.C. Hariharan, W.S. Koski, Joyce J. Kaufman, A.H. Lowrey and R.S. Miller. Int. J. Quantum Chem. S17, 375 (1983)

16. C. Petrongolo, H.J.T. Preston and Joyce J. Kaufman, Int. J. Quantum Chem. 13, 457 (1978).

17. Joyce J. Kaufman, Nora M. Semo, and W.S. Koski, J. Med. Chem. 18, 647 (1975).

18. D. Carlstrom, Acta Chem. Scand. 20, 1240 (1966).

19. M. Martinez-Ripoll and H.P. Lorenz, Acta Cryst. B32, 2322 (1976).

20. M. Martinez-Ripoll and H.P. Lorenz, Acta Cryst. B32, 2325 (1976).

21. T.H. Lu, T.J. Lee, C. Wong, K.T. Kuo, J. Chinese Chem. Soc. 26, 53 (1979).

22. W. Van Havere, A.T.H. Lenstra and H.J. Geise, Acta Cryst. B38, 2516 (1982).

23. S. Ginsburg and J.B. Wilson, J. Am. Chem. Soc. 79, 481 (1957).

24. S.F. Mason, J. Chem. Soc. 22 (1960).

25. J. Hagedorn, J. Stark and H.P. Lorenz, Angen. Chem. Int. Ed. 11, 307 (1972).

Chapter 14

METAL CLUSTER TOPOLOGY: APPLICATIONS TO GOLD AND PLATINUM CLUSTERS

R.B. King
Department of Chemistry, University of Georgia

ABSTRACT

The graph-theory derived approach for metal cluster bonding is extended to gold and platinum clusters including spherical and toroidal centered gold clusters and stacked triangle platinum clusters; the latter appear to be novel examples of Möbius systems.

INTRODUCTION

In 1977 we published a graph-theoretical interpretation of the bonding topology in delocalized inorganic polyhedral molecules [1]. Our initial treatment [1] focussed on polyhedral boranes, carboranes, and metal clusters. Subsequent work [2] extended these methods to bare metal clusters of post-transition elements such as tin, lead, and bismuth. Further details of our methods are given in a recent book chapter [3]. In general the results of the graph-theory derived methods, insofar as a comparison is possible, are consistent with other approaches to metal cluster bonding by workers such as Mingos [4,5], Stone [6,7], and Teo [8,9,10].

This paper extends our graph-theory derived approach for metal cluster bonding to gold and platinum clusters, which require a variety of new concepts. Mingos [11,12,13] has extended his methods to the treatment of gold clusters but relatively little success has been achieved until now in the understanding of the bonding in platinum clusters. For example, Teo's methods [9] do not give exact electron counts for some of the most common types of platinum carbonyl clusters.

BACKGROUND

The atoms at the vertices of polyhedral cluster compounds may be light atoms using only s and p orbitals for chemical bonding (e.g., boron or carbon) or heavy atoms using s, p, and d orbitals for chemical bonding (e.g., transition metals or post-transition elements). If these vertex atoms are normal, they use three valence orbitals for intrapolyhedral bonding leaving one or six external orbitals in the case of light or heavy atoms, respectively. The single external orbital of a light vertex atom such as boron or carbon normally bonds to a single monovalent external group (hydrogen, halogen, alkyl, aryl, nitro, cyano, etc.). The six external orbitals of a heavy vertex atom such as a transition metal may be used for a much greater variety of purposes including the following: (1) A single external orbital bonding to a carbonyl, phosphine, or isocyanide ligand; (2) Three external orbitals bonding to a benzene or cyclopentadienyl ring; (3) A single external orbital containing a non-bonding lone electron pair (common for post-transition element vertices).

An important question in polyhedral cluster compounds is whether their chemical bonding is localized along the edges of the polyhedron or delocalized in the surface and volume of the polyhedron. Delocalized bonding occurs when there is a mismatch between the vertex degree of the polyhedron (i.e., number of edges meeting at the vertex) and the number of internal orbitals from the vertex atom. For normal vertex atoms using three internal orbitals there are the following three fundamental cases:
(A) Planar polygons (all vertices of degree two): Mismatch $(3 \neq 2)$ leading to delocalized bonding in planar polygonal aromatic systems such as benzene and cyclopentadienide.
(B) Simple [14] polyhedra (all vertices of degree three): Match (3=3) leading to localized bonding such as in polyhedranes (e.g., cubane, dodecahedrane, etc.).
(C) Deltahedra (all triangular faces) having no tetrahedral chambers (i.e., all vertices of degree four or greater): Mismatch $(3 \neq 4,5,6,...)$ leading to delocalized bonding in three-dimensional aromatic systems such as polyhedral borane anions, carboranes, and many metal clusters. The last case, of course, is the one of greatest interest in the context of this paper.

The three internal orbitals of normal vertex atoms in delocalized polygons or polyhedra can be partitioned into two types: (1) Twin internal orbitals (sp^2 hybrids or p orbitals in a light vertex atom polygon (Case A) or polyhedron (Case C), respectively); (2) Unique internal orbital (p orbital or an sp hybrid in a light vertex atom polygon (Case A) or polyhedron (Case C), respectively). The intrapolyhedral bonding in delocalized deltahedra without tetrahedral chambers and having n vertices requires 2n + 2 skeletal electrons arising from the following sources:
(A) Surface bonding (2n skeletal electrons) arising from pairwise overlap (i.e., n K_2 graphs) of the vertex atom twin internal orbitals

in the polyhedral surface to give n bonding and n antibonding orbitals. (B) Core bonding (2 skeletal electrons) arising from n-center overlap (i.e., a single K_n graph) of the vertex atom unique internal orbitals at the polyhedral center to give a single bonding orbital and n-1 antibonding orbitals.

Electron-rich delocalized polyhedra having more than 2n + 2 skeletal electrons form polyhedra having one or more non-triangular faces whereas electron-poor delocalized polyhedra having less than 2n + 2 skeletal electrons form deltahedra having one or more tetrahedral chambers. A more detailed discussion of bonding models for these systems is given in the previous papers [1,3].

GOLD CLUSTERS

The vertex atoms in the polyhedral clusters treated in our previous papers [1,2,3] use a spherical bonding orbital manifold (sp^3 for light vertex atoms and d^5sp^3 for heavy vertex atoms) having equal extent in all three dimensions leading to the 8-electron (for light atoms) or 18-electron (for heavy atoms) configurations of the next rare gas. However, in some systems containing the late 5d transition and post-transition metals including gold, one or two of the outer p orbitals are raised to antibonding energy levels leading to toroidal $(d^5)sp^2$ or cylindrical $(d^5)sp$ bonding orbital manifolds, respectively. The $(d^5)sp$ toroidal bonding orbital manifold can bond only in the two dimensions of the plane of the ring of the torus leading, for example, to 16-electron square planar complexes of d^8 late transition metals such as Rh(I), Ir(I), Ni(II), Pd(II), Pt(II), and Au(III). Similarly, the $(d^5)sp$ cylindrical bonding orbital manifold can bond only in a single (axial) dimension leading, for example, to 14-electron linear complexes of d^{10} metals such as Pt(O), Ag(I), Au(I), Hg(II), and TI(III). The p orbitals raised to antibonding energy levels can participate in $d\sigma \rightarrow p\sigma^*$ or $d\pi \rightarrow p\pi^*$ back-bonding with filled d orbitals in adjacent atoms as noted by Dedieu and Hoffman [15] from extended Hückel calculations on Pt(O)-Pt(O) dimers. The raising of one or two outer p orbitals to antibonding levels in heavy late transition metal and post-transition metal complexes has been attributed to relativistic effects [16].

The gold clusters of particular interest [17,18] consist of a center gold atom surrounded by a puckered polygonal belt of peripheral gold atoms generally with one or more additional peripheral gold atoms in distal positions above and/or below the belt. The peripheral gold atoms in such clusters use a 7-orbital d^5sp cylindrical bonding orbital manifold, but their residual two orthogonal antibonding p orbitals can receive electron density from the filled d orbitals of adjacent peripheral gold atoms leading to bonding distances between adjacent peripheral gold atoms. Centered gold clusters can be classified as either spherical or toroidal clusters [19] depending upon whether the center gold atom uses a 9-orbital d^5sp^3 spherical bonding orbital manifold or an 8-orbital d^5sp^2 toroidal bonding orbital manifold, respectively. The topology of the core bonding in the centered gold clusters is generally not that of the

K_n complete graph found in the delocalized deltahedral clusters discussed above but instead corresponds to the topology of the polyhedron formed by the peripheral gold atoms. This apparently is a consequence of the poor lateral overlap of the cylindrical d^5sp manifolds of the peripheral gold atoms. Also the volume of the polyhedron of peripheral gold atoms must be large enough to contain the center atom. Thus the icosahedron formed by the twelve peripheral gold atoms in $Au_{13}Cl_2[P(CH_3)_2C_6H_5]_{10}^{3+}$ is regular [20] whereas the cube formed by the eight peripheral gold atoms in $Au_9[P(C_6H_5)_3]_8^+$ is distorted from O_h to D_3 symmetry [21]. This arises from the fact that the internal volume of an Au_{12} icosahedron is large enough to accomodate the center gold atom whereas the internal volume of an Au_8 cube is too small to accomodate the center gold atom. The resulting swelling of the Au_8 cube leads to the observed symmetry reduction.

In the electron counting of centered gold clusters the (neutral) center gold atom is a donor of one skeletal electron, i.e., 11 valence electrons minus the 10 electrons needed to fill its five d orbitals. A toroidal centered gold cluster requires 6 skeletal electrons whereas a spherical centered gold cluster requires 8 skeletal electrons. These numbers are fully consistent with the 12p + 16 total (skeletal plus external) electron rule for toroidal centered gold clusters and 12p + 18 total electron rule for spherical centered gold clusters (p is the number of peripheral gold atoms) used by Mingos and co-workers [19]. Such electron counting leads to the general formulas $Au_nL_yX_{n-1-y}^{(y-5)+}$ for toroidal centered gold clusters and $Au_nL_yX_{n-1-y}^{(y-7)+}$ for spherical centered gold clusters where L is a two-electron donor ligand such as phosphine or isocyanide and X is a halide or pseudohalide. Examples of well-characterized toroidal clusters conforming to the $Au_nL_yX_{n-1-y}^{(y-5)+}$ general formula include $Au_8[P(C_6H_5)_3]_7^{2+}$ [22], $Au_9[P(C_6H_5)_3]_8^{3+}$ [23], $Au_9(SCN)_3[P(c-C_6H_{11})_3]_5$ [24], $Au_{10}Cl_3[P(c-C_6H_{11})_2C_6H_5]_6$ [19], and $Au_9[P(C_6H_4OCH_3-p)_3]_8^{3+}$ [25]. Examples of well-characterized spherical clusters conforming to the $Au_nL_yX_{n-1-y}^{(y-7)+}$ general formula include $Au_9[P(C_6H_5)_3]_8^+$ [21], $Au_{11}I_3[P(C_6H_5)_3]_7$ [26], and $Au_{13}Cl_2[P(CH_3)_2C_6H_5]_{10}$ [20].

PLATINUM CARBONYL CLUSTERS

The well characterized large platinum carbonyl clusters fall into the following two categories (Figure 1):
(A) Stacked $Pt_3(CO)_6$ triangles leading to the dianions $Pt_{3k}(CO)_{6k}^{2-}$ (k = 2,3,4,5) [27].
(B) Three stacked Pt_5 pentagons (BDB in Figure 1) having a Pt_4 chain (ACCA in Figure 1) inside the stack thereby leading to the $Pt_{19}(CO)_{22}^{4-}$ cluster [28].
A common feature of both of these types of systems is the stacking of Pt_n polygons leading to a system containing a principal C_n axis on which none of the polygon platinum atoms are located. In the case of the stacked triangle $Pt_{3k}(CO)_{6k}^{2-}$ clusters having such a

C_3 axis, the number of electrons arising from the vertex atoms must be a multiple of 3 or the total number of skeletal electrons must be 2 (mod 3) after allowing for the -2 charge. This requirement alone leads to the $2n + 2$ ($n = 3k$ in this case) skeletal electrons required for delocalized deltahedra but this is not consistent with the observed stacked triangle geometry of the $Pt_{3k}(CO)_{6k}^{2-}$ clusters. Furthermore the height of some of these stacks (i.e., $k = 5$ is known [27]) prevents the unique internal orbitals of all 3k vertex platinum atoms from overlapping at the core of the stack in a 3k-center bond having the topology of a K_{3k} complete graph.

A bonding model for the stacked triangle $Pt_{3k}(CO)_{6k}^{2-}$ clusters based on the observed geometries and electron counts can incorporate the following ideas:

(A) The $Pt(CO)_2$ vertices are anomalous using four internal orbitals rather than the normal three. They therefore have five external orbitals and are donors of four skeletal electrons each.

(B) The vertices of the <u>interior</u> triangles in the $Pt_{3k}(CO)_{6k}^{2-}$ stack have degree four so that the internal orbitals from these $Pt(CO)_2$ vertices match the corresponding vertex degrees in accord with expectations for edge-localized bonding.

(C) The vertices of the two <u>exterior</u> triangles in the $Pt_{3k}(CO)_{6k}^{2-}$ stack have degree three. After using three internal orbitals of these $Pt(CO)_2$ vertices for edge-localized bonding, there remains one internal orbital from each of the six platinum atoms of the two exterior triangles for further skeletal bonding. Let us call these "extra" internal orbitals on each vertex atom of the exterior triangles the <u>Möbius orbitals</u>.

(D) Edge-localized bonding in each of the 6k - 3 edges of the $Pt_{3k}(CO)_{6k}^{2-}$ stack requires 12k - 6 skeletal electrons. Since there are a total of 12k + 2 skeletal electrons, eight skeletal electrons are left for the two groups of three Möbius orbitals at the top and bottom of the triangle stack. The symmetry of the C_2 axes of the D_{3h} $Pt_{3k}(CO)_{6k}^{2-}$ stacks forces equal allocation of these eight electrons to the top and bottom of the stack. This means that at each end of the $Pt_{3k}(CO)_{6k}^{2-}$ stack there are four electrons for the molecular orbitals formed by the three triangularly situated Möbius orbitals. This electron count suggests that at the top and bottom triangles of the $Pt_{3k}(CO)_{6k}^{2-}$ stacks, there is 4m electron (m is an integer, namely one in this case) Möbius overlap involving a twisted ring (Möbius strip) of the three relevant orbitals rather than 4m + 2 electron untwisted Hückel overlap found in planar aromatic hydrocarbons such as benzene [29]. If the Möbius orbitals are d orbitals, then twisted Möbius overlap is possible for an <u>odd</u> number of metal atoms (e.g., a triangle or pentagon, but <u>not</u> a quadrilateral) since d orbitals change phase (i.e., "twist") at each metal cluster.

This bonding model for the $Pt_{3k}(CO)_{6k}^{2-}$ stacked triangle clusters suggests edge-localized bonding along the 6k - 3 edges of the stack coupled with delocalized Möbius triangles at both the top and bottom of the stack. Thus the edge-localized bonding in the $Pt_{3k}(CO)_{6k}^{2-}$ clusters corresponds to the edge-localized carbon-carbon σ-bonding in benzene whereas the Möbius bonding at the top and the bottom of the $Pt_{3k}(CO)_{6k}^{2-}$ stack corresponds to the Hückel π-bonding in benzene.

The structure of the threaded tubular platinum carbonyl cluster $Pt_{19}(CO)_{22}^{4-}$ (Figure 1) can be built as follows:
(1) Three Pt_5 pentagons (BDB in Figure 1) are stacked on top of each other forming two pentagonal prismatic chambers sharing a pentagonal face.
(2) A linear Pt_4 chain (ACCA in Figure 1) is placed on the C_5 axis of the stacked pentagons so that the two end members of the Pt_4 chain are the apices of pentagonal pyramids at the top and bottom of the pentagonal stack and the two central members of the Pt_4 chain are located in the centers of the two pentagonal prismatic chambers noted above.

In this structure of $Pt_{19}(CO)_{22}^{4-}$ the internal orbitals from the platinum atoms are used as follows:
(A) End platinum atoms of the Pt_4 chain (two platinum atoms): Three internal orbitals are used for a delocalized pentagonal pyramid and the fourth internal orbital is used for a localized bond to the nearest interstitial platinum atom also in the Pt_4 chain.
(B) Platinum atoms of the top and bottom Pt_5 pentagons (ten platinum atoms): Three internal orbitals are used for a delocalized pentagonal pyramid and the fourth internal orbital is used for a localized bond to the nearest platinum atom of the middle Pt_5 pentagon.
(C) Interstitial platinum atoms (the two center platinum atoms of the Pt_4 chain): All nine platinum valence orbitals are internal orbitals so that all of the ten valence electrons of each interstitial platinum atom become skeletal electrons.
(D) Platinum atoms of the middle Pt_5 pentagon (five platinum atoms): All four internal orbitals are used for edge-localized bonds to neighboring platinum atoms.
This allocation of platinum internal orbitals leads to the following electron counting scheme for $Pt_{19}(CO)_{22}^{4-}$:

Source of skeletal electrons:
    17 PtCO vertices using 4 internal orbitals: (17)(2) =   34 electrons
    5 "extra" CO groups: (5)(2) =   10 electrons
    2 interstitial platinum atoms: (2)(10) =   20 electrons
    -4 negative charge on anion   4 electrons
Total available skeletal electrons   68 electrons

Use of skeletal electrons:
    Edge-localized bonding in the Pt_{15} tube: 25 edges =   50 electrons
    Edge-localized bonding in the Pt_4 chain: 3 edges =   6 electrons
    Incremental electrons for the two delocalized
        pentagonal pyramidal chambers: (2)[(2)(6)+4-10] =   12 electrons
Total skeletal electrons required   68 electrons

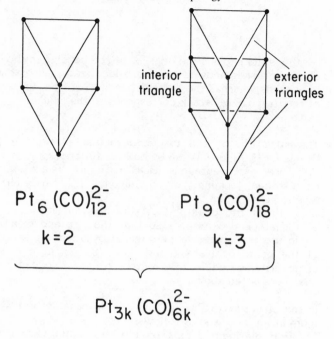

$$Pt_6(CO)_{12}^{2-}$$
$$k=2$$

$$Pt_9(CO)_{18}^{2-}$$
$$k=3$$

$$Pt_{3k}(CO)_{6k}^{2-}$$

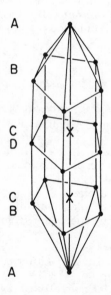

$$Pt_{19}(CO)_{22}^{4-}$$

Figure 1. Schematic diagrams of the platinum carbonyl clusters discussed in this paper.

This indicates that the anion $Pt_{19}(CO)_{22}^{4-}$ has exactly the number of electrons required for the above bonding model. Furthermore, the existence of a homologous series of threaded tubular clusters is predicted having the general formula $Pt_{6n+1}(CO)_{6n+4}^{4-}$.

SUMMARY

This paper shows how our graph-theory derived approach for metal cluster bonding can be extended to gold and platinum clusters exhibiting new structural features. For the treatment of these systems important ideas new to this theory need to be introduced, notably non-spherical (i.e., toroidal and cylindrical) bonding manifolds for the gold clusters and Möbius bonding for the stacked triangle $Pt_{3k}(CO)_{6k}^{2-}$ clusters.

ACKNOWLEDGEMENT

We are indebted to the U.S. Office of Naval Research for support of this research.

REFERENCES

[1] King, R.B.; Rouvray, D.H. J. Am. Chem. Soc. **1977**, 99, 7834.
[2] King, R.B. Inorg. Chim. Acta **1982**, 57, 79.
[3] King, R.B. in "Chemical Applications of Topology and Graph Theory," King, R.B., Ed; Elsevier: Amsterdam, 1983, pp. 99-123.
[4] Mingos, D.M.P. Nature (London), Phys. Sci. **1972**, 236, 99.
[5] McPartlin, M.; Mingos, D.M.P. Polyhedron **1984**, 3, 1321.
[6] Stone, A.J. Inorg. Chem. **1981**, 20, 563.
[7] Stone, A.J. Polyhedron **1984**, 3, 1299.
[8] Teo, B.K. Inorg. Chem. **1984**, 23, 1251.
[9] Teo, B.K.; Longoni, G.; Chung, F.R.K. Inorg. Chem. **1984**, 23, 1257.
[10] Teo, B.K. Inorg. Chem. **1985**, 24, 1627.
[11] Mingos, D.M.P. J. Chem. Soc. Dalton **1976**, 1163.
[12] Hall, K.P.; Gilmour, D.I.; Mingos, D.M.P. J. Organometal. Chem. **1984**, 268, 275.
[13] Mingos, D.M.P. Polyhedron **1984**, 3, 1289.
[14] Grünbaum, B. "Convex Polytopes," Interscience: New York, 1967, pp. 57-58.
[15] Dedieu, A.; Hoffmann, R. J. Am. Chem. Soc. **1978**, 100, 2074.
[16] Pyykkö, P.; Desclaux, J.-P. Accts. Chem. Res. **1979**, 12, 276.
[17] Schmidbaur, H.; Dash, K.C. Adv. Inorg. Chem. Radiochem. **1982**, 25, 243-249.
[18] Steggerda, J.J.; Bour, J.J.; van der Velden, J.W.A. Recl. Trav. Chim. Pays-Bas **1982**, 101, 164.
[19] Briant, C.E.; Hall, K.P.; Wheeler, A.C.; Mingos, D.M.P. J. Chem. Soc. Chem. Comm. **1984**, 248.
[20] Briant, C.E.; Theobald, B.R.C.; White, J.W.; Bell, L.K.; Mingos, D.M.P; Welch, A.J. J. Chem. Soc. Chem. Comm. **1981**, 201.
[21] van der Linden, J.G.M.; Paulissen, M.L.H.; Schmitz, J.E.J. J. Am. Chem. Soc. **1983**, 105, 1903.

[22] Vollenbroek, F.A.; Bour, J.J.; van der Velden, J.W.A. Recl. Trav. Chim. Pays-Bas **1980**, 99, 137.

[23] Bellon, P.L.; Cariati, F.; Manassero, M.; Naldini, L.; Sansoni, M. J. Chem. Soc. Chem. Comm. **1971**, 1423.

[24] Cooper, M.K.; Dennis, G.R.; Henrick, K.; McPartlin, M. Inorg. Chim. Acta **1980**, 45, L151.

[25] Hall, K.P.; Theobald, B.R.C.; Gilmous, D.I.; Mingos, D.M.P.; Welch, A.J. J. Chem. Soc. Chem. Comm. **1982**, 528.

[26] Bellon, P.L.; Manassero, M.; Sansoni, M. J. Chem. Soc. Dalton **1972**, 1481.

[27] Calabrese, J.C.; Dahl, L.F.; Chini, P.; Longoni, G.; Martinengo, S. J. Am. Chem. Soc. **1974**, 96, 2614.

[28] Washecheck, D.M.; Wucherer, E.J.; Dahl. L.F.; Ceriotti, A.; Longoni, G.; Manassero, M.; Sansoni, M; Chini, P. J. Am. Chem. Soc. **1979**, 101, 6110.

[29] Zimmerman, H.E. Accts. Chem. Res. **1971**, 4, 272.

Chapter 15

IONIZATION ENERGY AS PARAMETER FOR OXIDATIVE ELECTRON TRANSFER PROCESSES IN ORGANIC MOLECULES

Leo Klasinc[1,3], Hans Güsten[2] and Sean P. McGlynn[3]

[1]The Rugjer Bošković Institute, Zagreb, Croatia, Yugoslavia,
[2]Institut für Radiochemie, Kernforschungszentrum Karlsruhe, FR Germany,
[3]Department of Chemistry, Louisiana State University, Baton Rouge, Louisiana, USA

ABSTRACT

There are numerous examples of chemical properties, phenomena and processes which correlate surprisingly well with ionization energies of its subjects. However, in many cases such correlations lack a sound physical basis. Electrochemical oxidation of organic molecules and the one occurring in a noninteracting solution and/or gas phase by means of a strong oxidant can be considered, as processes with electron transfer taking place in the rate determining step. Since the latter processes are of great environmental significance the predictive power of an ionization energy reaction rate correlation would be of great use and importance. Examples leading to prediction of abiotic degradability of atmospheric pollutants and ozonization rates in drinking water treatment are presented.

INTRODUCTION

The primary concern of chemistry is compounds: their formation, properties and reactions. The obvious investigative route for satisfying these concerns is experiment, namely observation and measurement; the other is the development of models and the construction of theories. The focus of this work concerns the latter (i.e., models and theories). However, since we posit a chemocentric view, we assume that our results are also of fundamental significance to biology, physics, ecology, etc.

Chemistry assumes that electrons moving in the potential field of nuclei are crucial to the existence and development of those chemical forces that define the formation, properties and reactions of chemical compounds. One of the theoretical approaches that adopts this view is the orbital (atomic, AO, or molecular, MO) concept which describes the motion of an electron (or electrons) in the average field of a complicated system and assumes that the result bears on reality. The success of this method is well documented and the difficulties it encounters, particularly when confronted by spectroscopy [1], have been elaborated.

In this work, we will advocate the use of ionization energies as a basic parameter for the description of certain chemical processes. There are two main reasons for such an advocacy: (i), the ionization energy is an experimental quantity that is readily determined by numerous methods based on the release of electrons from chemical compounds; and (ii), a direct relationship of the ionization energy and the orbital model exists, although it is not always as straightforward as Koopmans' theorem [2] would suggest.

PHOTOIONIZATION, IONIZATION ENERGY AND KOOPMANS' THEOREM

The release of an electron from any chemical system is an ionization. We will confine ourselves here to the ionization of free atoms and molecules in the gas phase caused by electromagnetic radiation (photons). Such events are known as "photoionization" processes, the released electrons are termed "photoelectrons" and the technique used to determine their excess kinetic energy is referred to as "photoelectron spectroscopy." If the photon energy, as obtained, say, by monochromatized synchrotron radiation or tunable laser radiation, is exactly equal to the ionization energy, the photoelectrons will escape with zero kinetic energy. This type of electron spectroscopy is known as "threshold photoelectron spectroscopy" (TPS). If the photon energy exceeds the ionization energy, the surplus excitation is transferred to the ejected electron as a kinetic energy. The technologies that measure this excess energy are referred to as "X-ray or uv-photoelectron spectroscopy" or, for short, XPS and UPS, respectively. When the excess kinetic energy, E_k, of the photoelectrons ejected from a sample at fixed photon energy is plotted against their number, the result is a "photoelectron spectrum." The corresponding ionization energies, E_i, (or ionization potentials I, or electron binding energies BE) follow from

$$E_{i,j} = h\upsilon - E_k; \quad j = 1,2,...w; \quad E_{i,w} \leq h\upsilon$$

In the MO picture, the photoelectrons originate in individual electronic orbitals of the molecular ground state. If spinorbit coupling is small, each nondegenerate orbital is occupied by two electrons of opposite spin (Figure 1). However, the photoionization spectrum need not consist of the set of single-event processes that supposedly describes the UPS and XPS processes. Indeed, electron excitation can accompany electron ejection ("shake up"); two electrons can be ejected simultaneously ("shake off"); or a second electron can be subsequently ejected from the original highly excited ion produced in XPS (Auger process).

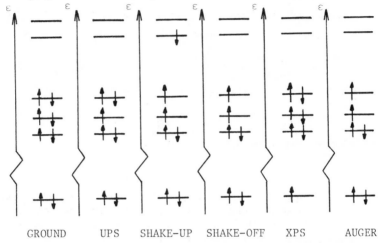

GROUND   UPS   SHAKE-UP   SHAKE-OFF   XPS   AUGER

Figure 1. Ionization processes in the single configuration approximation.

The connection of theory and experiment is given by Koopmans' theorem [2] which states that the electronic wavefunction of a singly ionized state is adequately described by Slater determinants based on the set of N-1 ground state self-consistent field (SCF) molecular spin orbitals (MSO). That is, if,

$$\Psi_0^N = | \phi_a (1) \phi_b (2) \ldots \phi_n (N) |$$

is a good descriptor for the ground state, then

$$\Psi_{-n}^{N-1} = | \phi_a (1) \phi_b (2) \text{----} \phi_m (N\text{-}1)|$$

is a good descriptor of the singly ionized state. The ionization energy, then is

$$E_{i,n} = E(\Psi_{-n}^{N-1}) - E(\Psi_0^N) = <\Psi_{-n}^{N-1}|H^{N-1}|\Psi_{-n}^{N-1}> - <\Psi_0^N|H^N|\Psi_0^N> = -\varepsilon_{nn}$$

where H^{N-1} and H^N are the SCF hamiltonians for the N-1 and N electron systems, respectively. This statement, namely that the ionization energy equals the negative of the orbital energy of the ejected electron, $-\varepsilon_{nn}$, is the first part of Koopmans' theorem.

Now, the function ψ_{-n}^{N-1} is by no means optimal. The optimal function may be written as the CI (configuration interaction) expansion

$$\psi^{N-1} = \sum_{k\varepsilon\delta cc} \psi_{-k}^u C_k^u + \sum_{\substack{k\varepsilon\ell\varrho cc \\ u\varepsilon\delta cc}} \psi_{-k\ell}^u C_{k\ell}^u$$

where, for example, $\psi_{-k\ell}^\mu$ denotes a determinant Ψ_0 in which spin-orbital ϕ_ℓ has been replaced by ϕ_μ; and where we have dropped the N-1 superscripting to avoid crowding. The function $\psi_{-k\ell}^\mu$, as is obvious, is a shake-up configuration (Figure 1). Simplification of ψ^{N-1} might consist of truncation to

$$\psi^{N-1} = \sum_{k\varepsilon\delta cc} \Psi_{-k} C_k$$

However, what we desire is really

$$\psi^{N-1} = \Psi_{-k}$$

This gross simplification is equivalent to the demand that we find an orthogonal transformation of the set of Hartree-Fock MSO's so that the cationic state can be represented by one single determinant constituted from this set, namely Ψ_{-k}, and the neutral ground state can be represented by one single determinant constituted from the same set, namely by Ψ_0. Koopmans' theorem asserts this possibility and it identifies the appropriate MSO set as the canonical Hartree-Fock set.

This latter assertion is the second and more important part of Koopmans' theorem. It may be rephrased alternatively: The only allowed ionizations are those which remove an electron from an MSO [or shake-up and shake-off transitions are forbidden]. If spin-orbit coupling is small (≤ 20 meV), a further restatement becomes possible: The only allowed ionizations are those which remove one electron from an MO.

Koopmans' theorem provides a salient experiment/ theory interface. Consequently, it is well to specify the approximations inherent in its derivation. These are: (i) Fixed-Nuclei Approximation - It is the Born-Oppenheimer approximation which permits the notion of "molecular geometry." Thus, in addition to this approximation, it is also understood that the cationic N-1 electron system which is the immediately terminal state of the process

N-electron system + $h\nu$ = (N-1)-electron system + e^-

is identical in all geometric detail to the initial state of the N-electron system. This, of course, is the Franck-Condon approximation. Consequently, Koopmans' theorem applies only to vertical ionization events.

(ii) The Correlation Energy - The neglect of correlation energy is intrinsic to the Hartree-Fock approximation. The correlation energy is caused by the fact that electrons adjust their motions to the instantaneous charge distribution, and not to an average charge distribution (as is assumed in the Hartree-Fock equations). In fact, the correlation energy is the difference between the correct energy and the Hartree-Fock energy associated with any given Hamilton operator. If relativistic effects are small, the latter is well known, and the "correct energy" is equivalent to the experimental energy. However, electrons of opposite spin usually tend to stay considerably further apart (i.e., correlate their motions better) than a single determinantal wavefunction will allow and, as a result, correlation energies can be quite substantial. Nonetheless, while large for any one state, it is only differences between two states, namely between the initial N- and terminal (N-1)-electron states, which is of significance to photoelectron spectroscopy. This difference may well be small. Koopmans' theorem implies that it is zero.

(iii) The Relaxation Energy - The same set of spinorbitals is used to construct the Slater determinants for the N- and (N-1)-electron systems. This supposition implies that the electrons of the cation do not adjust in any way to the reduction of interelectronic repulsions which must characterize the (N-1)-electron system. This supposition is known as the "frozen-core" or "frozen orbital" (fc) approximation.

(iv) The Non-Relativistic Approximation - This approximation is not a consequence of the functional nature of the wavefunctions; it is, rather, a defect caused by the omission of relativistic terms from the Hamilton operator. We have omitted these terms solely for convenience. The various relativistic terms -- for example, spin-orbit or spin-spin interactions -- might have been included in the Fock operator in a way which would not have altered any of our conclusions. In fact, in his original paper, Koopmans included relativistic effects explicitly -- and to no ill effects whatsoever.

(v) Restriction to Closed-Shell Systems - Koopmans' theorem is restricted to closed-shell N-electron systems. Thus, at least in the form expressed here, it is specifically inapplicable to non-closed-shell systems (e.g., many transition metal complexes).

The relationships between the experimental ionization energy, the MO energy and the Hartree-Fock ionization
energy are schematized in Figure 2. The correlation energy
is always negative and is shown to be slightly larger for
the system with the larger number of electrons.

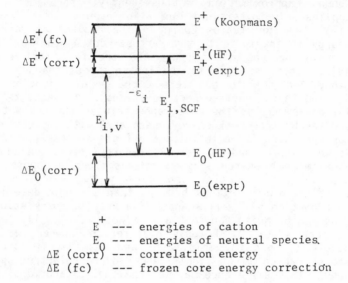

E^+ --- energies of cation
E_0 --- energies of neutral species.
ΔE (corr) --- correlation energy
ΔE (fc) --- frozen core energy correction

Figure 2. Relationship between experimental ionization energy, E_i, the
Koopmans' MO energy and the Hartree-Fock ionization energy, $E_{i,SCF}$

The reorganization energy for the cation [i.e., ΔE^+(fc) $\equiv$
[E^+(Koopmans) - E^+(HF)] is always positive. Hence, there is
a tendency for $|\Delta E^+$(fc)| + $|\Delta E^+$(corr)| to be approximately
equal to $|\Delta E^0$(corr)| and it is this tendency which is
responsible for the moderate success of Koopmans' theorem.
    Clearly, the higher the ionization energy (that is,
the deeper the MO from which electron ejection occurs) the
greater is the chance for breakdown of this simple one-
electron picture.
    Thus, our advocacy of the parametric use of experi-
mental ionization energies refers only to the lowest (or, in
some special cases, the next to lowest) I. The restrictions
to experimental values is not particularly stringent in view
of the numerous examples of correlations that exist between
calculated and experimental ionization energies (e.g.,
Schmidt [3]). Extensive correlations of experimental ion-

ization energies with various chemical properties are also available but, while often surprisingly good, they are usually difficult to justify theoretically. Such correlations can be very useful: for example, they may permit estimates of otherwise unattainable information. The following examples, all of which are based on processes in which organic compounds transfer an electron to a partner (electrode or reagent) and become oxidized, will now be discussed.

RELATIONSHIP OF THE STANDARD OXIDATION POTENTIAL TO THE IONIZATION ENERGY OF AROMATIC HYDROCARBONS [4]

Electron transfer to and from aromatic hydrocarbons plays an important role in various reductive and oxidative processes [5], [6]. The parameter of choice for the assessment of the ease of electron transfer during the reactions of these compounds with electrophiles and oxidants is the standard oxidation potential, E°. The most direct access to oxidation potentials is provided by electrochemical methods, among which cyclic voltammetry (CV) is undoubtedly the most appropriate for organic solvents. Unfortunately, values of E° for benzene derivatives were in short supply: the brief lifetime of arene cation radicals produces hysteresis in the cyclic voltammograms. However, the recent development of microvoltammetric electrodes [7], which makes it possible to record voltammograms at sweep rates larger than 10000V/s, has produced copious data for alkyl and polyalkylbenzenes [4]. Concurrently, we have measured the vertical ionization energies, $E_{i,v}$, using the HeI UPS technique [8-10]. Since $E_{i,v}$ is a gas phase value, any correlation with E° may permit estimation of solvation effects [11]. Similar, though different information may be obtained from i), the correlation of E° with the anodic peak potentials, E_p^a, obtained under irreversible CV conditions or from ii), the correlation of E° with the oxidation potentials of aromatic hydrocarbons when these are π-complexed to metal centers such as $Cr(CO)_3$.

The correlation of E° (volts relative to the normal hydrogen electrode, NHE) for 27 aromatic hydrocarbons in trifluoroacetic acid with $E_{i,v}$(eV) for the gas phase is found to be excellent. It is given by
$$E^\circ = 0.71 \, E_{i,v} - 3.68$$
with a correlation coefficient r = 9.98 (i.e., confidence level > 99.9%) The slope is considerably less than unity, indicative of the fact that the energetics in the gas and solution phases, while related, are not identical. In fact,

the solvation contribution is included in $E°$ whereas the reorganization energy of the arene cation radical, which undergoes Jahn-Teller distortion [12, 13], is not included in $E_{i,v}$. The difference in free energy, $\Delta G_s°$, associated with solution of the aromatic hydrocarbon Ar is

$$\Delta G_s° = [(G_{Ar}°{}^+)_s - (G_{Ar}°)_s] - [(G_{Ar}°{}^+)_g - (G_{Ar}°)_g]$$

where g and s refer to gas phase and solvation states. The standard oxidation potential

$$E° = (1/F) \left\{ \Delta G_s° + [(G_{Ar}°{}^+)_g - (G_{Ar}°)_g] \right\} + C$$

where F is the Faraday constant and C is a constant for a given electrode system. Since the vertical ionization energy is

$$E_{i,v} = (1/F) [(G_{Ar}°{}^+)_g - (G_{Ar}°)_g]$$

where $(G_{Ar}°{}^+)_g$ is the free energy of formation of the unrelaxed arene cation radical (Franck-Condon transition). Substitution in $E°$ yields

$$E° = E_{i,v} + [(\Delta G_r° + \Delta G_s°) /F] + C$$

where $\Delta G_r°$, the reorganization energy of Ar^+, represents the difference $(G_{Ar}°{}^+)_g - (G_{Ar}°{}^+)$. Comparison with the correlation line then yields

$$\Delta\Delta G° = \Delta G_r° + \Delta G_s° = -0.29F \cdot E_{i,v} + C'$$

$C' = 4.1 + C$, indicating that $\Delta\Delta G°$ is of the order of 7 kcal/mol for these compounds. Since the solvation energies of the neutral species are small, these effects must be attributed to the radical cations. It also appears that the deviation of the slope from unity represents variations of $\Delta G_s°$, because, among certain sterically-related compounds, the polyalkylbenzenes for example, the slope does approximate unity.

The correlation of $E°$ with the anodic peak potentials E_p obtained from irreversible CV in an acetonitrile medium yields

$$E° = 1.01 E_p - 0.001$$

where both potentials are expressed in volts referenced to NHE. This surprisingly good correlation (n = 0.98 for 18 compounds, slope ~1, intercept ~0) leads to a more general relationship that also includes many polycyclic aromatic hydrocarbons (PAH) [11]. This correlation is shown in Figure 3 and is given by

$$E_p = 0.63 E_{i,v} - 3.01$$

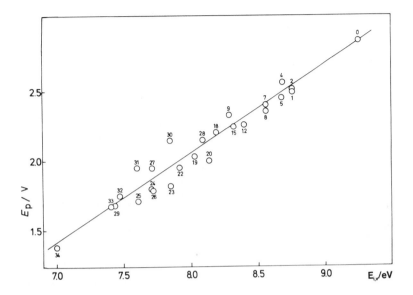

Figure 3. Correlation of the irreversible CV peak potentials, ep, of various aromatics in acetonitrile with the vertical ionization energy, $E_{i,v}$ numbers refer to benzene (0) and alkylbenzenes from [4] and naphthalene (28), anthracene (25), phenanthrene (30), chrysene (31), of benzanthracene (32), pyrene (33) and perylene (34) from [11].

where E_p is expressed in volts relative to NHE and $E_{i,v}$ is cited in eV. (r = 0.97 for 26 compounds). This linear correlation resembles ones reported by Miller et al. [14], Neikam et al. [15] and Pysh and Yang [16] for the polarographic oxidation potentials, $E_{\frac{1}{2}}$, of a great variety of organic compounds and for which the slopes $E_{\frac{1}{2}}/E_{i,v}$ = 0.89, 0.83 and 0.68, respectively. Thus, regardless of the variation of the free energy ΔG_r^o and the kinetic contributions from the follow-on reactions of the radical cation, remarkable correlation quality was obtained, indicating that these variations/contributions are either small and constant and/or that there is much fortuitous cancellation between them. We assume that the former is the more likely.

We now investigate our ability to correlate some important electron transfer processes in both the gas and the solution phases to ionization energies.

PREDICTION OF THE ABIOTIC DEGRADABILITY OF ORGANIC COMPOUNDS IN THE TROPOSPHERE [17]

Numerous oxidation reactions occur in the troposphere. Many of these lead to degradation of natural gaseous organic compounds as well as anthropogenic pollutants. These reactions are part of the self-purifying tropospheric process and they belong to the global carbon cycle. It is now well established that certain photochemically-produced, short-lived radicals are responsible for these oxidative, tropospheric reactions [18]. Among these, the extremely reactive OH radical is surely determinative of the lifetime and the distance a pollutant can travel in the course of tropospheric transport. Indeed, even ten years ago [19], the relative reactivity O:OH was known to be ~1:100.

The absolute rate, k_{OH}, for the reaction of OH with a gaseous chemical compound sets the upper limit for the troposphere lifetime of this compound. Indeed this reaction follows the pseudo-first order equation

$$\tau_{\frac{1}{2}} = \ell n2/k_{OH}[OH]$$

where $\tau_{\frac{1}{2}}$ is the tropospheric lifetime and [OH] is the mean annual concentration of tropospheric OH, for which Crutzen [20] cites the value 5×10^5 OH radicals/cm^3. Unfortunately, this equation is only approximate: other processes such as direct photolytic degradation by sunlight, the adsorption onto soil and the diffusion into the stratosphere can shorten the lifetime considerably. Nonetheless, this equation does provide a secure prediction for the maximum persistence time of chemicals that enter the atmosphere.

Measurements of OH reactivity are complicated and restricted to readily vaporizable compounds. Consequently, a vital need for a predictive capability of "hydroxyl reactivity" exists. Such a predictive capability, were it to exist, would also be of use in the liquid phase: Güsten et al. [21], for example, have shown that a statistically significant correlation of OH reactivity in water and OH reactivity in the gas phase does exist. Consequently, a great deal of correlative effort has been expended in this area: For H-atom abstraction from alkanes by OH-radicals, a correlation of the reaction rate with the bond energy has been uncovered [22-25]. Zetzsch [26] used structure reactivity relationships such as the Hammett equation to predict OH reactivities; and Gaffney and Levine [27] showed that the rate constant for the reaction of OH with alkenes and dienes correlates linearly with the first ionization energy.

We have recently constructed a large data base of k_{OH} and $E_{i,v}$ values [17] with the express purpose of studying their use in predicting abiotic degradibility and tropo-

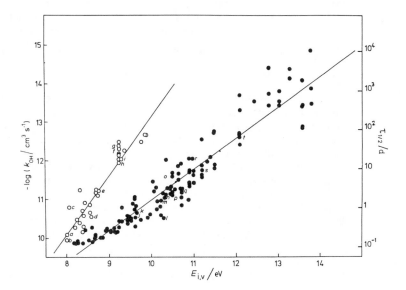

Figure 4. Correlation of – $\log k_{OH}$ vs. $E_{i,v}$ for 161 organic compounds in the gas phase at 300K. The text should be consulted for further elaboration. Open circles refer to aromatics and closed circles refer to aliphatics. The lettered datapoints, a-t, indicate reference chemicals [28].

Table 1. Predicted OH rate constants and mean tropospheric lifetimes of some chemicals at 300K.

Compound	$E_{i,v}$/eV	$-\log(k_{OH/cm^3s^{-1}})$	$\tau_{1/2}$/days
Dichlorodiphenylethylene (DDE)	8.23*	10.4	0.40
Trichlorobiphenyl (TCB)	8.34*	10.6	0.64
DDT	8.62*	11.0	1.6
Methylisocyanate (MIC)	10.6	11.4	5
p-Nitrophenol	9.38	12.1	25
2,6-Dichloro-benzonitril	9.79*	12.8	100

*See text for approximations

spheric lifetimes. This repository contains 361 k_{OH} data points for 161 different compounds and a considerably larger number of ionization energies. No simple correlation of $-\log k_{OH}$ vs. E_i was found to exist. However, knowing that $-\log k_{OH}$ for substituted benzenes does correlate with the corresponding ionization energies [29], it was decided to treat the aromatics and the aliphatics separately. This correlative attempt was successful, the only sour note being the discordant behavior exhibited by ketones, carboxylic esters, epoxides and halides (fluorides excepted). All these latter compounds have one property in common: their lowest energy ionization is associated with removal of an electron that is localized on the characteristic group (or atom). Consequently, it appears that the electrophilic OH radical does not attack organic molecules at the lone pair centers. Therefore, it seemed proper to substitute the deviant ionization energies by those for substitute molecules which were identical in all regards except for exclusion of the possibility of lone-pair ionization. These substitute molecules are:

$R-CO-R' \rightarrow RCH_2R'$; $RCOOR \rightarrow RCH_2R'$; $RCH-CH-R' \rightarrow RCH_2CH_2R'$
                                        (with O bridge)

$RX \ (\ X=C\ell, \ Br, \ I) \rightarrow RH$ or, better, RF

The correlative result using this tactic and embracing all 161 compounds is presented in Figure 4. Two clearly distinct, excellent correlations exist:

- for aromatics, with n = 32, r = 0.95, s = 0.29 and t = 16, we find

$$-\log (k_{OH}/cm^3 s^{-1}) = (1.52 \pm 0.10) \ E_{i,v}/eV - (2.06 \pm 0.84)$$

- for aliphatics, with n = 129, r = 0.95, s = 0.36 and t = 36, we find

$$-\log (k_{OH}/cm^3 s^{-1}) = (0.79 \pm 0.02) \ E_{i,v}/eV + (3.06 \pm 0.24)$$

where n is the number of molecules in the class; r is the correlation coefficient; s is the standard deviation; and t is the Student's t function. Both equations yield a predictive capability for k_{OH} that is accurate to one order of magnitude or better, the probability being about 90% ($\pm$ 1.5s). These two linear equations obviously reflect alternative reaction paths of the OH radical for aromatics and aliphatics at 300K, a view that is supported by the kinetic data of Rinke and Zetzsch [29] and Lorenz and Zellner [30]. In any event, for aromatics, the temperature dependency of k_{OH} indicates that OH addition is the dominant reaction path at room temperature.

On the basis of the regression line behavior we have predicted the reaction rate constants and the mean trospospheric lifetimes of a few chemicals of environmental significance (Table 1). One of these, methylisocyanate (MIC), is the compound responsible for the recent Bhopal (India) tragedy which killed or injured thousands. The tropospheric lifetime of this aliphatic compound was unknown but, from the published ionization energy [33], it was easily estimated as $\tau_{\frac{1}{2}} = 5d$.

PREDICTION OF LIFETIME OF TRACE ORGANICS IN DRINKING WATER TREATED WITH OZONE

The growing pollution of surface and ground water by chemical compounds increases the possibility that these will

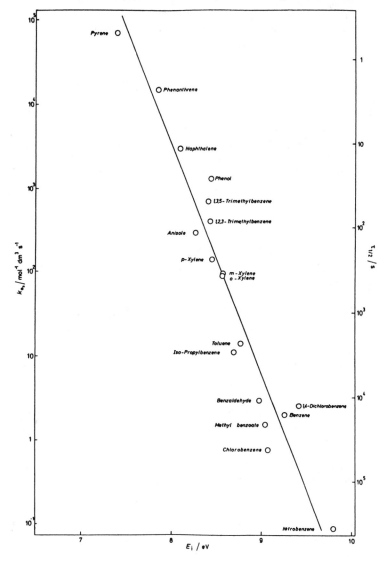

Figure 5. Correlation of log k_{03} vs. $E_{j,v}$ [10] for organic compounds in water at 25°C. The right ordinate gives half-lifetimes, $t_{1/2}(s)$, for an ozone concentration in water of 1mg/l.

enter the final stages of drinking water treatment. Among other methods for the final treatment of drinking water, the use of ozone, in particular, has a long tradition in Europe especially for removing bacteria and odor and improving taste. Thus, more than 150 ozone purification plants exist in Switzerland alone. These, operating with a 4-19 m contact with 0.4-10 mg/ℓ ozone, achieve 99% reduction in bacterial count [36-36]. Hoigné and Bader [34-36] have performed extensive measurements of reaction rates with ozone for numerous organic compounds in water, their aim being to determine pollutant lifetimes and to optimize the ozonation process. These authors found that the reaction was first order in both ozone and the organic, indicative of a situation similar to that for OH radicals. Thus, they found

$$\tau_{\frac{1}{2}} = \ell n2/k_{03}[O_3]$$

where k_{03} is the reaction rate constant and $[O_3]$ is ozone concentration in the water system. Thus regardless of the actual reaction mechanism(s), it is almost certain that the rate determining step is an oxidative electron transfer from the dissolved substrate. If so, a linear correlation of log k_{03} with ionization energies is to be expected and a predictivity of $\tau_{\frac{1}{2}}$ for trace organic compounds in water subjected to ozonation would appear to be at hand.

The reaction rate constants k_{03} of some substituted aromatics [38, 39] are plotted against lowest ionization energies in Figure 5.

The correlation is excellent. This correlation permits the prediction, for given E_i and $[O_3]$, of $\tau_{\frac{1}{2}}$ or, equivalently, predicts the appropriate contact time and $[O_3]$ for the reduction of contaminant concentration to an acceptable value. The right-hand ordinate of Figure 5 shows the value of $\tau_{\frac{1}{2}}$ at an ozone concentration of 1 mg/ℓ (or $2\times10-5\,M$).

Within the 20m time used for bacterial desinfection at this ozone concentration, it is seen that all compounds higher than toluene are oxidized. Benzene and the chlorinated benzenes, however, require contact times of 3h under these same conditions. Thus, in order to achieve oxidation of benzene or the chlorinated benzenes in times under 20 minutes, an ozone concentration of 10 mg/ℓ ($2\times10^{-4}M$) is necessary.

The higher aromatic compounds are oxidized within a few seconds which, in view of their carcinogenicity, is an important finding. Indeed, until recently [38] polycyclic hydrocarbons were thought to be highly resistive to ozonation [40, 41]. It is now clear that this is not so.

Correlations such as those of Figure 5 can be used to predict the persistence of an organic to ozone if its ionization energy is known. Although k_{O_3} is not as difficult to determine as k_{OH}, such prediction are very useful indeed.

ACKNOWLEDGEMENT

Support of this work by the Republic Council for Science of SR Croatia (SIZ-2),the Bundesministerium fur Forschung and Technologie (BMFT),the US Department of Energy and the National Institutes of Health is gratefully acknowledged.

REFERENCES

1. K. Wittel and S. P. McGlynn,Chem. Rev. 77 (1977) 745.
2. T. Koopmans, Physica 1 (1933) 104.
3. W. Schmidt, J. Chem. Phys. 66 (1977) 828.
4. T. O. Howell, J. M. Goncalves, C. Amatore, L. Klasinc, R. M. Wightman, and J. K. Kochi, J. Am. Chem. Soc. 106 (1984) 3968.
5. I. P. Beletskaya and D. I. Makhon'kov, Russ. Chem. Rev. (Engl. Transl.) 50 (1981) 1007.
6. M. Julliard and M. Chanon, Chem. Rev. 83 (1983) 425.
7. R. M. Wightman, Anal. Chem. 53 (1981) 1125A.
8. L. Klasinc, B. Kovač,and B. Ruščić, Kem. Ind. (Zagreb) 23 (1974) 569.
9. L. Klasinc, Pure Appl. Chem. 52 (1980) 1509.
10. L. Klasinc,B. Kovač, and H. Güsten, Pure Appl. Chem. 55 (1983) 289; L. Klasinc, unpublished results.
11. V. D. Parker, J. Am. Chem. Soc. 98 (1976) 98.
12. T. Nakajima, A. Toyota, and M. Kataoka, J. Am. Chem. Soc. 104 (1982) 5610.
13. M. Iwasaki, K. Toriyama, and K. Nunome, J. Chem. Soc. Chem. Comm. 1983, 320.
14. L. L. Miller, G. D. Nordblom, and E. A. Mayeda, J. Org. Chem. 37 (1972) 916.
15. W. C. Neikam, G. R. Dimeler, and M. M. Desmond, J. Electrochem. Soc. 111 (1964) 1190.
16. E. S. Pysh,and N. C. Yang, J. Am. Chem. Soc. 85 (1963) 2124.
17. H. Güsten, L. Klasinc, and D. Marić, J. Atmos. Chem. 2 (1984) 83.
18. T. E. Graedel, "Chemical Compounds in the Atmosphere", Academic Press, NewYork 1978.
19. H. Güsten,and R. D. Penzhorn, Naturwiss. Rundschau 27 (1974) 56.
20. P. J. Crutzen in "Atmospheric Chemistry" (E. D. Goldberg, Ed.) Springer-Verlag, Berlin-Heidelberg-New York, 1983, p. 313.

21. H. Güsten, W. G. Filby, and S. Schoof, Atmos. Environ. 15 (1981) 1763.
22. N. R. Greiner, J. Chem. Phys. 53 (1970) 1070.
23. K. R. Darnall, R. Atkinson, and J. N. Pitts, Jr., J. Phys. Chem. 82 (1978) 1581.
24. R. Atkinson, Int. J. Chem. Kinet. 11 (1980) 1197.
25. J. Heicklen, Int. J. Chem. Kinet. 13 (1981) 651.
26. C. Zetzsch, Conference on Chemicals in the Environment, Copenhagen, Oct. 18-20, 1982. pp. 302-312.
27. J. S. Gaffney and S. Z. Levine, Int. J. Chem. Kinet. 11 (1979) 1197.
28. The list of the compounds which were compiled by the European authorities (EC and OECD) is published by: B. Scheele, Chemosphere 9 (1980) 293.
29. M. Rinke and C. Zetzsch, Ber. Bunseng. Phys. Chem. 88 (1984) 55.
30. K. Lorenz and R. Zellner, Ber. Bunseng. Phys. Chem. 87 (1983) 629.
31. R. Atkinson, K. R. Darnall, A. C. Loyd, A. M. Winer and J. N. Pitts, Jr., Adv. Photochem. 11 (1979) 375.
32. H. Niki, P. D. Maker, C. M. Savage and L. P. Breitenbach, J. Phys. Chem. 82 (1978) 132.
33. J. H. D. Eland, Phil. Trans. Roy. Soc. (London) A268 (1970) 87.
34. C. J. Gomella, J. Amer. Water Works Assoc. 64 (1972) 39.
35. A. Netzer, A. Lugowski and S. Beszedits, Ozone: Sci. Engng. 1 (1979) 281.
36. J. C. Joret, J. C. Block, Ph. Hartemann and Y. Richard, Ozone:Sci. Engng. 4 (1982) 91.
37. J. Hoigné and H. Bader, Ozone:Sci. Engng, 1 (1979) 73.
38. J. Hoigné and H. Bader, Water Res. 17 (1983) 173.
39. V. Butković, L. Klasinc, M. Orhanović, J. Turk, and H. Güsten, Environ. Sci. Technol. 17 (1983) 546.
40. J. B. Andelman and J. E. Snodgrass, CRC Crit. Rev. Environ. Controll 4 (1974) 69.
41. R. M. Harrison, R. Perry and R. A. Wellings, Water Res. 9 (1975) 331.

Chapter 16

CHEMICAL GRAPH-THEORETIC CLUSTER EXPANSIONS

D.J. Klein
Department of Marine Sciences
Texas A&M University at Galveston

ABSTRACT

A general unifying computationally-amenable chemico-graph-theoretic cluster expansion method is offered as a paradigm for graph-theoretic applications in chemistry. The scheme is outlined and some of the multitude of possible applications are briefly indicated.

INTRODUCTION

There is a vast range of problems for which simple chemical structure ideas are relevant, and as a consequence graph-theoretic methods are expected to aid in formalizing, quantifying, and extending these ideas. Such problems range from the empirical to semiempirical and on to more purely theoretical realms. Some corresponding example problems include: the organization of the variations of a biological activity for a range of related molecules or for a similar organization of chromatographic separation coefficients (in the empirical realm); component (or cluster) expansions of the ground-state electronic energy of individual molecules or of molecular magnetic susceptibilities (in the semiempirical area); and the construction of size-consistent ab initio wavefunctions or the computation of statistical mechanical partition functions (in the theoretical regime). Because simple chemical bonding ideas are used in conjunction with such a wide range of diverse problems there arises the fundamental question: "Is there an underlying unifying computationally-amendable chemico-graph-theoretic paradigm applicable over this wide range of problems?" Here we suggest a form for such a paradigm so as to indicate an affirmative answer to this question.

There has been previous work toward such a paradigm. Notable efforts in this area are by Gordon, Kennedy and coworkers [1,2,3] primarily for empirical applications and by Domb [4] for statistical mechanical problems. In addition to indicating a broader range of application we extend the paradigm to encompass a wider family of mathematical approaches. For instance, no limitation of the choice for the cluster function to the so-called zeta function is made here. The present lifting of earlier restrictions to connected subgraphs permits the paradigm to extend to new (so-called multiplicative and derivative) classes of quantities, including partition functions and wavefunctions [5]. Much research [6-16] concerning so-called "topological indices" or graph-theoretic invariants for correlation with molecular properties can be viewed as making special choices for low- (and often fixed-) order cluster expansions of the type described here. A step further "back" is an immense quantity of work tabulating numerical results for particular properties via what may often be interpreted as fixed low-order graph-theoretic cluster expansions. For such work concerning thermochemical properties of organics see ref. [17-20]. Curiously, related developments in statistical mechanics seem to have taken place quite independently: earlier work dates back to the thirties [21-23]; the relevance of formal graph theory was emphasized in the fifties [24,25]; and Möbius inversion was used in the sixties [26-28]. In the quantum chemical regime it turns out, for instance, that the (clearly graph-theoretically related) valence-bond type wavefunctions [29-31] can be viewed [32] within the present context; still the ideas apply over a much broader range [5] of wavefunctions, as well as, to matrix element evaluation.

GRAPH-THEORETIC BACKGROUND

As usual a <u>graph</u> G is identified in terms of first a set V(G) of <u>sites</u> or vertices and second a set E(G) of <u>edges</u> consisting of (unordered) pairs of sites. There are many possibilities for what the sites may represent. Some examples include:

(a)　electrons or orbitals in atoms or molecules;
(b)　chemical bonds or bond- and lone-pair geminals in a molecule;
(c)　atoms or atomic ions in molecules or crystals;
(d)　functional groups in molecules or polymers;
(e)　molecules in an aggregate, liquid, or crystal.
The corresponding different choices for edges of G are associated to near-lying or more directly interacting pairs of sites. The usual molecular graphs of elementary chemistry are but one example of graphs.

The nomenclature here for subgraphs is standard [33]. Connected subgraphs are used to describe systems with no isolated (or noninteracting) pieces (or subsystems). A spanning subgraph G' of the graph G is such that $V(G')=V(G)$. As indicated later these subgraphs are useful in describing many-body ("multiplicative") global quantities. The subgraph partial ordering relation $G' \subseteq G$ is defined to mean that $V(G') \subseteq V(G)$ and $E(G') \subseteq E(G)$. The components of a graph are the maximal connected subgraphs. A selected set of connected subgraphs is denoted by $C(G)$, and the corresponding set of spanning subgraphs each component of which is a member of $C(G)$ is denoted by $C^x(G)$.

A useful type of function mapping graphs onto the real numbers consists of size functions $s(\cdot)$. By definition

$$s(G) \geqslant 0$$

$$G' \subseteq G \implies s(G') \leqslant s(G)$$

Given such a size function, say $c(\cdot)$, on connected graphs a corresponding size function $c^x(\cdot)$ on spanning graphs is to be such that $c^x(G)$ is the maximum of the sizes of its (disjoint) components as measured by $c(\cdot)$. One choice for the size $s(G)$ is the number $|V(G)|$ of vertices of G. Another choice for $c(G)$ is the (connected) graph's diameter, as measured by the maximum graph-theoretic distance between two vertices of G.

CLUSTER EXPANSIONS

Consider a general "property" X and its realizations $X(G)$ for systems labelled by graphs G. The associated cluster expansion is

$$X(G) = \sum_{G'} f(G,G')\, x(f,G')$$

where the $x(f,G')$ are f-irreducible cluster quantities corresponding to X and to the cluster function $f(\cdot,\cdot)$ which maps a pair of graphs onto the real numbers such that

$$f(G,G') \begin{cases} = 0 & , \; G \not\supseteq G' \\ \neq 0 & , \; G = G' \\ = ? & , \; G \supset G' \end{cases}$$

For example, for expansions of heats of combustion of hydrocarbons [7,17-20] in terms of connected-subgraph cluster expansions the $x(f,G')$ would be site energies, bond energies, etc. as G' increases from 1 site, to 2 sites, etc. A common choice for $f(\cdot,\cdot)$ is the zeta function

$$\zeta(G,G') = \begin{cases} 0 & , \; G \not\supseteq G' \\ 1 & , \; G \supseteq G' \end{cases}$$

with G, G' restricted to a set S of graphs such as $C(G)$ or $C^{\times}(G)$. Other common choices for $f(\cdot,\cdot)$ retain an "independence" of the property X and take the function values to be nonnegative and are such that $f(G,G')$ depend only upon portions of G that are "close" to G'. Another example is

$$f(G,G') = \begin{cases} \displaystyle\sum_{u \in V(G')} d(u,G)^{-\frac{1}{2}} & , \; G \supseteq G' \\ 0 & , \; \text{otherwise} \end{cases}$$

where $d(u,G)$ is the "degree" of u in G and the sub-graphs are restricted to a set $S = C(G)$. This set introduced by Randić [12] has been extensively applied [13] to a great variety of biological activities.

The cluster expansions may be truncated to yield a sequence of approximants, the nth order one being

$$X(f,n;G) = \sum_{G'}^{s(G') \leqslant n} f(G,G') \; x(f,G')$$

A significant feature here is that low-order approximants often yield very accurate estimates, as is commonly individually noted for a vast number of particular properties. (See, e.g., refs. [6-20].) A second feature is that in principle a whole hierarchy of approximants is possible. Different choices for X, $f(\cdot,\cdot)$, associated set S, and $s(\cdot)$ lead to different cluster expansion approximants. A crucial point concerns the manner of determination of the $x(f,G')$.

Perhaps the most common approach to estimate the $x(f,G')$ is to restrict attention to a single low-order approximant (with say $s(G') \leqslant n=2$) and choose the associated $x(f,G')$ so that $X(f,n;G)$ least-squares fit the experimental $X(G)$ for some chosen data set of systems G. Gordon and Kennedy [1] note that this least-square fitting approach has the undesired feature of dependence upon the data set chosen. Further extensions to higher order would entail a (typically) ever more rapidly increasing number of empirical parameters. Thence the physical-chemical meaning of these parameters becomes clouded and the possibility of multiple solutions to the least-squares optimization increases.

A second approach is to determine the $x(f,G)$ by "inversion" (from small subsystem data). That is, the f-irreducible cluster quantities are obtained from $X(G'')$ for smaller subsystems G'',

$$x(f,G') = \sum_{G''} f^{-1}(G,G'') \, X(G'')$$

where $f^{-1}(\cdot,\cdot)$ is the inverse cluster function such that

$$\sum_{G'} f(G,G')f^{-1}(G',G'') = \delta(G,G'')$$

Thus $f^{-1}(\cdot,\cdot)$ for reasonable size graphs is available via standard matrix inversion or via the recursion relation

$$f^{-1}(G,G) = 1/f(G,G)$$

$$f^{-1}(G,G'') = -\frac{1}{f(G,G)} \sum_{G' \subset G} f(G,G')f^{-1}(G',G'')$$

This relation is entirely analogous to that for the Möbius function [34], which is the inverse of the zeta function, and which is already explicitly given [2-4,28] for several choices of the graph set S. The inversion approach for the $x(f,G')$ has several desirable features: first, the $x(f,G')$ take values dependent only on the $X(G'')$ for $G'' \subseteq G'$, so that as data for larger graphs is added (or improved) the $x(f,G')$ do not change; and, second, no optimization is required but only a guaranteed well-defined computationally tractable inversion problem independent of the property X.

Yet a third or fourth approach is sometimes possible.

For instance, for wavefunctions the Rayleigh-Ritz variational principle might be used to make an optimal choice for the irreducible cluster wavefunctions. Treatment of many theoretical quantities is feasible via perturbation expansions.

An illuminating analogy comparing the first two approaches is possible. The optimization method (in the nth order) is analogous to curve fitting an nth degree polynomial to a data set of points. On the other hand the inversion method is analogous to a Taylor-like series expansion with the $x(f,G')$ corresponding to derivatives and the $f(G,G')$ to the monomials; more generally the $f(G,G')$ correspond to (perhaps orthogonal) polynomials of increasing degrees with the $x(f,G)$ the associated expansion coefficients.

CATEGORIZATION OF PROPERTIES

A quantity X may be categorized in terms of its behavior in the limit that a system $G = A\lor B$ breaks up into two separate noninteracting subsystems A and B. The four considered possibilities

$$X(A\lor B) \to X(A) + X(B)$$

$$X(A\lor B) \to X(A) \text{ or } X(B)$$

$$X(A\lor B) \to X(A)\cdot X(B)$$

$$\partial X(A\lor B) \to \partial X(A)\cdot X(B) + X(A)\cdot \partial X(B)$$

are here termed additive, constantive, multiplicative, and derivative, respectively. In the last case the derivative quantity ∂X is associated with a corresponding multiplicative property X.

Very frequently attention has in the past been restricted to additive quantities. Representative examples include:
(a) various energies (free, combustion, ground-state, etc.);
(b) entropy;
(c) magnetic susceptibility;
(d) melting and boiling points;
(e) optical refractivity;
(f) chromatographic separation coefficients;
(g) biological activities;
(h) model Hamiltonians; and
(i) 1- and 2-particle Green's functions.
Actually most of the previous work [1-20] overlooks nonscalar examples such as the last two mentioned;

these ideas, however, have actually been implemented
[35-37] for such cases.

Constantive quantities are in fact just particular
cases of additive quantities, where when the system
separates one of the component subsystem quantities
is zero. But there are many examples realizing this
special circumstance:
(a)   Many molecular excitation energies;
(b)   ionization potentials;
(c)   infra-red vibrational frequencies;
(d)   NMR chemical shifts;
(e)   dissociation constants; and
(f)   model Hamitonians for an impurity center in a
      crystal.
Again nonscalar examples occur.

There are many examples of multiplicative quantities:
(a)   wavefunctions;
(b)   overlap matrices over bases of Slater deter-
      minants with nonorthogonal localized orbitals;
(c)   statistical mechanical partition functions;
(d)   many-bond density operators;
(e)   characteristic polynomials for adjacency
      matrices;
(f)   matching polynomials; and
(g)   possibly irreducible cluster quantities x(f,G).
Also if X is additive, then exp{X} is multiplicative
and may be used in making cluster expansions even
though X might be the actual quantity of interest.
For instance, Hosoya et al. [10] treatment of boiling
points X(G) actually is of the form of a low-order
cluster expansion for exp{X(G)}; similar comments
apply to one scheme [38] for the treatment of
resonance energies.

There are many examples of derivative quantities
also:
(a)   Hamiltonian matrix elements for cluster expanded
      wavefunctions;
(b)   Hamiltonian matrices over bases of Slater deter-
      minants with nonorthogonal localized orbitals;
(c)   the statistical mechanical trace of
      $H \cdot \exp\{-H/kT\}$; and
(d)   ordinary (first) derivatives of many
      multiplicative quantities.
There also occur second derivative quantities both in
statistical mechanical and quantum chemical
applications.

A key point is that each category entails its own
general type of choice for the set S of subgraphs to
be summed over in a cluster expansion. For the

additive case S should be of the type $C(G)$, consisting of connected subgraphs; arguments supporting this contention for the circumstance of size-extensive quantities are found elsewhere [4,27]. For the constantive case the relevant S consists of those $G' \varepsilon C(G)$ such that G' contains the smallest subgraph G'' associated to a sole nonzero component in the limit where G'' is separated off as a noninteracting subsystem. For the multiplicative and derivative cases S should be of the type $C^x(G)$, consisting of spanning subgraphs. Under weak conditions on $f(\cdot,\cdot)$ it turns out that the irreducible cluster quantities $x(f,G)$ factor

$$x(f, A\cup B) = x(f,A) \cdot x(f,B)$$

whenever $G = A\cup B$ factors (i.e., when $A\cap B = 0$).

Finally there are several interrelations amongst the four categories of quantities and their associated cluster-expansion approximants. If X is multiplicative, then typically $\ln\{X(G)\}$ and $\partial X(G)/X(G)$ are additive as well as $\ln\{X(f,n;G)\}$, $\partial X(f,n;G)/X(f,n;G)$, $\{\ln X\}(f,n;G)$, and $\{\partial X/X\}(f,n;G)$. When X is additive, then X and its approximants typically are "size-extensive" in the sense that $X(G)/|V(G)|$ and $X(f,n;G)/|V(G)|$ are bounded as $|V(G)| \to \infty$, and often their limits exist. When X is multiplicative similar size-extensivity comments apply to $\ln\{X\}$, $\partial X/X$ and associated cluster-expansion approximants. All this indicates the correct qualitative behavior for cluster expansions as the system size changes, even in approaching the infinite extreme. That is, for instance, such wavefunction expansions [5] are "size-consistent" in a sense that many wavefunction calculations are not [39-42].

CONCLUSION

This brief presentation and overview of the chemico-graph-theoretic method of approximation has sought several goals. First, we have noted that many special treatments can be recognized as (often quite successful) low-order cluster expansions of our general type. Second, different attitudes toward and methodologies for the determination of the f-irreducible cluster quantities were indicated. Third, some ideas about the categorization of quantities and associated effects in the cluster expansions and their approximants were hinted at. Fourth, the range of problems to which these techniques was indicated to be much wider than often thought. As such it is suggested that the overall view that emerges may represent a powerful and wide-ranging new paradigm.

The author wishes to thank W. A. Seitz, T. G. Schmalz, R. D. Poshusta, and M. Randić for helpful discussion concerning the subject matter. Also research support by The Robert A. Welch Foundation of Houston, Texas is acknowledged.

REFERENCES

1. Gordon, M., and Kennedy, J.W., 1973, J. Chem. Soc. Faraday II 69, 484.
2. Essam, J.W., Kennedy, J.W., Gordon, M., and Whittle, P., 1977, J. Chem. Soc. Faraday II 73, 1289.
3. Kennedy, J.W., and Gordon, M., 1979, Ann. N.Y. Acad. Sci. 319, 331.
4. Domb, C., in Phase Transitions and Critical Phenomena III, 1974, pages 1-94, ed. C. Domb and M.S. Green (Academic Press, N.Y.).
5. Klein, D.J., 1977, Intl. J. Quantum Chem. 11, 255.
6. Redgrove, H.S., 1917, Chem. News 116, 37.
7. Franklin, J.L., 1949, Ind. Eng. Chem. 41, 1070.
8. Platt, J.R., 1952, J. Phys. Chem. 56, 328.
9. Smolenski, E.A., Russ. J. Phys. Chem. 38, 700.
10. Hosoya, H., Kowasaki, K., and Mizutani, K., 1972, Bull. Chem. Soc. Japan 45, 3415.
11. Rouvary, D.H., 1973, Am. Sci. 61, 729.
12. Randić, M., 1975, J. Am. Chem. Soc. 97, 6609.
13. Kier, L.B., and Hall, L.H., 1976, Molecular Connectivity in Chemistry and Drug Research (Academic Press, N.Y.).
14. Seydel, J.K., and Schaper, K.J., 1979, Chemische Struktur und Biologische Aktivitat von Wirkstoffen, Methoden der Quantitativen Strktur-Wirkung Analyse (Verlag Chemie, Weinheim).
15. Randić, M., and Wilkins, C.M., 1979, J. Phys. Chem. 83, 1525.
16. Mekenyan, O., Bonchev, D., and Trinajstic, N., 1980, Intl. J. Quantum Chem. 18, 369.
17. Tatevski, V.M., Benderskii, V.A., and Yarovoi, S.S., 1961, Rules and Methods for Calculating the Physico-Chemical Properties of Paraffinic Hydrocarbons (Pergamon Press, New York).
18. Janz, G.J., Thermodynamic Properties of Organic Compounds, 1967 (Academy Press, N.Y.).
19. Cox, J.D., and Pilcher, G., 1970, Thermochemistry of Organic and Organometallic Compounds (Academic Press, N.Y.).

20. Lyman, W.J., Reehl, W.F., and Rosenblatt, D.H.,
    1982, Handbook of Chemical Property
    Estimation Methods (McGraw-Hill, New York).
21. Ursell, H.D., 1927, Proc. Comb. Phil. Soc. 23,
    685.
22. Mayer, J.E., 1937, J. Chem. Phys. 5, 399.
23. Mayer, J.E., and Montroll, E.W., 1941, J. Chem.
    Phys. 9, 2.
24. Riddell, R.J., and Uhlenbeck, G.E., 1953, J.
    Chem. Phys. 21, 2056.
25. Uhlenbeck, G.E., and Ford, G.W., 1962, in
    Studies in Statistical Mechanics I, pages
    123-211, ed. J. deBoer and G.E. Uhlenbeck
    (North-Holland, Amsterdam).
26. Sherman, S., 1965, J. Math. Phys. 6, 1189.
27. Sykes, M.F., Essam, J.W., Heap, B.R., 1966, J.
    Math. Phys. 7, 1557.
28. Essam, J.W., 1971, Disc. Math. 1, 83.
29. Rumer, G., 1932, Nachr. Ges. Wiss. Gottingen,
    337.
30. Pauling, L., 1933, J. Chem. Phys. 1, 280.
31. Herndon, W.C., and Ellzey, M.L. Jr., 1974, J. Am.
    Chem. Soc. 16, 6631.
32. Klein, D.J., 1979, Phys. Rev. 19B, 870.
33. Trinajstić, N., 1983, Chemical Graph Theory
    (CRC Publishers, Boca Raton, Florida).
34. Rota, G.C., 1964, Zeit. Wahr. Verw. Geb. 2, 340.
35. Poshusta, R.D., and Klein, D.J., 1982, Phys. Rev.
    Lett. 48, 1555.
36. Malrieu, J.P., Maynau, D., and Daudey, J.P.,
    1984, Phys. Rev. B30, 1817.
37. Schmalz, T.G., Poshusta, R.D., and Klein, D.J.,
    1985, in preparation.
38. Swinbourne-Sheldrake, R., Herndon, W.C., and
    Gutman, I., 1975, Tetra. Lett. 755.
39. Langhoff, S.R., and Davidson, E.R., 1974, Intl.
    J. Quantum Chem. 8, 61.
40. F. Sasaki, 1977, Intl. J. Quantum Chem. 11S, 125.
41. Malrieu, J.P., 1979, J. Chem. Phys. 70, 4405.
42. Meunier, A., and Levy, B., 1979, Intl. J.
    Quantum Chem. 16, 955.

Chapter 17

ARE ATOMS DESTROYED BY FORMATION OF THE CHEMICAL BONDS?

Z.B. Maksić

Theoretical Chemistry Group, The "Rudjer Bošković" Institute, 41001 Zagreb, Yugoslavia and Faculty of Natural Sciences and Mathematics, University of Zagreb, Marulićev trg 19, 41000 Zagreb, Yugoslavia

ABSTRACT

Extensive evidence is provided which shows that atoms retain to a large extent their identity within molecules. It appears that a number of molecular properties can be rationalized by the model of perturbed or modified atoms in a molecule (MAM). Atomic modifications upon formation of chemical bonds can be classified as isotropic and anisotropic changes. The former is given by the atomic monopole which is a consequence of the intramolecular charge migration. It successfully reproduces diamagnetic shielding of the nuclei, diamagnetic susceptibility of molecules and ESCA chemical shifts. The anisotropic part of the electron charge distribution of an electron in a chemical environment is most easily described by the hybridization concept. It interprets directional features of covalent bonds and some energetic properties of molecules. In conclusion, the present results offer an intuitively appealing picture of molecules consisting of charged atomic cores immersed in a shallow "sea" of the mixed electron density.

INTRODUCTION

There are several hints which indicate that atomic bricks are identifiable in molecular buildings. One clue supporting the notion of a modified atom in a molecule is provided by molecular binding energies which are by two orders of magnitude smaller than a sum of total energies of atoms forming chemical bonds. The other is given by the X-ray deformation density maps which convincingly show that the electron charge redistribution accompanying the molecular formation is very small /1/. It is, therefore, not surprising that a large variety of molecular properties can be expressed as sums of atomic like entities. We shall discuss first the simplest model which belongs to the MAM category. This is the so called promolecule model where spherical and neutral atoms are situated at the equilibrium positions tacitly assuming that their mutual interactions equal zero. This obvious idealization reproduces quite closely some diamagnetic properties. Then we shall remove the electroneutrality constraint allowing for the charge migration. This is of crucial importance in describing molecular properties involving atoms with pronounced difference in electronegativity. Next we shall consider asymmetry of the local atomic charge distribution by using Pauling's hybridization concept /2/ neglecting at the same time the intramolecular charge transfer. This type of approach is appropriate, e.g., in hydrocarbons where the charge drift can be abandoned to the first approximation. It appears that hybridization has a very rich chemical content yielding useful information about molecular shape and size, local molecular properties like bond energies, angular strain, spin-spin coupling constants across one bond, C-H stretching frequencies and the like. Finally, the importance of the physical concepts in interpretational quantum chemistry will be briefly discussed.

THE MODIFIED ATOM IN A MOLECULE (MAM) MODEL

The idea of the distorted atom in a molecule was put forward by Moffitt /3/ in early fifties. More manageable variants of this approach can be found in recent works of Balint-Kurti and Karplus /4/, Goddard et al. /5/ and others /6/. An interesting attempt to define an atom in a chemical environment is provided by the virial partitioning of the electron charge distribution /7/ offering deep insight into some bonding phenomena /8/. We shall adopt a simple and pragmatic working hypothesis which is based in the first place on the assumption that molecules have a definite geometric structure defined within the Born-Oppenheimer clamped nuclei approximation. Then it will be supposed that there is a good bona fide partitioning of the mixed electron charge. The Mulliken population analysis will serve the purpose in our model calculations. This step in the MAM model building can be easily refined in the later stage if necessary. In particular, the population analysis which preserves higher atomic multipoles could be devised if desired. Finally, the electron charge distributions in molecules will be produced by the semiempirical IEHT method, which in turn appears to be one of the most reliable semiempirical schemes /9/. The local atomic anisotropies can be represented by hybrid atomic orbitals (HAO's). The latter can be generated in a number of different ways. Since hybridization has not an absolute meaning, each theoretical scheme defines its own scale for this useful bonding parameter. Hence, if the hybridization is to be studied in a large variety of sizeable molecules, a simple and efficient criterion is desirable. The most economical procedure in a sense of the Ockham's razor principle is provided by the maximum overlap method /10-12/. We shall employ the iterative maximum overlap (IMO) method which is capable to give a good description of the molecular shapes and sizes particularly in hydrocarbons /13/. The hybridization parameters produced by the IMO method can be favourably compared with indices calculated by the ab initio methods /14/.

CALCULATION OF MOLECULAR PROPERTIES BY THE MAM MODEL

Electric monopoles of atoms and magnetic properties

One of the most important parameters characterizing a modified atom in a·molecule is the formal atomic charge. Some magnetic properties can be estimated by the atomic point charge approximation to a very good accuracy. One of them is the diamagnetic part of the temperature independent magnetic susceptibility

$$\chi_{aa}^{d} = K \left[\langle b \rangle_e^2 + \langle c \rangle_e^2 \right] \tag{1}$$

Here a, b and c denote the inertial coordinates, $K = Ne^2/4m_e c_0^2$ where N is the Avogadro constant and c_0 is the velocity of light in the vacuum. We have shown that the average values of the electronic second moments $\langle a^2 \rangle_e$, $\langle b^2 \rangle_e$ and $\langle c^2 \rangle_e$ can be readily obtained once the atomic charges and atomic coordinates are known /15,16/:

$$\langle a^2 \rangle_e = \sum_A Q_A a_A^2 + \sum_p n_p k_p \tag{2}$$

where sum over A goes over all atoms. The first term gives a dominating contribution being a quadratic function of atomic coordinates a_A. Q_A is the total electron charge ascribed to the atom A. The second term is a relatively small correction to the first. It is isotropic and arises from the spatial extension of the atomic orbitals. Interestingly enough, the isotropic contribution is constant for all atoms belonging to the same p-th period of the Mendeleev system of elements. Therefore n_p is the number of atoms belonging to the p-th row and k_p are the corresponding empirical parameters (Table 1).

Table 1. Empirical K_p parameters for the calculation of molecular second moments (in 10^{-16} cm$_2$)

row-p	0	1	2	3	4
k_p	0.2	1	2.5	3.5	5.5

The latter correspond quite closely to the free-atom Hartree-Fock $\langle (r^2/3) \rangle$ values /17/ averaged over the row of the periodic system. Consequently, by using these ab initio results, the equation (2) can be written in a form free of adjustable parameters. It turns out that the molecular second moments are not very sensitive to the intramolecular charge drift. Hence, the promolecule model involving $Q_A = Z_A$ (notice that the minus sign of the second moments is dropped for the sake of simplicity) has a very good performance in most cases /15,16/. Then the formula (2) takes a pocket-calculator form

$$\langle a^2 \rangle_e = \sum_A Z_A a_A^2 + \sum_p n_p k_p \tag{3}$$

The charge transfer is important if atoms exhibiting widely different electronegativities are involved /18/. Alkali halides deserve a special attention in this respect. We assumed that 100 % ionic bond takes place in this family of compounds involving transfer of one valence electron of the alkali metal to the halide atom. Denoting alkali and halide atoms by M and X, respectively, the expression (3) reads as follows:

$$\langle a^2 \rangle_e = (Z_M - 1) a_M^2 + (Z_X + 1) a_X^2 + k_{p_M - 1} + k_{p_X} \tag{4}$$

This formula gives results in good accordance with the available ab initio data /19/ yielding at the same time a transparent explanation of the ab initio finding that the second moments in NaF and KCl are almost invariant to the shift of the coordinate system from the alkali, atom to the halide atom /20/.

Hence, the ionic $M^{+1}X^{-1}$ model works here very well.

Summarizing our extensive calculations of the diamagnetic (Langevin) susceptibilities of molecules /9,15,16,18,19,21/ we can say that formulas (2-4) yield results of the good quality which can be favourably compared with ab initio and (or) experimental findings. In some cases they have led to detection of errors in the literature data. In particular, they are helpful in determining sign of the molecular g-tensor which in turn is experimentally estimated only up to the sign /22/. In conclusion, a brief comment on the Flygare's et al. /23/ method for the calculation of second moments will be made. Their formula reads:

$$\langle a^2 \rangle = \sum_A Z_A a_A^2 + \sum_A 2a_A \langle a_A' \rangle + \sum_A \langle (a_A')^2 \rangle \qquad (5)$$

There is an apparent similarity to the formula (3) because the first and the last terms in expressions (3) and (5) closely correspond to each other. Flygare's approach has an additional term which involves atomic dipole components $\langle a_A' \rangle$. The latter are extracted from a large number of molecular dipole and quadrupole moments under the tacit assumption that atomic monopole contributions to these entities can be neglected. This is, however, a wrong hypothesis /9/ and consequently the Flygare et al. scheme should be revised accordingly.

Diamagnetic shielding of the nuclei (Lamb's shift) is another property which can be conveniently decomposed into atomic components. It was shown by Ramsey /24/ that the following approximate expression should hold:

$$\sigma_{av.}^d(A) = \sigma_{av.}^d(FA) + (e^2/3mc^2)\sum_B^i Z_B/R_{AB} \qquad (6)$$

where $\sigma_{av.}^d(A)$ refers to the free-atom value of atom A. One immediately observes that Ramsey's formula (6) is consistent with the promolecule picture. It yields reasonable estimates of the average diamagnetic shielding /25-28/. Allowing for the intramolecular charge transfer and including some adjustable constants in order to improve the performance, a semiempirical formula of the form:

$$\sigma_{av.}^d(A) = K_{A1}\sum_\mu^A (\xi_{A\mu} Q_\mu^A/n_{A\mu}) + K_{A2}\sum_B^i Q_B/R_{AB} + K_{A3} \qquad (7)$$

is obtained /29/. Here Q_μ^A and Q_B are gross orbital and gross atomic electron populations, respectively, $\zeta_{A\mu}$ is the AO's screening constant and $n_{A\mu}$ is the corresponding principal quantum number. It should be mentioned that the non-parametrized form $K_{A1}=K_{A2}=1$ and $K_{A3}=0$ yields also good results. The formula (7) is more general than the Ramsey's expression (6) because it encompasses charged species where (6) is not applicable. Test calculations have shown that IEHT--MO electron populations in conjunction with the formula (7) give reliable diamagnetic shieldings in radicals and cations /9/. Less satisfactory results can be expected in anions because the used IEHT method has then subminimal basis set. If the charge migration is highly pronounced like in alkali halides, then the screening constants are functions of the orbital populations, i.e., $\zeta_{A\mu}=f(Q_\mu^A)$ which should be explicitly taken into account /30/. It should be also pointed out that the atomic dipole moment method for the calculation of diamagnetic shieldings of Gierke and Flygare /31/ suffers the same conceptual drawback as in the case of diamagnetic susceptibilities (vide supra).

Electric monopoles of atoms and energetic properties

The total molecular SCF energy is roughly given by a sum of potentials V_A exerted on the nuclei /32,33/

$$E_{SCF} = \sum_A k_A\, Z_A\, V_A \tag{8}$$

where Z_A are atomic numbers and k_A are weighting factors which depend only on the nature of the atom A. The formula is able to recover about 99.5 % of the total energy. Since the potentials at the nuclei can be calculated with a satisfactory accuracy by using the atomic monopole approximation /34/, the formula (8) takes a transparent form

$$E_{SCF} = \sum_A k_A Z_A \left\{ - \sum_\mu^A (\zeta_\mu Q_\mu^A / n_{A\mu}) + \sum_B^i (Z_B - Q_B)/R_{AB} \right\} \tag{9}$$

involving orbital and atomic electron populations. The IEHT charge distributions yield total SCF energies exhibiting standard deviation of 0.1 a.u. /35/. This is not too bad in view of the simplicity of the model. It is worthwhile to mention that adjustable k_A parameters are close to 0.5 value, which is required by the virial theorem.

The dissociation energies to the ionic limit of the alkali halides are reasonably well reproduced by the ionic $M^{+1}X^{-1}$ point charge model /9/. Trends of changes are particularly well described. For example, the energy decreases along the series MX for a fixed M and X=F,Cl,Br and I. Similarly, if the halogen atom is kept fixed, the bond strength decreases along the MX family (M=Li,Na,K,Rb, and Cs). For large interionic distances the ionic $M^{+1}X^{-1}$ model yields almost quantitative estimates of the dissociation energies.

The atomic monopoles in conjunction with the electrostatic approximation are extremely useful in rationalizing the ESCA chemical shifts. As it is well known, X-ray photoelectron spectroscopy (XPS or ESCA) is a powerful tool for studying the charge distribution in molecules and crystals /36,37/. This finding is based on the fact that binding energies (BE) of the localized inner core electrons exhibit strong dependence on chemical environment. Another interesting feature is that binding energy shifts (Δ BE) parallel the changes in electrostatic potentials exerted on the nucleus of the ionized atom /36,38,39/. We have shown in a number of papers /40-43/ that, at the semiempirical level, considerable gain in accuracy is obtained if the IEHT method is employed. The basic formula for the ground state potential approach (GPM) reads

$$\Delta BE_A = k_1 Q_A + k_3 M_A + k_4 \qquad (10)$$

where M_A denotes the so called Madelung potential. The weighting parameters k_1 and k_3 absorb a good deal of the relaxation energy. However, generally better results are obtained if the reorganization energy of the valence electron cloud due to the creation of the positive hole is explicitly taken into account. This can be achieved by two models. The first invokes the equivalent core concept /44,45/ leading to the expression

$$\Delta BE_A = k_1 (\xi_A Q_A + \xi_{\bar{A}} Q_{\bar{A}}) + k_3 (M_A + M_{\bar{A}}) + k_4 \qquad (11)$$

if the atomic monopoles are used. Here the bar denotes the equivalent atom possessing the equivalent core. An alternative approach is provided by the pseudo-atom concept which simulates the transition potential describing the ionization process /46/. The corresponding formula is of the form

$$\Delta BE_A = k_1 Q_A^{TP} + k_3 M_A^{TP} + k_4 \qquad (12)$$

where the superscripts TP refers to pseudo-atom entities. Formulas (11) and (12) belong to the transition potential formalism (TPM) for an obvious reason. Our extensive IEHT calculations have conclusively shown that the electrostatic (monopole) approximation successfully describes main features of ESCA shifts in gaseous state and in molecular solids /40--43/. Performance of the IEHT method is illustrated by Table 2. Perusal of the data indicates that the

Table 2.   Standard Deviations of the ESCA Chemical Shifts as Calculated by the SCC-MO Wavefunctions employing Atomic Monopole Approximation (in eV).

Atom	GPM	RPM
B	0.3	-
C	0.6	0.3
N	0.4	0.4
O	0.6	0.6
F	0.2	-
Si	1.1	0.2
S	0.5	0.2
Ge	1.2	0.4

RPM approach is definitely better for heavier atoms like Si, S, Ge. The largest standard deviation is found for oxygen chemical shifts. It should be mentioned, however, that the IEHT method has appreciably better performance than other semiempirical methods. This is remarkable because the IEHT method cen be easily applied to large molecules involving heavy atoms. It proved very useful in discussing charge distribution in biologically important purines and pyrimidines /42,43/, sydnones, ylides, keto-enol tautomerism of ascorbic acid /47/ etc.

Atomic multipole moments and one-electron properties

Electric dipole and quadrupole moments of molecules can be reduced to atomic monopoles and dipoles /48/. One can quite generally say that higher molecular multipoles can be expressed by atomic multipole moments /49/. Another property of interest is extramolecular electrostatic potential (EP) which provides an indicator of chemical reactivity /50/. It can be satisfactorily reproduced by using polycentric expansion of the 1/r operator by using local atomic multipole expansion /49/ which can be extended to encompass overlap charges /51/. It should be strongly pointed out that atomic monopoles do not suffice and vice versa, the formal atomic charges derived from molecular EP's /52/ are unrealistic /9/.

HYBRIDIZATION AND GLOBAL MOLECULAR PROPERTIES

Hybrid orbitals /2,6,9/ conform to the local site symmetry of an atom in a molecular environment. Therefore hybrids represent chemically adapted atomic orbitals being particularly useful in describing localized (Lewis) covalent bonds. The approximate maximum overlap calculations are well documented /6,9,11--14/. Consequently, we shall briefly discuss some of the main results. The local hybrid orbitals reproduce the salient features of the electron charge distribution in small strained rings offering a simple and natural explaination of the bent bond phenomenon /53/. Spatial characteristics of hydrocarbons are well reproduced. Bond angles are predicted with an accuracy of a few degrees. Larger errors occur sometimes in dihedral angles. This is, however, not unexpected because the nonbonding repulsions are not explicitly considered and dihedral angles are easily deformed as a rule (the barrier is typically a couple of kcal/mole). The IMO method has a high predictive power in estimating interatomic distances. In several interesting cases the predicted geometry was correct as confirmed later by experimental measurements and ab initio calculations /9/. The bond overlap integrals can be successfully correlated with the heats of formation. The latter yield reasonable heats of hydrogenation and strain energies defined by a scale provided by the corresponding homodesmotic reactions /6,9/. In addition to these gross molecular properties, hybrid orbitals give a simple interpretation of local bond properties like the spin-spin coupling constants, C-H stretching frequencies, proton acidity and last but not least - bond energies. Since hybridization is a physical model, hybrid orbitals offer probably the best possible basis sets for approximate (semiempirical) methods. This conjecture follows directly from the fact that hybrids are symmetry adapted zero-th order local wavefunctions.Hence hybridization concept is a golden mine which is not fully exhausted as yet.

CONCLUSION

Several general conclusions can be drawn from the presented material. Atoms combine in myriads of ways to form molecules exhibiting different propeties. A remarkable finding is that electronic structure of atoms is not scrambled by the formation of chemical bonds. Instead, a picture of atomic cores embedded in the mixed –electron density seems to be essentially correct. It should be strongly pointed out that atomic cores involve not only the inner-shell electrons, but also a good deal of valence electron density. This is reflected in a number of molecular properties which can be concomitantly expressed as sums of atomic-like entities. On the other hand, shared (mixed) electron densities are well described by the perturbed atom model, as revealed by the astonishing success of the hybridization concept. The latter rationalizes inter alia the main facet of covalent bonding - spatial arrangement of chemical bonds - in an amazingly simple and elegant way. Of course, there are fine details and subtle molecular properties which can not be explained by the elementary models discussed above. They require closer scrutiny and more involved methods. Nevertheless, results presented here indicate rather strongly that chemistry is relatively simple although we don't know it in most cases. Another point of interest is that proper physical models usually require simple mathematics. All results given above can be obtained by the use of a mini-computer or even by a pocket-calculator. This is a very general feature of the quantum chemical methods. For example, the Hartree-Fock theory is numerically feasible because it is based on the polycentric LCAO basis set expansion. This is compatible with the empirical idea about the atomic structure of molecules. Hence the contemporary ab initio methods are not of an a priori type as some people seem to think. As a counter example one can mention the so called one-center method where the total molecular electron density is obtained by an expansion at the single (heavy atom) point.This approach was a failure in spite of the sophisticated mathematics. Its poor performance can be traced down to the underlying basic assumption which is conceptually wrong. It follows that the choice of a proper physical model can considerably simplify the necessary mathematical procedures, leading to the results in a most economical and meaningful way.

Acknowledgement: I would like to thank the Alexander von Humboldt Stiftung for a generous support in scientific books.

REFERENCES

1. P.Coppens and M.B.Hall,eds.,Electron Distribution and the Chemical Bond,Plenum Press,New York (1982).

2. L.Pauling,Proc.Nat.Acad.Sci. 14,359(1928); J.Am. Chem.Soc. 53,1367(1931).

3. W.Moffitt,Proc.Roy.Soc.London,Ser.A 210,224,245 (1951).

4. G.G.Balint-Kurti and M.Karplus,in Orbital Theories of Molecules and Solids,H.March,ed.,Clarendon Press,Oxford (1974).

5. W.A.Goddard,T.H.Dunning Jr.,W.J.Hunt and P.J.Hay, Acc.Chem.Res. 6,368(1973).

6. Z.B.Maksić,Pure Appl.Chem. 55,307(1983).

7. R.F.W.Bader and T.T.Nguyen-Dang,Adv.Quant.Chem. 14,63(1981).

8. D.Cremer and E.Kraka,Croat.Chem.Acta 57,1259(1984).

9. Z.B.Maksić,M.Eckert-Maksić and K.Rupnik,Croat. Chem Acta 57,1295(1984) and the references therein.

10.C.A.Coulson and T.H.Goodwin,J.Chem.Soc.2851(1962).

11.M.Randić and Z.B.Maksić,Theoret.Chim.Acta 3,59 (1965).

12.Z.B.Maksić,L.Klasinc and M.Randić,Theoret.Chim.Acta 4,273(1966).

13.K.Kovačević and Z.B.Maksić,J.Org.Chem.39,539(1974).

14.K.Kovačević and Z.B.Maksić, to be published.

15.Z.B.Maksić and J.E.Bloor,Chem.Phys.Letters 13,571 (1972); J.Phys.Chem. 77,1520(1973).

16.Z.B.Maksić,Croat.Chem.Acta 45,431(1973); J.Mol. Structure 20,41(1974); Croat.Chem.Acta 48,309(1976)

17.C.F.Fischer-Froese,The Hartree-Fock Method for Atoms,John Wiley and Sons,New York (1977).

18.Z.B.Maksić and N.Mikac,Chem.Phys.Letters 56,363 (1978); Mol.Phys. 40,455(1980).

19.Z.B.Maksić and N.Mikac,J.Mol.Structure 44,255(1978)

20.R.L.Matcha,J.Chem.Phys. 47,4995(1967) and the series of papers in this journal.

21.A.Graovac,Z.B.Maksić,K.Rupnik and A.Veseli,Croat. Chem.Acta 49,695(1977).

22.W.H.Flygare,Molecular Structure and Dynamics,Prentice Hall Inc.,Englewood Cliffs (1978).

23.T.D.Gierke,H.L.Tigelaar and W.H.Flygare,J.Am.Chem. Soc. 94,330(1972);W.H.Flygare,Chem.Rev. 74,653(1974).

24. N.F.Ramsey,Am.Sci. 49,509(1961).

25.S.I.Chan and A.S.Dubin,J.Chem.Phys. 46,1745(1967).

26. W.H.Flygare and J.Goodisman,J.Chem.Phys. $\underline{49}$,3122 (1968).
27. C.Deverell,Mol.Phys. $\underline{17}$,551(1969);$\underline{18}$,319(1970).
28. Z.B.Maksić and K.Rupnik,Croat.Chem.Acta $\underline{56}$,461 (1983).
29. Z.B.Maksić and K.Rupnik,Theoret.Chim.Acta $\underline{62}$,397 (1983).
30. Z.B.Maksić and K.Rupnik, to be published.
31. T.D.Gierke and W.H.Flygare,J.Am.Chem.Soc. $\underline{94}$,7277 (1972).
32. P.Politzer and R.G.Parr,J.Chem.Phys. $\underline{61}$,4258(1974).
33. P.Politzer,Isr.J.Chem. $\underline{19}$,224(1980).
34. Z.B.Maksić and K.Rupnik,Z.Naturforsch. $\underline{38a}$,308 (19837.
35. Z.B.Maksić and K.Rupnik,Theoret.Chim.Acta $\underline{62}$,219 (1983).
36. K.Siegbahn et al.,ESCA Applied to Free Molecules, North Holland,Amsterdam (1969).
37. U.Gelius,P.F.Heden,J.Hedman,B.J.Lindberg,R.Manne, R.Nordberg,C.Nordling and K.Siegbahn,Phys.Scr. $\underline{2}$, $\underline{70}$(1970).
38. H.Basch,Chem.Phys,Letters $\underline{5}$,337(1970).
39. M.E.Schwartz,Chem.Phys.Letters $\underline{6}$,631(1970).
40. Z.B.Maksić and K.Rupnik,Theoret.Chim.Acta $\underline{54}$,145 (1980);Croat.Chem.Acta $\underline{50}$,307(1977);Z.Naturforsch. $\underline{35a}$,988(1980).
41. Z.B.Maksić,K.Rupnik and N.Mileusnić,J.Organomet. Chem. $\underline{219}$,21(1981).
42. Z.B.Maksić and K.Rupnik,Nouv.J.Chim. $\underline{5}$,515(1981).
43. Z.B.Maksić,K.Rupnik and A.Veseli,Z.Naturfrosch. $\underline{38a}$,866(1983).
44. W.L.Jolly and D.N.Hendrickson,J.Am.Chem.Soc. $\underline{92}$, 1863(1970).
45. D.W.Davis and D.A.Shirley,Chem.Phys.Letters $\underline{15}$,185 (1972).
46. O.Goscinski,B.T.Pickup and G.Purvis,Chem.Phys.Letters $\underline{22}$,167(1973).
47. M.Eckert-Maksić,Z.B.Maksić and K.Rupnik,in progress.
48. F.L.Hirshfeld,Theoret.Chim.Acta $\underline{44}$,129(1977).
49. J.Bentley,in Chemical Applications of Atomic and Molecular Electrostatic Potentials,P.Politzer and D.G.Truhlar,eds.,Plenum Press,New York (1981).
50. E.Scrocco and J.Tomasi,Top.Curr.Chem.$\underline{42}$,95(1973).
51. A.J.Stone,Chem.Phys.Letters $\underline{83}$,233(1981).
52. F.A.Momany,J.Phys.Chem. $\underline{82}$,592(1978);S.R.Cox and D.E.Williams,J.Comp.Chem. $\underline{2}$,304(1981).
53. Lj.Vujisić,D.Lj.Vučković and Z.B.Maksić.J.Mol. Structure Theochem $\underline{106}$,323(1984).

Chapter 18

GIANT ATOMS AND MOLECULES

S.P. McGlynn, L. Klasinc, D. Kumar, P. Clancy, S.W. Felps and J. Dagata
Chemistry Department, The Louisiana State University, Baton Rouge, LA 70803, USA

ABSTRACT

Recent advances in the theory of atomic rydberg states have led to a reexamination of previous attitudes and to new modes of description for the highly-excited states of polyatomic molecules. This paper attempts to clarify some of these notions and to apply them, in order of increasing complexity, to the vacuum ultraviolet spectroscopy of a set of simple molecules.

With the demonstration by Seaton of a simple connection between the phase shift of scattering theory and the quantum defect of a rydberg equation, the whole of scattering theory infused the interpretation of rydberg spectra with new powerful technologies. It is the aim of this work to apply these technologies to a discussion of (i), single-channel; (ii), multi-channel quantum defect theory, particularly the Lu-Fano modification of Seaton's work, and its utility in treating perturbed spectra; and (iii), the final elaboration, generalized quantum defect (GQDT) theory, in which both the electron coordinate r and the nuclear coordinate R are both variable and both productive of continua at either r or R equal to infinity. GQDT provides a simple means of introducing channel rydberg and discrete valence state interactions. These interactions, depending on the energy of the rydberg state (i.e., pre or post the first ionization limit) and the nature of the valence states (i.e., dissociative or non-dissociative) can lead to autoionizations, predissociations or complicated energy/intensity/bandshape behavior.

INTRODUCTION

The purpose of this paper is to discuss the new field of giant atoms and molecules [1]. It begins with the n-dependency of atom size and proceeds from there to discuss the ways in which the physics and chemistry of these "giants" differ from the ordinary. In doing so, we have chosen to be eclectic: that is, we use the opportunity to present odds and ends of hitherto unpublished material from our own laboratories.

We also attempt to present a brief outline of single-channel, multi-channel and generalized quantum defect theory [2,3,4]. We do so with intent, because we are convinced that this approach provides the method of choice for theoretical investigations of molecular electronic spectroscopy and molecular dynamics. Unfortunately, limitations of space dictate a certain brevity and our discussion is mainly aimed at demonstrating simplicity and pertinence.

Many varieties of "giant" molecule exist and many of these have nothing at all to do with Rydberg states. For example, the delocalized excitations so common to the ordered solid state (e.g., excitons) [5] could well be considered to represent "giant" excitations, at least in a spatial sense. We will not consider such excitations in this work. However, we do consider one sort of delocalized excitation, namely a resonance phenomenon that typifies the spectroscopy of negative ions in aqueous phases. This phenomenon, commonly termed "charge transfer to solvent (CTTS) [6], is included here for two reasons: firstly, we believe the resonance description to be appropriate and we demonstrate this by specific consideration of the hydroxyl ion, OH⁻; secondly, such resonances may represent the norm, rather than the exception for small negative ions, ones which may not possess any bound rydberg states whatsoever. But, all in all, the inclusion of this topic is another example of the eclecticism that infuses this work.

THE SIZE OF RYDBERG ATOMS/MOLECULES

Atoms or molecules in highly-excited Rydberg states (i.e., states for which the principal quantum number n is big) can be exceedingly large. Indeed, since they can be considerably larger than either polymers or macromolecules, it is not improper to refer to them as "giant atoms" or "giant molecules" ... a terminology that is now generally accepted and, in our opinion, proper.

The best way to appreciate just how large a highly-excited Rydberg entity can be is to direct attention on the hydrogen atom and the dependency of its physical properties on n. These are listed in Table 1. In order to emphasize size, we now direct attention to the n^2 dependency of the electronic "radius" r. We also choose two very commonplace molecules, namely benzene and methyl iodide, in order to make our conclusions concrete:

Table 1.
Dependence of Some Physical Properties of Rydberg State of a Hydrogenic
System on the Principal Quantum Number n.

PROPERTY	n-DEPENDENCY
Radius	n^2
Cross-sectional area	n^4
Volume	n^6
F(Adjacents)^a	n
F(Series)^b	n^{-3}
ΔE(adjacents)^c	n^{-3}
ρ(E) δE^d	n^5
Electrostatic binding energy	n^{-2}
Electric polarizability	n^6
Diamagnetic susceptibility	n^4
Spin-orbit coupling	n^{-3}
Exchange energy	$n^{-x}(2<x<7)$

a) Transition probability between adjacent rydberg levels n ↔ n+1

b) Transition probability in the individual bands of a rydberg series m → n, where n = m+1, m+2, m+3, etc.

c) Energy separation between adjacent rydberg levels n ↔ n+1

d) Density of rydberg levels of a given series in the energy interval δE

---Benzene: The circle which just encompasses the benzene hexagon has r = 1.4 x 10^{-8}cm. Thus, when n = 30, r = 3.2 x 10^{-6}cm (320Å); and when n = 100, r = 3.5 x 10^{-5}cm (3500Å). It is important to stress that rydberg states of benzene for which n > 30 have been observed (i.e., that "benzene" with r > 320Å does exist) [7].

---Methyl iodide: Methyl iodide is not spherical, the C-H bond distance being 1.09Å and the C-I distance being 2.0Å. If we assume this C_{3v} molecule to be roughly spherical, we find r = 1.55 x 10^{-8}cm and, therefore, its size for n = 30 and 100 should be 5.6×10^{-7}cm (56Å) and 6.2×10^{-6} (620Å), respectively. Rydberg states of methyl iodide for which n > 35 have been observed. That is, CH_3I molecules for which r > 56Å exist [8].

These sizes appear in context when we note that a standard cell-wall protein, for which the number of amino acid residues is ~100, has a root mean square radius r_{rms} = 110Å; and that one of the very largest polymers, say polystyrene of molecular weight 10^6, exhibits an r_{rms} ~ 400Å. That is, known rydberg states exceed macromolecules in size.

Finally, the use of two color laser spectroscopy has led to the detection of molecular rydberg states for which n = 65 [9]. Indeed, in principle anyway, there exists no reason why appropriate molecules should not exhibit discrete rydberg states for which n > 100. That is, molecular rydberg states of radius greater than the largest known polymer have been detected already and ones that are at least ten times larger will be detected shortly.

Another way of looking at size, one with important consequences for the question of existence criteria for rydberg states in condensed media, queries the number of solvent molecules sampled by the rydberg electron of a giant molecule that is embedded in the solvent. If we take the n = 30 rydberg state of benzene and if we assume a close-packed solvent structure in which each solvent molecule is spherical and possesses a radius r = 1.75Å, we conclude that the rydberg electron of benzene samples 10^6-10^7 solvent entities during the course of one orbit. That is a very large number, indeed. It suggests that a), such a rydberg state cannot exist in a condensed medium; b), if anything resembling a rydberg state exists at all then it surely must behave as a Wannier exciton; and c), an exceedingly high probability exists for a trapping of the rydberg electron at some impurity or transient defect site (i.e., charge transfer to impurity or to solvent [CTTI] or [CTTS]) in the solvent medium.

It is clear, then, that the rydberg states of atoms and molecules can readily surpass in size any known macromolecule and that the terminology "giant" is an apt descriptor.

CONSEQUENCES OF SIZE

One consequence of size, namely existence criteria for discrete molecular rydberg states in high pressure or condensed media, has been alluded to already and will be discussed in some detail later. For now, we content ourselves with a few, terse, important observations, some of which will be elaborated in the next section.

(i) Diamagnetic susceptibility, with its cross-sectional n^4 dependence, will swamp any paramagnetism of the molecular system. That is, a linear Zeeman effect will be supplanted, at moderate n, say n $\geq$ 8, by a quadratic field dependence, and previously parity-forbidden transitions will begin to acquire intensity. At higher n values, the situation will be complicated further by progressive onsets of ℓ-mixing, n-mixing and various types of Landau resonances.

(ii) As with magnetic field effects, so also with electric field behavior. The electric polarizability, with its volume dependency, will completely dominate static dipole field effects, even at moderate n. Indeed, even the induced dipoles are not guaranteed to lie either parallel or anti-parallel to the zero-field static dipole moment, should one exist, of the ground state entity. And, as in (i), the induction of n- and ℓ-mixing will cause the appearance of many previously forbidden electronic transitions.

(iii) Spin-orbit coupling, ζ, and core/rydberg exchange, K, energies drop off rapidly with increasing n. However, the exchange energy drops precipitously (See Table 1), the result being that spin-orbit coupling dominates even at quite low n. Thus, in the atomic case, $\zeta/K \gg 1$, for $n \geq 5$ and (j,j)- coupling obtains. In the molecular case, (Ω, ω)-coupling will dominate the affairs of molecular rydberg states.

(iv) The density of rydberg levels at high n, $\rho(E)\delta E$, becomes high (See Table 1) and the spacing between them, ΔE, becomes small. In fact, many level separations fall in the microwave region and, since the energy of these levels is very sensitive to applied electric fields (see item ii above), such a system may have manifest technological importance. Similarly, since the ionization potential refers to $n = \infty$, the gross sensitivity to electric field, implied in the n^6-dependence of polarizability, may also provide certain technological advantages. The two technological possibilities hinted at here will be elaborated later. However, it is emphasized that these are randomly chosen and only exemplify the myriad device possibilities inherent in the area of giant atoms/molecules.

(v) The large size of the rydberg orbital suggests that the emphasis should lie on large r rather than small r behavior. In turn, this suggests that a scattering-theory approach or some variant thereof might be the approach of preference even for reasonably tightly-bound rydberg electrons. Such a variant exists and is known as quantum defect theory (QDT). Its basic premise consists of the assumption that the motion of the rydberg electron (or electrons) when it is <u>outside</u> the atomic molecular core may be treated differently from its motion when it is <u>inside</u> that core. The advantage of this dual view is that the properties of the separate motions, when appropriately parametrized and when matched at the core boundary, lead to relatively simple analytical formulations for energies, transition probabilities, band shapes, angular distributions of ionized electrons, etc. We will discuss certain of these formulations in some detail.

(vi) It is well to emphasize that some expectedly small rydberg states (i.e., ones of low n) may acquire "giant" character by virtue of interaction with ionization continua. A case in point is the phenomenon of autoionization in those low-n rydbergs of xenon that terminate on the 2nd ionization limit, I $(^2P_{\frac{1}{2}})$. These are usually referred to as Beutler-Fano resonances [3]. Similar effects also occur in molecules, as for example in the low-n rydbergs that terminate on the $^2E_{\frac{1}{2}}$ ionization limit of gaseous methyl iodide [10]. However, there exists another type of autoionization, one that occurs in condensed media and which is commonly referred to as "charge transfer to solvent" (CTTS) [6]. It is our contention that the CTTS process is actually a resonance phenomenon and that it does confer a form of giganticism on the molecule in which it occurs. However, this form of giganticism may have little or nothing to do with rydberg states. In view of that, a seeking for balance suggests that we discuss it in some detail. That we will do later.

- -

Finally, we emphasize that we will not discuss any form of exciton behavior (which, of course, is also a form of orbital giganticism). This topic has been addressed by others and we refer the interested reader to those sources.

MANIFESTATIONS OF SIZE

We now discuss items (i)-(vi) individually, the primary emphasis being on items (v) and (vi), with the others receiving terse but illustrative treatment.

(i) DIAMAGNETISM: The diamagnetic magnitude [11] is

given by

$$H_D = (e^2B^2/8m)n^4a_o{}^2$$

where B is the magnetic field and a_o is the Bohr radius, and the n^4-dependence is explicitly stated. Relative to H_D we may now formulate the various onsets for different magnetic perturbations. These are:

ℓ-Mixing: Occurs when $2R/n^3 \sim H_D$, R being an electron coordinate in a center of mass frame. Total angular momentum loses its "good" quantum number characteristic in this regime whereas n remains unsullied.

n-Mixing: Occurs when $2R/n^3 \gg H_D$. Adjacent rydberg manifolds merge under these conditions and the spectrum may become "non-rydberg."

Strong Mixing: Occurs when $2R/n^3 \sim \hbar\omega_c/2\pi$ where ω_c is the cyclotron frequency $\omega_c = -eB/m$. At this point, coulomb and external magnetic field forces are effectively comparable.

Landau Region: Occurs when $2R/n^3 << h\omega_c$. The dominant force is now magnetic in nature and the result is an oscillatory spectrum reminiscent of cyclotron resonance behavior.

(ii) POLARIZABILITY: The extreme sensitivity of the n^6 behavior is best illustrated by Figure 1. This figure compares

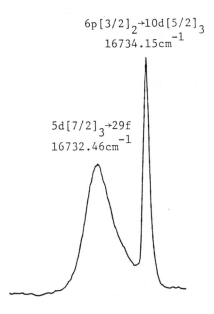

$$6p[3/2]_2 \to 10d[5/2]_3$$
$$16734.15 cm^{-1}$$

$$5d[7/2]_3 \to 29f$$
$$16732.46 cm^{-1}$$

Figure 1. A comparison of the effect of an rf field on bandshapes of two rydberg transitions, one to low n (n = 10) and one to high n (n = 29), in gaseous xenon. The broadening of the $16732.46 cm^{-1}$ band is attributable to extensive ℓ-mixing with the nominal $\ell = 3$ terminal orbital for this transition. This spectrum was obtained by optogalvanic sensing [12].

the bandshapes of two adjacent rydberg transitions, one of relatively high n, the other of relatively low n. The large half bandwidth of the high n transition is the result of extensive ℓ-mixing produced by an rf field, whereas the low n transition is unaffected by that same field.

(iii) SPIN-ORBIT COUPLING: This topic has been discussed in some depth [10]. Consequently, it will suffice here to provide one very illustrative, hitherto unpublished set of observations. Of the four distinct transitions that result from the $5p \to 6s$ rydberg excitation of CH_3I, the oscillator strength ratio to the two terminals Π_1 states [13]

$$F(^1\Pi_1)/F(^3\Pi_1) = [K/\zeta + (1 + K^2/\zeta^2)^{\frac{1}{2}}]^2$$

turns out to be a very sensitive function of K/ζ. A plot of this function is given in Figure 2 and superposed on it are the experimental data points for $5p \to s6$ (HI), $4p \to 5s$ (HBr) and $3p \to 4s$ (HCl). Obviously, the fit of experiment and theory is excellent. Furthermore, it is quite clear that HI, even in its lowest energy rydberg states, represents an (Ω, ω)- coupling situation.

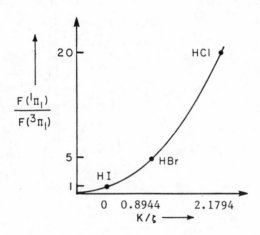

Figure 2. A plot of the observed oscillator strength ration $F(^1\Pi_1)/F(^3\Pi_1)$ for the hydrogen halides versus the exchange/spin orbit ratio K/ξ. The solid curve is also the theoretical plot.

(Ω, ω)- coupling in the higher states of these molecules, even in HF, can be taken to be complete.

(iv) TECHNICAL DEVICES: The large rydberg densities at high-n imply small energy separations and the n^6-dependence of the polarizability implies the ability to alter these level separations. The conjunction of these two characteristics suggests the use of high-n rydberg systems as broad-band microwave detector/amplifiers. Finally, the ease of field-ionization of rydberg systems for which $n \geq 100$ and the ready detectability of the field electrons suggests the use of these systems as weak-field detectors. A schematic of both these devices is given in Figure 3.

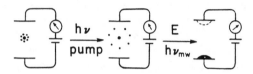

Figure 3. A schematization of a broad band microwave detector/amplifier and a weak electrical field sensor. The atomic core is denoted by a large solid point, the optical electron by smaller dots. The large size of the rydberg atom is shown in the middle circuit and the complete breakdown into core and electron, and their attachments to the appropriate electrodes, is shown on the extreme right.

The reader may choose to "invent" other uses as his/her mood dictates.

(v) QUANTUM DEFECT THEORY: An overview of this topic is available in Rau [2], Lu [3] and Greene [4], and the interested reader should consult these authors for details. We shall develop the topic in stages, starting with the simplest problem, namely the one-electron Coulomb problem.

--- One-electron Coulomb problem: We begin with the hydrogenic atom. The radial Schroedinger equation (in Rydberg units) is

$$[\frac{d^2}{dr^2} - \frac{\ell(\ell+1)}{r^2} + \frac{2Z_c}{r} + 2\varepsilon_j] \, f_j(r) = 0$$

where

$$\varepsilon = -Z_c^2/k^2; \quad k = \nu_j \text{ for } \varepsilon < 0; \quad k = i\gamma \text{ for } \varepsilon > 0$$

The bound states, $\varepsilon < 0$, are given by the regular Coulomb function f, which asymptotically is

$$f = u(\nu,r) \sin \pi\nu - v(\nu,r) \cos \pi\nu$$

where u is a rising and v a falling exponential in r. Since bound functions must vanish at $r=\infty$, we find

$$\sin \pi v = 0$$

and, consequently, v must equal an integer n and the energy becomes

$$\varepsilon_n = -z_c^2/n^2$$

We now remove the hydrogen-like restriction by introducing another potential term V such that the radial equation, still spherical, contains the added term $-2V$. We will also suppose that V is short-ranged so that at distances greater than some cut-off distance, say r_o, $V=0$ and the hydrogenic Schroedinger equation is recovered. That is, an excited electron in any other atom than hydrogen sees a complicated potential, one that is usually stronger than $-1/r$, at $r<r_o$ and a Coulombic potential at $r>r_o$. The boundary condition at r_o, namely that there be a smooth joining of the interior and exterior solutions to the Schroedinger equation containing V, yields for $r \geq r_o$

$$F_{j\ell}(r) = f_j(r)\cos \pi\mu_\ell - g_j(r)\sin \pi\mu_\ell$$

where f and g are the regular and irregular Coulomb functions. In the absence of a core (i.e., $V=0$), the solution reverts to f simply because the boundary r_o retreats to $r=0$. That is, $\mu \to o$ as $r_o \to o$.

Using appropriate asymptotic forms for f_j and g_j, we can rewrite

$$F_{j\ell}(r)^2 \cong u_j \sin \pi(v+\mu_\ell) - v_j \cos \pi(v+u_\ell)$$

whence if follows that

$$\sin \pi(v+u_\ell) = 0$$

and

$$\varepsilon_{n\ell} = -z_c^2/(n-\mu_\ell)^2$$

In this way, the Rydberg equation has been established. It has also been determined that n is an integer and that μ_ℓ, the ℓ-dependent non-integer part, is descriptive of the core effects. The quantity μ_ℓ is known as the quantum defect and its physical interpretation is straightforward: If V is attractive, the radial wavefunction is pulled into the core whereas, if V is repulsive, the radial wavefunction is repelled by the core; this attraction/repulsion is measured, in terms of its effects on the hydrogenic energy, by μ_ℓ and, in terms of its effects on the hydrogenic wavefunction, by the phase shift $\delta \equiv \pi\mu_\ell$. That is, $\pi\mu_\ell$ is the phase difference of the functions F and f.

---Single-channel quantum defect theory (SQDT): The treatment just given involves a single ionization limit and it treats only those levels that converge on that limit. It is known as SQDT. It has the advantage of physical simplicity. It also places levels that lie above or below that limit on much the same footing (i.e., it does not impose any artificial limit on the size of the basis but includes all levels, discrete and continuum, into the same complete basis set).

---Multi-channel quantum defect theory (MQDT): MQDT recognizes that most systems possess multiple ionization limits and that it is not a simple matter to assign a quantum defect to a given Rydberg level. Thus, if the system possess two distinct limits I_1 and I_2, it is not clear that the quantum defect for a Rydberg level of energy ε should be given by $\varepsilon = I_1 - Z_c^2/(n-\mu)^2$ or from $\varepsilon = I_2 - Z_c^2/(n'-\mu')^2$. This inherent ambiguity in the assignment of a specific level to a specific channel is merely an admission that considerable channel mixing may occur. Thus, at the energy ε, the eigenfunction may be written as a superposition of the various channels. One such superposition, for $r > r_0$, is

$$\Psi = \sum_i \phi_i \, [f(\nu_i,r) \sum_\alpha U_{i\alpha}\cos \pi\mu_\alpha A_\alpha$$

$$-g(\nu_i,r) \sum_\alpha U_{i\alpha}\sin \pi\mu_\alpha A_\alpha]$$

where ϕ_i represents the wavefunction for the ith ion core, for spin and for angular momentum coupling in the ith channel. That is, only the radial part of the outer electron is excluded from ϕ_i. This wavefunction has a rather simple interpretation. Once one speaks of a core state such as ϕ_i, one is essentially speaking of a fragmentation channel in which the core and certain of the optical electron characteristics are held fixed; in other words, one is describing a situation that obtains only at large r. In the case of an atom this would be a (J_c,j)-coupled situation. At small r, on the other hand, core-electron interactions may be so large as to dwarf the I_i-threshold separations and one must speak of compound states or compound channels. These are best described, particularly in lighter atoms, by LS coupling. They are, in fact, referred to as α-channels. In this context then, $U_{i\alpha}$ is merely an element of the matrix that projects i onto α (i.e., of the matrix that transforms the (J_c,j)-set into the (L,S)-set). $U_{i\alpha}$ is termed the frame transformation and $\pi\mu_\alpha$, as previously, is the phase shift induced in the large-r single-electron function caused by scattering from the core.

Facile ways of using MQDT have been devised by Lu [3] for atoms and Dagata [8] for molecules.

---Generalized Quantum Defect Theory (GQDT): GQDT is necessary for the extension from atoms to molecules, where discrete and continuum states may also be built on excitations of the atomic framework. That is, we must admit the possibility of rotational and vibrational motions and, hence, of dissociative molecular events. To be precise, for the molecule, we may write V as

$$V = V_{ex} + V_{so} + V_{ss} + V_{Js} + F_{JT}$$

where the exchange, spin-orbit and spin-spin parts are common to both the atom and the molecule but where representative electronic-nuclear coupling terms (for example, spin uncoupling F_{Js} and dynamic Jahn-Teller V_{JT} terms) now mix the i/α channels with framework channels. This sort of coupling may have two important consequences: it may lead to Born-Oppenheimer (BO) breakdown and, in fact, provide a way of handling the non-BO situation; and it implies that the frame transformation matrix must be generalized to cover body frame/laboratory frame basis set superpositions [14].

Such GQDT approaches are in process of refinement. The reader is referred to Greene [4], and Dagata and McGlynn [14] for further reading in an area that is destined for importance.

(vi) CHARGE TRANSFER TO SOLVENT (CTTS): The topic of rydberg states of neutral molecules in condensed media has been discussed by Jortner. We will not engage this topic further. Instead, we will broach the topic of charge transfer to solvent (CTTS) transitions. Such CTTS transitions are common to small inorganic anions such as I^-, CN^-, SO_3^-, OH^-, etc. It has been assumed that the excess negative charge on these ions causes them to have low ionization potentials and, hence, low-lying rydberg states; and that these rydberg states, because of a presumed big-orbit character and a low energy, are the natural precursors to CTTS and the production of trapped electrons. This, no doubt, is an interesting suggestion but one which may be flawed:

Firstly, if the ionization potential is very low, the potential surface on which the most loosely-bound electron moves may well be supportive of only one bound state, namely the ground state. That is, it is quite possible that no bound excited states, rydberg or otherwise, exist in many such ions. In fact, in an atomic negative ion, since the core possesses zero charge, rydberg states cannot exist.

Secondly, if CTTS is a very probable and very fast process, then this process is best thought of as a resonance. That is, CTTS may be akin to autoionization in a condensed medium. As such, the resonance "state" may be described by a binodal wavefunction, a part from the ion and a part from the surrounding solvent medium. Since that from the medium may well be dominant, it is this part that may be determinative of energy, symmetry, etc. with the ion part important only in that it confers transition probability on the optical excitation process. In that sense, the ion part of the binodal wavefunction need bear no relationship to any state or type of state, bound or continuum, of the gaseous ion.

The topic of CTTS, being related to big-orbit states, will now be discussed, the hydroxyl ion, OH^-, being taken as exemplar. However, we reemphasize that neither this implied big-orbit nature nor our discussion of CTTS in this work suggests any necessary connectedness to molecular rydberg states.

The ionization energy of OH_g^-, where g denotes gas phase, is only 1.8eV[15]. The ground state of OH_g^- is well characterized. However, no bound excited state of OH_g^- has ever been detected and none is to be expected. Thus, with reasonable assurance, we may accept that OH_g^- possess only one bound state, namely the ground state.

When OH_g^- is introduced into water, a diffuse absorption band onsets at ~2500Å (~5eV). That is, aquation of OH_g^- to yield OH_{aq}^- leads to absorption in a region in which OH_g^- does not absorb. The resolution of this rather puzzling result is provided by the further observation that the 5eV optical excitation leads to production of $e_{aq, trap}$ (as verified by near infrared absorption) and hydroxyl radicals, OH_{aq} (as verified by ready detection of the characteristic Gaydon emission system at 3300Å). As far as is known, the evolution of OH_{aq} and e^-_{aq} are esentially coincident with the absorptive act. Thus, the CTTS event

$$OH_{aq}^- + h\nu = OH_{aq} + e_{aq}^-$$

is almost surely a resonance, an autoionization that evolves into the cited products.

A considerable amount of thermodynamic data is available for hydroxyl radical, hydroxyl ion and the electron, in both gaseous and aquated states [16]. These data are used to construct Figure 4. The known data points in Figure 4 are all incorporated in the energies of the horizontal lines, the only one that is indeterminate being that for $OH_{aq} + e^-_{aq, trap}$ which may lie anywhere between -2.2 and-3.8 eV, dependent on trap type, water purity, etc. The potential energy curves are purely schematic, the upper referring to the gaseous system and the lower to the aquated situation. The dominant conjecture contained in this schematization is the supposition that the CTTS event leading to $OH_{aq} + e^-_{aq}$ proceeds through an intervening $OH_{aq} + e$ set of intermediate products and may also eventuate in an $OH_{aq} + e^-_{aq, trap}$ set of final products.

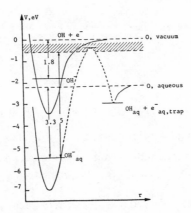

Figure 4. A schematization of potential energy curves for OH^- in the gaseous (upper) and aqueous (lower) phases. The CTTS transition is denoted by the 5eV, arrow which terminates in an effective continuum. The 1.8eV arrow denotes the ionization potential of OH^-_g and the 3.3eV arrow denotes the "barrier-free" ionization potential of OH^-_{aq}. The barrier in the lower curve occurs at -0.4eV and corresponds to $OH_{aq}+e^-$. The dashed line at -2.2eV corresponds to $OH_{aq}+e^-_{aq,free}$. The symbol r denotes an electron coordinate.

The resonance, now placed at 5eV, is also shown on Figure 4, the termini of the optical absorption being located in a continuum that barely dips below the potential maximum at -0.4eV. In view of this, the inner region (i.e., low r) part of the binodal function descriptive of the resonance "state" can and surely does possess characteristics of virtually any excited state (i.e., rydberg, valence, dissociative, etc.) conceivable for an OH^-_{aq} system. Nonetheless, the absorption is surely one that leads to the initial production of a giant molecule and to the rapid evolution of new sets of products that more aptly describe the resonance that does any supposed connection to states of the OH^-_g entity.

CONCLUSION

The purpose of this work is straightforward: To point out that attitudes and techniques convenient for the study of low-energy, intravalence, molecular states may have little or no bearing on the types of theory and experiment required to provide insights into vuv spectroscopy. A secondary objective of course consists of the demonstration that vuv spectroscopy has much to say about electronic structure and dynamics and that it will be very surprizing indeed if it does not alter our conceptual approaches to both structure and dynamics in very fundamental ways.

REFERENCES

1. S. P. McGlynn and G. L. Findley, J. Photochem. $\underline{17}$, 461 (1981).
2. A. R. P. Rau, p. 191-215 in "Photophysics and Photochemistry in the Vacuum Ultraviolet" (eds: S. P. McGlynn, G. L. Findley and R. H. Huebner), D. Reidel Publishing Co., Dordrecht (Holland), 1985.
3. K. T. Lu, p. 217-243, ibid.
4. C. H. Greene, p. 245-259, ibid.
5. J. Jortner, E. E. Koch and N. Schwentner, p. 515-578, ibid.
6. M. Fox, p. 333 in "Concepts of Inorganic Photochemistry" (eds: A. W. Adamson and P. D. Fleischer), Wiley-Interscience, New York, 1975.
7. J. P. Connerade, M. A. Baig and S. P. McGlynn, J. Phys. B: At Mol. Phys. $\underline{14}$, L67 (1981).
8. M. A. Baig, J. P. Connerade, J. Dagata and S. P. McGlynn, J. Phys. B: At Mol. Phys. $\underline{14}$, L25 (1981).
9. M. B. Robin (Bell Telephone Laboratories, Murray Hill, N. J.) unpublished results.
10. H. T. Wang, W. S. Felps, G. L. Findley, A. R. P. Rau and S. P. McGlynn, J. Chem. Phys. $\underline{67}$, 3940 (1977).
11. J. C. Gay, p. 631-706 in "Photophysics and Photochemistry in the Vacuum Ultraviolet" (eds: S. P. McGlynn, G. L. Findley and R. H. Huebner), D. Reidel Publishing Co., Dordrecht (Holland), 1985.
12. D. Kumar, L. Klasinc, P. L. Clancy, R. V. Nauman and S. P. McGlynn, "Pulsed Laser Optogalvanic Spectroscopy of Xenon in RF Discharge", Int. J. Quant. Chem., Proceedings of the 1985 Sanibel Symposia, in press.
13. W. S. Felps, G. L. Findley and S. P. McGlynn, Chem. Phys. Letters, $\underline{81}$, 490 (1981).
14. J. Dagata and S. P. McGlynn, J. Chem. Phys., in press.
15. K. P. Huber and G. Herzberg, "Molecular Spectra and Molecular Structure. IV. Constants of Diatomic Molecules", Van Nostrand Reinhold Co., New York, 1979, p. 516 [See Hotop, Patterson and Lineberger, J. Chem. Phys. $\underline{60}$, 1806 (1974)].
16. P. B. Merkel and W. H. Hamill, J. Chem. Phys. $\underline{54}$, 1695 (1971); $\underline{55}$, 2174 (1971).

Chapter 19

DIFFERENTIAL AND ALGEBRAIC TOPOLOGY OF CHEMICAL POTENTIAL SURFACES

Paul G. Mezey
Department of Chemistry, University of Saskatchewan,
Saskatoon, Saskatchewan, Canada S7N OWO

ABSTRACT

Differential and algebraic topology are exceptionally suitable mathematical tools for the local and global description of molecular structures and reaction mechanisms. In Reaction Topology geometrical concepts, such as nuclear position are replaced by topological concepts, such as open sets of wave packets. With respect to potential energy hypersurfaces, topology offers a new approach to the concept of quantum chemical reaction mechanisms. Whereas <u>chemical species</u> are represented by <u>catchment regions</u> of potential energy hypersurfaces, <u>reaction mechanisms</u> are represented by <u>homotopy equivalence classes</u> of reaction paths. These reaction mechanisms form a group, the one dimensional homotopy group of the low energy regions of the potential energy hypersurface. This group, referred to as the <u>fundamental group of reaction mechanisms</u>, serves as an algebraic framework for computer-based quantum chemical synthesis planning and molecular design.

INTRODUCTION

The energy content of a given collection of reacting molecules depends on their mutual arrangements in the ordinary, three dimensional space. In a semiclassical model of chemical reactions these arrangements can be described in terms of some internal coordinates, e.g. internuclear distances, bond angles, torsion angles, etc. When studying many chemical problems one may disregard the translation and rotation of the entire molecular system as a whole within a laboratory frame coordinate system. In such a case the mutual molecular arrangements (nuclear configurations) can be described within a 3N-6 dimensional, "reduced" nuclear configuration space M, that is a <u>metric</u> space (here N is the number of nuclei, and we assume that N>2). One can represent the energy of the reacting system of molecules in some specified electronic state as a hypersurface E(K) defined over M. The outcome of chemical reactions is, of course, strongly influenced by this energy hypersurface. In general, a chemical reaction can be thought of as a change in the nuclear configuration that corresponds to a formal displacement from one multidimensional basin of E(K) to another. These basins and analogous, lower dimensional point sets on the energy hypersurface are called the "catchment regions" of E(K).

Potential energy curves, surfaces and hypersurfaces have been studied for many reactions. Some (usually rather limited) parts of many hypersurfaces have been calculated and analyzed using quantum chemical methods [for recent reviews see e.g. 1-7] and also using the methods of molecular mechanics (empirical force fields) [8]. Most studies on the local properties of potential surfaces have been based on the semiclassical <u>geometric model</u> of molecules. Nonetheless, the fact that molecules are quantum mechanical entities is well recognized. Whereas the geometrical concepts of nuclear position and internuclear distance are satisfactory within a semiclassical model, alternative models are also explored, which are inherently more compatible with the probability density approach and the uncertainty relation of quantum mechanics. One such theoretical model, where geometrical concepts are replaced by <u>topological concepts</u>, is the subject of this review. In addition to a brief overview of some chemical applications of differential and algebraic topology, a simple method is given for the introduction of a fuzzy set structure and a fuzzy classification of chemical species within the differentiable manifold model of potential energy hypersurfaces.

DIFFERENTIABLE MANIFOLDS, MOLECULAR GEOMETRY AND MOLECULAR TOPOLOGY

The mass-weighted cartesian coordinates of N nuclei define a 3N dimensional nuclear configuration space ^{3N}R. The n=3N-6 dimensional reduced nuclear configuration space M is obtained from ^{3N}R as the quotient space with respect to the laboratory frame equivalence of nuclear configurations related to one another by rigid rotation and rigid translation [9,10]. The potential energy functional can be represented as a continuous hypersurface $E(K)$ over space M.

Various chemical species are represented by catchment regions [11,12] of $E(K)$. Each catchment region $C(\lambda,i)$ belongs to a critical point $K(\lambda,i)$, and is a collection of all points K of M from where a steepest descent leads to $K(\lambda,i)$. In the above notation λ is the critical point index of the i-th critical point $K(\lambda,i)$. For a stable chemical species $\lambda=0$, (that is, $K(\lambda,i)$ is a minimum), for a transition structure ("transition state") $\lambda=1$ (that is, $K(\lambda,i)$ is a simple saddle point) and for formal unstable chemical species represented by lower dimensional catchment regions $C(\lambda,i)$, $\lambda \geq 2$. Note, that in a complete catchment region partitioning of M these lower dimensional catchment regions (even zero dimensional, single point catchment regions $C(n,i)=\{K(n,i)\}$, representing a highly unstable, formal "chemical species") play an important mathematical role, although their direct chemical significance is minimal. (However, they represent constraints on the chemically important stable species and transition structures). Those catchment regions from where steepest descent paths lead to points where $E(K)$ (in fact, the original $E(\underline{r})$, $\underline{r} \in {}^{3N}R$) is not twice continuously differentiable, are denoted as $C(-1,i)$.

These catchment regions generate a complete partitioning of M:

$$M = \bigcup C(\lambda,i). \tag{1}$$

It is convenient to introduce local coordinate systems into these catchment regions. If T denotes the metric topology on M, then (M,T) is a <u>normal</u> topological space that fulfills the following separation axiom: if $\bar{C}(\lambda,i)$ and $\bar{C}(\lambda',j)$ denote the closures of catchment regions $C(\lambda,i)$ and $C(\lambda',j)$, respectively, in the metric of M, and if

$$\bar{C}(\lambda,i) \bigcap \bar{C}(\lambda',j) = \emptyset \tag{2}$$

then there exist T-open sets $G(i)$ and $G(j)$ such, that

$$C(\lambda,i) \subset G(i) \tag{3}$$

$$C(\lambda',j) \subset G(j) \tag{4}$$

and

$$G(i) \bigcap G(j) = \emptyset . \tag{5}$$

We may assume that an open set $G(i)$ is given for every

catchment region $C(\lambda,i)$ that satisfies the above conditions for all possible pairwise combinations of catchment regions. Evidently, such $G(i)$ open set exists, since, if $G_1(i)$, $G_2(i)$ ... $G_k(i)$ are sets chosen for a sequence 1, 2 ... k of pairs, involving a fixed catchment region $C(\lambda,i)$, then the set

$$G(i) = \bigcap_{\ell=1}^{k} G_\ell(i) \tag{6}$$

fulfills all k separation conditions.

We assume that a class $\{G(i)\}$ of such sets is given. These T-open $G(i)$ sets can then be used to define diffeomorphisms $\phi(i)$ between subsets of M and subsets of the n-dimensional Euclidean space ^nE. We shall assume that for every index i a diffeomorphism

$$\phi(i):G(i) \;\to\; H(i) \subset {}^nE \tag{7}$$

is given, where $H(i)$ is open in the usual metric of ^nE.

The Euclidean space ^nE is provided with the usual coordinate functions $\{u_k\}$, which are compatible with the usual metric, and are interpreted by

$$u_k(\underline{t}) = t_k \tag{8}$$

where

$$\underline{t} = (t_1, \ldots t_n) \in {}^nE. \tag{9}$$

The space (M,T) is a metric topological space, hence it is also a Hausdorff space. Function $\phi(i)$ is a diffeomorphism, hence it is also a homeomorphism form an open set of M to an open set of ^nE. These are precisely the conditions required for a coordinate system, consequently, $\phi(i)$ is an n-dimensional coordinate system in M. Since the catchment regions generate a partitioning of the nuclear configuration space M it follows that open sets $\{G(i)\}$ form an open cover of M:

$$M = \bigcup G(i) . \tag{10}$$

That is, the nuclear configuration space M is covered by domains of n-dimensional coordinate systems $\{\phi(i)\}$, consequently, M is an n-dimensional topological manifold [13,9].

The $G(i)$ domain of coordinate system $\phi(i)$ is called the coordinate neighbourhood of $\phi(i)$ and if $K \in G(i)$ then $\phi(i)$ is said to be the coordinate system at K.

The composition of function f and g, that is, g followed by f, is denoted by f.g,

$$f \cdot g: A \longrightarrow B \tag{11}$$

where

$$g: X \longrightarrow Y \tag{12}$$

$$f: Z \longrightarrow W \tag{13}$$

and

$$A = g^{-1}(Z) \cap X, \tag{14}$$

$$B \subset W. \tag{15}$$

The domain A of f.g may be empty. The composition of coordinate function u_k and diffeomorphism ϕ will be denoted by x_k,

$$x_k = u_k \cdot \phi . \tag{16}$$

Usually the same term, "coordinate system", is applied for both ϕ and the set of functions

$$(x_1, x_2 \ldots x_n) = (u_1 \cdot \phi , u_2 \cdot \phi , \ldots u_n \cdot \phi). \tag{17}$$

The concept of coordinate neighbourhoods allows one to introduce a fuzzy set [14] topology into M and to give a fuzzy classification of nuclear configurations within M. The family

$$G = \{G(i)\} \tag{18}$$

is a generating subbase for a topology T_G on M. One may define a family of membership functions $\{M_i\}$ fulfilling the following conditions:

$$M_i: M \longrightarrow I \tag{19}$$

$$\sum_i M_i (K) = 1, \forall K \in M \tag{20}$$

$$M_i(K) = 0 \text{ if } K \notin G(i) \tag{21}$$

where I = [0,1], the unit interval.
From eqs. (20) and (21) one obtains

$$M_i(K) = 1 \text{ if } K \notin \bigcup_{j \neq i} G(j). \tag{22}$$

The introduction of fuzzy membership functions relaxes the strict assignment of each point K(each nuclear configuration K) to one and only one formal chemical species $C(\lambda,i)$ as expressed by the "element of" relation $K \in C(\lambda,i)$. In the fuzzy set model where chemical species are represented by the $G(i)$ open sets, it is possible to regard a formal nuclear configuration K to belong to two or more chemical species, to a "degree" expressed by the magnitudes of the membership function values $\mathcal{M}_i(K)$.

A simple choice for the membership functions, that fulfills conditions (19)-(21) is as follows. Define a function

$$\nu : M \longrightarrow \{\text{positive integers}\} \tag{23}$$

$$\nu(K) = \text{number of } G(i) \text{ sets for which } K \in G(i). \tag{24}$$

Then the membership function defined as

$$\mathcal{M}_i(K) = \begin{cases} 0 \text{ if } K \notin G(i) \\ \\ 1/\nu(K) \text{ otherwise} \end{cases} \tag{25}$$

fulfills the above conditions. For some applications it is advantageous to define membership functions which vary continuously with K.

Two coordinate systems $\phi(i)$ and $\phi(j)$ are C^∞-related if both

$$\phi(i) \cdot (\phi(j))^{-1} \in C^\infty \tag{26}$$

and

$$\phi(j) \cdot (\phi(i))^{-1} \in C^\infty \tag{27}$$

meaning that the above compositions are infinitely differentiable.

The open sets $\{H(i)\}$ of Euclidean space ^nE are diffeomorphic images of the $\{G(i)\}$ supersets of catchment regions $C(\lambda,i)$, which represent chemical species in M. Consequently, structural relations of chemical species in a nuclear configuration space (which is in general non-Euclidean) can be studied in an Euclidean space ^nE.

Relations between sets $G(i)$ and $G(j)$, where

$$G(i) \cap G(j) \neq \emptyset \tag{28}$$

may be obtained by defining a homeomorphism

$$\phi(ij) = \phi(i) \cdot (\phi(j))^{-1} \tag{29}$$

where

$$\phi(ij) : \phi(j) \ (G(i) \cap G(j)) \longrightarrow \phi(i)(G(i) \cap G(j)). \quad (30)$$

If the family $\{G(i)\}$ of domains of coordinate systems is countable and if each $\phi(ij)$ is differentiable, then M is an n-dimensional differentiable manifold [13,9].

In a non-empty set of the form

$$G(i) \cap G(j) \subset M \quad\quad\quad\quad\quad\quad\quad\quad\quad\quad (31)$$

at least two coordinate systems, $\phi(i)$ and $\phi(j)$, are given. The Jacobian determinant for the corresponding coordinate transformation $\phi(i) \longrightarrow \phi(j)$ is defined by virtue of eq. (17) as

$$\det \left| \frac{\partial x_k(j)}{\partial x_\ell(i)} \right| \quad\quad\quad\quad\quad\quad\quad\quad\quad (32)$$

where k is the row index and ℓ is the column index of the determinant. In $G(i) \cap G(j)$ homeomorphism $\phi(ij)$ always has inverse, which is mapping $\phi(ji)$,

$$(\phi(ij))^{-1} = \phi(ji). \quad\quad\quad\quad\quad\quad\quad\quad (33)$$

It follows that the Jacobian determinant (32) is non-zero at all points of $G(i) \cap G(j)$.

Within the differentiable manifold model continuity and differentiation of functions defined over the nuclear configuration space M may be defined in terms of local coordinate systems of catchment regions, representing various chemical species. Properties of potential energy hypersurfaces can be analyzed in terms of local coordinate systems $\phi(i)$ over $G(i)$ supersets of catchment regions, using the coordinates in open subsets $H(i)$ of Euclidean space of E^n to label the corresponding points in $G(i)$.

Consider a general real valued function f defined over a T-open subset G of the nuclear configuration space M. Take a point $K \in G$ which is also an element of some sets $G(i)$ and $G(j)$,

$$K \in G \cap G(i) \cap G(j) . \quad\quad\quad\quad\quad\quad\quad (34)$$

Consider the composed functions $f \cdot (\phi(i))^{-1}$ and $f \cdot (\phi(j))^{-1}$, defined over $\phi(i) (G \cap G(i))$ and $\phi(j) (G \cap G(j))$, respectively. These functions may be thought of as function f expressed in terms of local coordinates set up in $G(i)$ and $G(j)$ by diffeomorphisms $\phi(i)$ and $\phi(j)$, respectively. A useful property of these functions is that $f \cdot (\phi(i))^{-1}$ is differentiable in a neighbourhood of $\phi(i)(K)$ if and only if $f \cdot (\phi(j))^{-1}$ is differentiable in a neighbourhood of $\phi(j)(K)$. This property follows from the identity

$$f \cdot (\phi(i))^{-1} = f \cdot (\phi(j))^{-1} \cdot \phi(j) \cdot (\phi(i))^{-1} =$$

$$= f \cdot (\phi(j))^{-1} \cdot \phi(ji). \tag{35}$$

If $f \cdot (\phi(j))^{-1}$ is differentiable, then $f \cdot (\phi(i))^{-1}$ is a composition of differentiable functions and itself must be differentiable.

As it has been pointed out, [13,9] the above property ensures that the Euclidean coordinate representations of functions defined over M, consequently, various applications, e.g. normal coordinate representations of vibrational potentials, are related to one another in a simple way within the overlapping regions of coordinate domains. When reference from one chemical species $C(\lambda,i)$ is switched to reference to another chemical species $C(\lambda',j)$, then the change of coordinate systems does not interfere with the continuity and differentiability properties of function f. Consequently, functions defined over the entire nuclear configuration space M, for example, the energy hypersurface E(K) itself, may be treated locally in coordinate neighbourhoods G(i) as functions defined on an ordinary Euclidean space. At the same time the global interpretation is also preserved by ensuring an "orderly" switch of coordinate systems. The differentiable manifold model of potential energy hypersurfaces combines some of the advantages of local and global representations.

It is convenient to choose the $\phi(i)$ coordinate systems in a special form, which assigns the origin of the Euclidean space ^nE,

$$(0,0, \ \dots \ 0) \in {}^nE \tag{36}$$

to the critical point $K(\lambda,i)$ in $C(\lambda,i)$.

Such a representation is not in general identical to an internal coordinate or normal coordinate representation of a vibrational problem, defined at an equilibrium point. However, by further restrictions on $\phi(i)$ the local coordinate system may be made equivalent to a set of normal coordinates, at least in the immediate vicinity of the critical point $K(\lambda,i)$. Note, however, that $\phi(i)$ is defined on a T-open set G(i) containing the entire catchment region $C(\lambda,i)$, within which large deviations from the harmonic approximation may be found. On the other hand, the large amplitude motion formalism shows some analogies with the manifold representation of M.

A global multidimensional model also have some disadvantages, associated primarily with computational difficulties. Even if the analysis is restricted to a few coordinate neighbourhoods G(i), the fact that the dimension n of the manifold may be very high, can render the analysis impractical. It is natural then to introduce further

restrictions by investigating a given subspace only. In
manifold theory the concept of subspace is much too general.
If one differentiable manifold is contained in another, it
is advantageous if the local coordinate systems and the
coordinate neighbourhoods in the two are related in some
simple way.   There exists such a simple relation for a
differentiable manifold and its submanifold.   If
$(x_1, x_2 \ldots x_n)$ are the local coordinates around point K of the
n-dimensional manifold M, then an m-dimensional submanifold
$^m M$ has the local equations

$$x_{m+1} = x_{m+2} = \ldots = x_n = 0, \tag{37}$$

and around point K the local coordinates are $(x_1, x_2 \ldots x_m)$ in
submanifold $^m M$. A submanifold of a manifold is a
generalization of a cross-section of a hypersurface.

FUNDAMENTAL GROUPOIDS AND FUNDAMENTAL GROUPS OF POTENTIAL SURFACES

Consider the level set

$$F^-(A) = \{K : E(K) < A\} \tag{38}$$

of space M with respect to energy hypersurface E(K) and
energy bound A. In a formal sense, $F^-(A)$ contains all
nuclear configurations where the energy is less than the
bound A.

A general reaction path p is a mapping [15-17]

$$p : I \longrightarrow F^-(A) \tag{39}$$

of the unit interval I to level set $F^-(A)$. A special path,
called a constant path, is one for which the entire image is
a single point $K \in F^-(A)$:

$$p(I) = K \in F^-(A) . \tag{40}$$

Let P denote the family of all paths within $F^-(A)$.

For each path $p \in P$ we define two mappings, L* and R*, as

$$L^* : P \longrightarrow P \tag{41}$$

$$R^* : P \longrightarrow P \tag{42}$$

$$L^*(p) = q \in P \tag{43}$$

where

$$q(I) = p(0) \in F^-(A) , \tag{44}$$

and

$$R^*(p) = q' \in P \tag{45}$$

where

$$q'(I) = p(1) \in F^-(A). \tag{46}$$

Mapping L^* (mapping R^*) assigns to each path $p \in P$ the constant path q at the origin $p(0)$ (the constant path q' at the extremity $p(1)$, respectively). Paths $L^*(p)$ and $R^*(p)$ are the left and right zero paths of reaction path p, respectively.

Evidently, for a closed path p, $p(0)=p(1)$, one has

$$L^*(p) = R^*(p). \tag{47}$$

The product $p_3=p_1p_2$ of paths p_1 and p_2 is defined as the continuation of p_1 by p_2:

$$p_3(u) = \begin{cases} p_1(2u) \text{ if } 0 \le u \le 1/2 \\ p_2(2u-1) \text{ if } 1/2 \le u \le 1 \end{cases} \tag{48}$$

that product exists if and only if

$$R^*(p_1) = L^*(p_2). \tag{49}$$

Two paths, p and p' are <u>homotopically equivalent</u> in $F^-(A)$,

$$p \sim p' \tag{50}$$

if they have common endpoints and if they can be continuously deformed into each other within $F^-(A)$.

The homotopy equivalence class to which path p belongs is denoted by $[p]$. Evidently,

$$L^*(p) = L^*(p') \tag{51}$$

$$R^*(p) = R^*(p')$$

for any two $p,p' \in [p]$.

Let us denote the family of all such equivalence classes by $\pi(F^-(A))$ or in short, by π :

$$\pi(F^-(A)) = \{[p_\alpha]\}. \tag{52}$$

This set π is simpler than set P of all reaction paths. We may define two mappings L and R on π as

$$L: \pi \to \pi \tag{53}$$

$$R: \pi \longrightarrow \pi \tag{54}$$

$$L([p_\alpha]) = [L^*(p_\alpha)] \tag{55}$$

$$R([p_\alpha]) = [R^*(p_\alpha)] \tag{56}$$

The condition

$$R([p_1)] = L([p_2)] \tag{57}$$

implies the existence of the product $[p_1][p_2]$ of equivalence classes $[p_1]$ and $[p_2]$ of reaction paths, defined as

$$[p_1][p_2] = [p_1 p_2]. \tag{58}$$

This product, if exists (i.e. if (57) is fulfilled), is unique and does not depend on the choice of reaction paths $p_1, p_2 \in P$, representing equivalence classes $[p_1], [p_2] \in \pi$.
The family π of all homotopy equivalence classes of the complete set P of all reaction paths within level set $F^-(A)$, is a $\underline{\text{groupoid}}$, which, together with mappings L and R, fulfills the following conditions:

i) $L \cdot L = L = R \cdot L$ $\qquad\qquad\qquad\qquad\qquad$ (59)

$\qquad L \cdot R = R = R \cdot R$. $\qquad\qquad\qquad\qquad$ (60)

ii) For any class $[p] \in \pi$ the products $L([p])[p]$ and $[p]R([p])$ exist and

$$L([p])[p] = [p] = [p]R([p]) \in \pi. \tag{61}$$

iii) The products $L([p])L([p])$ and $R([p])R([p])$ exist for each $[p] \in \pi$ and

$$L([p])L([p]) = L([p]) \in \pi \tag{62}$$

$$R([p])R([p]) = R([p]) \in \pi \tag{63}$$

i.e. $L([p])$ and $R([p])$ are idempotent.
iv) For any two $[p_1], [p_2] \in \pi$, fulfilling condition (57)

$$L([p_1][p_2]) = L([p_1 p_2]) = L([p_1]) \tag{64}$$

$$R([p_1][p_2]) = R([p_1 p_2]) = R([p_2]) \tag{65}$$

hence, if in addition to (57), the condition

$$L([p_3]) = R([p_2]) \tag{66}$$

is also valid for some $[p_3] \in \pi$, then the following

products also exist:

$$([p_1][p_2])[p_3] \in \pi \qquad\qquad\qquad (67)$$

$$[p_1]([p_2][p_3]) \in \pi . \qquad\qquad\qquad (68)$$

v) The product "$p_1 p_2 p_3$" of reaction paths, is <u>homotopically associative</u>, i.e.

$$(p_1 p_2)p_3 \sim p_1(p_2 p_3). \qquad\qquad\qquad (69)$$

Hence, if these products exist then for the products (67) and (68) of homotopy classes of reaction paths <u>associativity</u> is also assured:

$$([p_1][p_2])[p_3] = [p_1]([p_2][p_3]) \qquad\qquad (70)$$

and one may simply write $[p_1][p_2][p_3]$.

vi) There exists a unique inverse path p^{-1} for every $p \in P$, defined as

$$p^{-1}(u) = p(1-u). \qquad\qquad\qquad (71)$$

This implies the existence of a unique inverse

$$[p]^{-1} = [p^{-1}] \in \pi \qquad\qquad\qquad (72)$$

for every $[p] \in \pi$, for which $[p], [p]^{-1}$ pair

$$L([p]) = R([p]^{-1}) \qquad\qquad\qquad (73)$$

$$R([p]) = L([p]^{-1}). \qquad\qquad\qquad (74)$$

Family π is the <u>fundamental groupoid</u> of level set $F^-(A)$. Elements $[p]$ of π represent <u>configuration-to-configuration reaction mechanisms</u> on the potential energy hypersurface [17] subject to the energy bound A. This interpretation of $[p]$ equivalence classes is motivated by the actual chemical equivalence of all those reaction paths, which interconnect the same pair of nuclear configurations, and which paths are not separated by high energy domains above bound A.

For an arbitrary point $K_0 \in F^-(A)$ consider the following subset π_1 of groupoid π:

$$\pi_1(K_0) = \{[p]:p(0) = p(1) = K_0, \forall p \in [p], [p] \in \pi\}. \quad (75)$$

It follows, that

$$L([p]), R([p]) \in \pi_1(K_0) \qquad\qquad\qquad (76)$$

$$[p_1][p_2] \in \Pi_1(K_0) \tag{77}$$

and

$$[p_1]^{-1} = [p_1^{-1}] \in \Pi_1(K_0), \forall [p], [p_1], [p_2] \in \Pi_1(K_0). \tag{78}$$

Eqs. (76)-(78) are the conditions of stability for a subset of a groupoid, hence $\Pi_1(K_0)$ is stable within groupoid Π, for any choice $K_0 \in F^-(A)$. With mappings L and R restricted to subsets $\Pi_1(K_0)$, these stable subsets are indeed subgroupoids of groupoid Π.

Furthermore, definition (75) of $\Pi_1(K_0)$ implies that

$$L([p]) = R([p]) = [p_0] \tag{79}$$

for every $[p] \in \Pi_1(K_0)$, where equivalence class $[p_0]$ contains the element constant path p_0 at point K_0

$$p_0 \in [p_0] \tag{80}$$

$$p_0(I) = K_0. \tag{81}$$

Hence both L and R, when restricted to $\Pi_1(K_0)$, are constant maps. Consequently, there exists a unique neutral element and subgroupoid Π_1 is a __group,__ a subgroup of groupoid Π.

If $F^-(A)$ is arcwise connected, then these groups are isomorphic for any choice of point K_0, hence then the specification of K_0 is not essential. $\Pi_1(K_0)$ is the __fundamental group of reaction mechanisms__ [14,15] in a connected level set $F^-(A)$. These groups are dependent on the energy bound A, and are expected to find applications in theoretical, computer aided molecular design.

It has been shown that there exists a set of algebraic relations between the fundamental group of reaction mechanisms and the catchment region topology T_C of the energy hypersurface [15,18]. This connection can be utilized in the actual construction of these groups. The determination of catchment regions $C(\lambda, i)$ is equivalent to the determination of their boundaries, that can be reduced to boundary networks [19]. These boundaries and boundary networks fulfill certain symmetry constraints on the corresponding nuclear configurations, analogous to symmetry constraints on transition structures and minimum energy reaction paths.

ACKNOWLEDGEMENTS

This work has been supported by a grant from the Natural Sciences and Engineering Research Council of Canada.

REFERENCES

1. P. Pulay, Direct use of gradients for investigating molecular energy surfaces, in "Applications of electronic structure theory", Ed. H.F.Schaefer, Plenum press, New York, 1977.
2. P.G. Mezey, Analysis of Conformational Energy Hypersurfaces, in Progr. Theor. Org. Chem., 2, 127(1977).
3. G. Leroy, M. Sana, L.A. Burke, and M.-T. Nguyen, in "Quantum Theory of Chemical Reactions", Eds. R. Daudel, A. Pullman, L. Salem, and A. Veillard, Reidel, Dordrecht, 1979.
4. W.H. Miller, N.C. Handy, and J.E. Adams, J. Chem. Phys., 72, 99(1980).
5. K. Muller, Angewandte Chemie, Internat. Ed., 19, 1(1980).
6. K. Fukui, Accounts Chem. Res., 14, 363(1981).
7. D.G. Truhlar, B.C. Garrett and R.S. Grev, in "Potential energy surfaces and dynamics calculations", Ed. D.G. Truhlar, Plenum, New York, 1981.
8. N.L. Allinger, MM2 Molecular Mechanics Program, Quantum Chem. Progr. Exch. 12, 395(1980).
9. P.G. Mezey, Reaction Topology: Manifold Theory of Potential Surfaces and Quantum Chemical Synthesis Design, in "Chemical Applications of Topology and Graph Theory", Ed. R.B. King, Elsevier, Amsterdam, 1983, pp. 75-98.
10. P.G. Mezey, Reaction Topology, in "Applied Quantum Chemistry", Eds. V.H. Smith et al, Reidel Publ. Co., Dordrecht, 1985.
11. P.G. Mezey, Theor. Chim. Acta, 58, 309(1981).
12. P.G. Mezey, J. Chem. Phys., 78, 6182(1983).
13. P.G. Mezey, Int. J. Quantum Chem. Q.B. Symp. 8, 185(1981).
14. M.M. Gupta, R.K. Ragade and R.R. Yager, Eds., Advances in Fuzzy Set Theory and Applications, North Holland, Amsterdam, 1979.
15. P.G. Mezey, Internat. J. Quantum Chem. Symp., 18, 77(1984).
16. P.G. Mezey, Theor. Chim. Acta, 67, 43(1985).
17. P.G. Mezey, Theor. Chim. Acta, 67, 91(1985).
18. P.G. Mezey, Internat. J. Quantum Chem., 00, 000(1985).
19. P.G. Mezey, Internat. J. Quantum Chem Symp., 18, 675(1984).

Chapter 20

THREE-DIMENSIONAL STRUCTURE-ACTIVITY RELATIONSHIPS AND BIOLOGICAL RECEPTOR MAPPING

I. Motoc, R.A. Dammkoehler
Department of Computer Science, School of Engineering and Applied Science,
Washington University, St. Louis, MO 63130, USA
G.R. Marshall
Department of Pharmacology, School of Medicine, Washington University,
St. Louis, MO 63110, USA

Quantitative structure-activity relationships (QSAR) quantify bioactivity as a function of molecular structure. The topological nature of traditional approaches to QSAR (denoted here by 2D-QSAR, i.e., Hansch [1] and Free-Wilson [2] type models) restricts their applicability to essentially congeneric and conformationally rigid molecules.

Development of a conceptually and computationally integrated framework applicable to structurally diverse compounds requires a topographical basis for QSAR. The resulting three-dimensional quantitative structure-activity relationships (3D-QSAR) would encompass, within a single QSAR, congeneric and non-congeneric molecules and account explicitly for the conformational variable, at least as far as the drug is concerned, in order to gain insight into the topography of the receptor itself by inference.

We emphasize here the mathematical aspects of the paradigm for extracting three-dimensional information from activity data. The present work is complementary to our recent review [3], and is organized as follows: In Sect. 1 a method for systematic search of the conformational hyperspace available to a flexible molecule is presented; Sect. 2 summarizes some details of our force field molecular energy calculation and methods for comparison of either rigid or flexible molecules by geometrical congruence [4]. In Sect. 3 the pharmacophore concept [5] is reviewed and in Sect. 4 three-dimensional molecular shape descriptors (3D-MSD and SIBIS-type) are discussed.

THE CONFORMATIONAL HYPERSPACE

The goal of this Section is to present a method for systematic search of the conformational hyperspace available to a conformationally flexible molecule. This analysis greatly facilitates identification of the most stable conformation of a complex molecule, assessment of the validity of a pharmacophore hypothesis, and estimation of the probable binding mode of a substrate in the active site of a receptor. Subsections 2.1 and 2.2 summarize the necessary background concerning molecular topology and topography, and 2.3 describes the search method.

Molecular Topology

Formally, the topology of a molecule is fully described by a simple graph (i.e., a discrete topological space) $G = (V,E)$ associated with the chemical constitution of the molecule considered [6]. The finite non-empty set V of p vertices and the prescribed set E of q edges collect, respectively, the atoms and the bonds constituting the molecule. We formalize the relationships between these sets using simple binary relations.

A binary relation γ from a set A to a set B is a subset $R_\gamma \subseteq A{\times}B$; it is conveniently represented on a $n{\times}m$ array M_γ by marking the positions $(kl) \in M_\gamma$ with 1 if $a_k\gamma b_1$, $a_k \in A$, $b_1 \in B$, and 0 otherwise. Here, $n = |A|$, $m = |B|$, and $|\ |$ denotes cardinality.

Two binary relations γ and δ are composable if $R_\gamma \subseteq A{\times}B$ and $R_\delta \subseteq B{\times}C$. The composite $\gamma\delta$ is a binary relation $R_{\gamma\delta} \subseteq A{\times}C$ such that $a(\gamma\delta)c$ if, for some b, $a\gamma b$ and $b\delta c$, $a \in A$, $b \in B$, and $c \in C$. Further, the array $M_{\gamma\delta}$ representing the composite $\gamma\delta$ is $M_{\gamma\delta} = M_\gamma * M_\delta$. The product "$*$" is defined [7] by the following rule: the entry $(ik) \in M_{\gamma\delta}$ is 1 iff there is at least one j, $1 \le j \le m$, such that the j-th position of the i-th row in M_γ and of the k-th column in M_δ are simultaneously 1; otherwise, it is 0.

Consider now the binary relations

$$\alpha: \quad R_\alpha \subseteq V{\times}E \ , \tag{1}$$

$$\beta: \quad R_\beta \subseteq E{\times}V \ , \qquad (V,E) = G \ , \tag{2}$$

which assign vertices to edges and, respectively, edges to vertices. Evidently, β is the inverse of α, $\beta = \alpha^{-1}$, or $\alpha = \beta^{-1}$, and the array M_β is the transpose of the array M_α, $M_\beta = M_\alpha^T$, or $M_\alpha = M_\beta^T$.

The binary relations α and β are composable and the composites $\alpha\beta$ and $\beta\chi$ are the binary relations $R_{\alpha\beta} \subseteq V \times V$ and $R_{\beta\alpha} \subseteq E \times E$. $\alpha\beta = \alpha\alpha^{-1}$ represents a compatibility relation [7] on the set V, and $\beta\alpha = \beta\beta^{-1}$ is a compatibility relation on the set E; $\alpha\alpha^{-1}$ is called the connectivity relation (i.e., the off-diagonal entries of $M_{\alpha\alpha^{-1}}$ provide complete information regarding the vertex adjacencies in G), and $\beta\beta^{-1}$ is called the adjacency relation (i.e., the off-diagonal entries of $M_{\beta\beta^{-1}}$ provide complete information regarding the edge adjacencies in G).

Given a compatibility relation ε on A, a compatibility class induced by ε is a subset $D \subset A$ such that, for any d_1, $d_2 \in D$, $d_1 \varepsilon d_2$. A compatibility class which is not properly contained in any other compatibility class is called a maximal compatible [7].

Let $E_i \subset E$, $0 \le i \le k-1$, be the maximal compatibles induced by $\beta\beta^{-1}$ on E, and $V_i \subset V$ collect the end vertices of the edges contained in E_i. It follows that $\bigcup_i E_i = E$, $\bigcup_i V_i = V$, and for any E_i there is at least one E_j such that $E_i \cap E_j \ne \phi$, $V_i \cap V_j \ne \phi$. The subgraph $G_i = (V_i, E_i)$ is called an aggregate.

Two maximal compatibles E_i and E_j are composable if

$$E_i \cap E_j \ne \phi , \tag{3}$$

and the composite $E_{ij} \subset E$ is given by

$$E_{ij} = E_i \cup E_j . \tag{4}$$

Similarly, E_{ij} and E_g are composable if $E_{ij} \cap E_g \ne \phi$ and the composite $E_{ijg} \subset E$ is given by $E_{ijg} = E_{ij} \cup E_g$.

We restrict* the discussion to trees for which the following relations hold:

$$|E_i \cap E_j| \le 1, \text{ for any } 0 \le i, j \le k-1, i \ne j, \tag{5}$$

$$|E_{...ij} \cap E_g| = 1, \text{ for any } ...ij \text{ and at least one g.} \tag{6}$$

The concepts discussed above are illustrated in Figure 1.

The complete composition E_λ, $\lambda = (\lambda_0, \lambda_1, \ldots, \lambda_{k-1}) \in \times\{0,1,\ldots,k-1\}$ is called the topological specification of the molecule considered. The set

$$E' = \{ \bigcup_{i,j} (E_i \cap E_j) | 0 \le i,j \le k-1, i \ne j\} \tag{7}$$

---

*As it is shown below, using appropriate topographical constraints, cyclic molecules can be converted into acyclics (trees) which adequately mimic the former.

collects the edges shared by the k aggregates, and
$$|E'| = m \tag{8}$$
Note that for any $e' = (i,j) \in E'$ there is one and only one λ_x, and one and only one $\lambda_y, \lambda_x, \lambda_y \in X\{0,1,\ldots,k-1\}$, such that

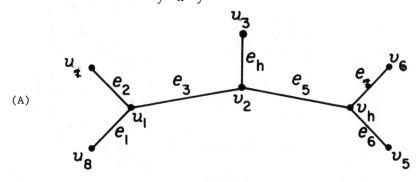

(A)

$$G = (V,E); \quad V = \{v_1,\ldots,v_8\} \quad \text{and} \quad E = \{e_1,\ldots,e_7\}$$

(B)

$M_\alpha =$

	e_1	e_2	e_3	e_4	e_5	e_6	e_7
v_1	1	1	1	0	0	0	0
v_2	0	0	1	1	1	0	0
v_3	0	0	0	1	0	0	0
v_4	0	0	0	0	1	1	1
v_5	0	0	0	0	0	1	0
v_6	0	0	0	0	0	0	1
v_7	0	1	0	0	0	0	0
v_8	1	0	0	0	0	0	0

(C)

$M_{\alpha\alpha^{-1}} =$

	v_1	v_2	v_3	v_4	v_5	v_6	v_7	v_8
v_1	1	1	0	0	0	0	1	1
v_2	1	1	1	1	0	0	0	0
v_3	0	1	1	0	0	0	0	0
v_4	0	1	0	1	1	1	0	0
v_5	0	0	0	1	1	0	0	0
v_6	0	0	0	1	0	1	0	0
v_7	1	0	0	0	0	0	1	0
v_8	1	0	0	0	0	0	0	1

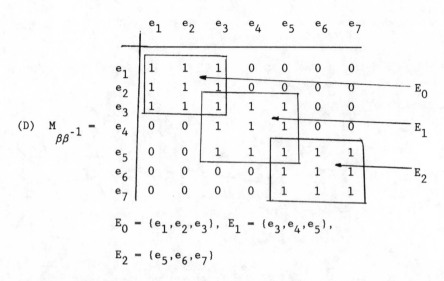

(D) $M_{\beta\beta^{-1}} =$

$$E_0 = \{e_1, e_2, e_3\}, \quad E_1 = \{e_3, e_4, e_5\},$$

$$E_2 = \{e_5, e_6, e_7\}$$

Figure 1. Graph G = (V,E) (A), array corresponding to binary relation α(B), array corresponding to compatibility relation $\alpha\alpha^{-1}$ (C), and array corresponding to compatibility relation $\beta\beta^{-1}$ and the maximal compatibles induced by $\beta\beta^{-1}$ on E(D).

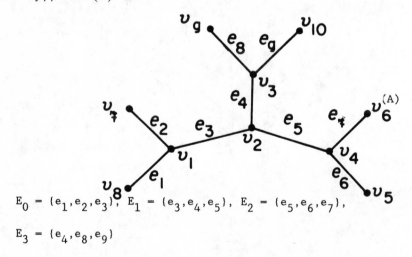

$$E_0 = \{e_1, e_2, e_3\}, \quad E_1 = \{e_3, e_4, e_5\}, \quad E_2 = \{e_5, e_6, e_7\},$$

$$E_3 = \{e_4, e_8, e_9\}$$

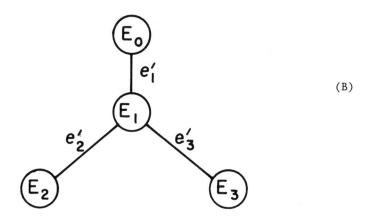

(B)

Figure 2. The maximal compatibles induced by $\beta\beta^{-1}$ on $\{e_1,e_2,...,e_9\}$ (A), and the graph of composability relation $e_1' = e_3 = (E_0,E)$, $e_2' = e_5 = (E_1,E_2)$, $e_3' = e_4 = (E_1,E_3)$. (B)

$E_{\lambda_x} \cap E_{\lambda_y} = (i,j)$. It follows that composability is a binary relation on the set $\{E_i\}$ induced by the set of subscripts assigned to the maximal compatibles and its graph coincides with E' (see Figure 2 for an illustrative example). Further, the isomorphism $(i,j) \simeq (\lambda_x,\lambda_y)$ allows systematic ennumeration of the paths in E'.

Molecular Topography

Given the topological specification E_λ of a molecule, the molecular topography of each conformation is completely determined by A, a set of cartesian coordinates, where $|A| = n$ is the number of constituent atoms, and $a_i \in A$ is the coordinates triple defining the equilibrium position of the atomic nucleus i, $1 \le i \le n$. For a conformationally flexible molecule there are an infinite number of coordinate sets each specifying the topography of a particular conformation. Within the rigid geometry approximation, we represent the infinite set of conformations of a molecule with m rotational degrees of freedom in a continuous m dimensional hyperspace, in which each dimension corresponds to a variable torsion angle ω_j, $1 \le j \le m$.

Let P be a point in the continuous hyperspace. The coordinates of P, $(p_1,p_2,...,p_m)$, $0 \le p_j \le 360°$, $1 \le j \le m$, are the values of torsion angles $(\omega_1,\omega_2,...,\omega_m)$ defining a particular conformation of the molecule. P represents a sterically allowed conformation if for all non-bonded atom pairs the inequality

$$d_{ij} - c_{ij} \geq 0, \ 1 \leq i \leq n-1, \ i < j \leq n \tag{9}$$

is satisfied. Here, d_{ij} is the Euclidian distance between the non-bonded atoms i and j, and c_{ij} is the sum of their van der Waals radii.

For practical reasons, we approximate the continuous hyperspace by a discrete topological space of the same dimensionality. A generic point in that space is defined by $P(n_1\Delta, n_2\Delta, \ldots, n_m\Delta)$, $0 \leq n_j < 360/\Delta$, $1 \leq j \leq m$. The use of a constant Δ produces an uniform sampling over all dimensions of the continuous space. The number of points (NP) in the discrete space is a function of m and Δ, and obtained by the formula

$$NP = r^m, \text{ where } r = 360/\Delta . \tag{10}$$

For each P there is a set of coordinates defining the topography of the corresponding conformation. These coordinate sets can be generated by performing a series of coordinate transformations on the set A. The coordinates $a_t, a_s, a_r \in A$ of three atoms, corresponding to vertices $v_t, v_s, v_r \in V_0$, $(V_0, E_0) = G_0$ and $(v_s, v_r) = e_1'$ are used to orient the initial conformation. A global translation and rotation are performed on the coordinates A so that a_s lies at the origin, a_r lies on the positive x axis, and a_t lies in the xy plane. The unit length vector $\underline{u} = (1,0,0)$ defines the direction of the axis of the first torsional rotation.

Methods for transforming the coordinates of a point as it rotates with respect to a fixed axis are conventionally given in matrix form [8]. An equivalent vector formulation for a torsional rotation can be obtained through algebraic analysis of the matrices [9], or derived from first principles [10] as shown below:

Let $\underline{v}$ be a displacement vector, $\underline{v} = a_j - a_r$, and $\underline{u}$ the unit length vector corresponding to the direction cosines of the axis of rotation; $\underline{v}$ can be resolved into three orthogonal component vector by the operations shown below

$$\underline{v}_3 = \underline{v} \times \underline{u}$$
$$\underline{v}_2 = \underline{u} \times \underline{v}_3$$
$$\underline{v}_1 = \underline{v} - \underline{v}_2 ,$$

The rotated displacement vector and the rotated coordinates are then given by

$$\underline{v}(\omega) = \underline{v}_1 + \underline{v}_2 \cos\omega + \underline{v}_3 \sin\omega \tag{11}$$

$$a_j(\omega) = a_r + \underline{v}(\omega) \tag{12}$$

Next, consider a fixed (non-rotating) atom a_i and the displacement vector $\underline{s} = a_i - a_r$. The square of the distance d_{ij} as a function of ω is given by direct application of the cosine law,

$$d_{ij}^2(\omega) = |\underline{s}|^2 + |\underline{v}|^2 - 2|\underline{s} \cdot \underline{v}(\omega)| \tag{13}$$

A more compact form may be obtained by observing that the sum of the constant terms in (13) defines a point a_p which is the projection of a_j onto the axis of rotation. One may therefore rewrite (11) and (12) as

$$\underline{v}(\omega) = \underline{v}_2 \cos\omega + \underline{v}_3 \sin\omega \tag{14}$$

$$a_j(\omega) = a_p + \underline{v}(\omega), \text{ where } a_p = a_r + \underline{v}_1 . \tag{15}$$

Finally, redefining $\underline{s}$ as the displacement vector from a_p to a_i, one obtains an equation with scalar coefficients,

$$d_{ij}^2(\omega) = d_1 + d_2 \cos\omega + d_3 \sin\omega , \tag{16}$$

where

$$d_1 = |\underline{s}|^2 + |\underline{v}|^2; \quad d_2 = -2(\underline{s} \cdot \underline{v}_2); \quad d_3 = -2(\underline{s} \cdot \underline{v}_3) .$$

The equation above can be converted to a useful quadratic form by a substitution of variables, $x = \tan\frac{\omega}{2}$. Then, $\sin\omega = 2x/(1+x^2)$ and $\cos\omega = (1-x^2)/1+x^2)$. Following algebraic manipulation, one obtains

$$d_{ij}^2(x) = (ax^2+bx+c)/(1+x^2) , \tag{17}$$

where

$$a = d_1 - d_2; \quad b = 2d_3; \quad c = d_1 + d_2.$$

The relation shown below is called the <u>differential distance function</u>. Observe that the

$$\delta_{ij}(\omega) = d_{ij}^2(\omega) - c_{ij}^2 \tag{18}$$

function is positive when $d_{ij}^2(\omega) \geq c_{ij}^2$ and negative otherwise. By converting (18) to quadratic form, i.e., first inserting (16) into (18), we can evaluate the discriminant, b^2-4ac, to determine whether there are any values of ω ($\omega = 2$ arctan x) for which $\delta_{ij}(\omega) = 0$. If the discriminant is strictly positive, δ_{ij} has real roots indicating the van der Waals constraint (9) will be violated for some range of values of ω. If the discriminant is negative, δ_{ij} has complex or real double degenerate roots indicating that $\delta_{ij}(\omega)$ will be positive or, respectively, negative for all values of ω, depending whether $c - c_{ij}^2 \geq 0$, or, respectively, $c - c_{ij}^2 < 0$.

Systematic Search of the Conformational Hyperspace

Systematic Search is a computational procedure which generates and records all points in discrete hyperspace which correspond to sterically allowed conformations. The procedure also provides the capability for measuring and recording geometric parameters derived from each allowed conformation; typical parameters are: interatomic distances between selected functional groups, coordinates defining the locii of one or more specified atoms with respect to a fixed reference frame, and vectors specifying the relative orientation of pairs of atoms in each conformation.

A topological specification, $E_{\lambda_0 \lambda_1 \ldots \lambda_m}$, provides the information required to decompose an initial sterically allowed conformation, represented by A, into substructures, A_j, $0 \leq j \leq m$, where $A_j \subset A$ determines the topography of the j-th aggregate. Consistent with the assumptions of rigid geometry, the substructure represented by A_j has no internal degrees of freedom. More importantly, the distances between all atom pairs, $(a_r, a_s) \in A_j$, are invariant with respect to the torsional rotations around an axis defined by any pair $(a_r, a_s) \in A_j$. As a result, while the coordinates of A_j are transformed by a torsional rotation, there is no requirement to verify that the re-oriented substructure is sterically allowed.

The topological specification of the molecule also provides an order in which the substructures may be combined to form complete conformations of the molecule, i.e., recall the isomorphism $(i,j) \simeq (\lambda_x, \lambda_y)$. Each edge of the graph of the composability relation induced by the labelling of the aggregates is interpreted as an axis of a torsional rotation. The two atoms corresponding to the end vertices of the edge shared by two adjacent aggregates, G_i and G_j, $i < j$, are assigned to the substructure A_i. As a result, $A_i \cap A_j = \emptyset$.

procedure $search(dim)$
        begin
                $m := dim$
                $A_m := update(LA_m, LD_m)$
                $LP_m := copy(LT_m)$
                        if$(LC_m \neq \phi)$ **then** $LP_m := screen(LC_m, LP_m)$
                $LP_m := validate(LA_m, LP_m)$
                        for $n = 1$ *to sizeof* (LP_m)
                        begin $p_m := $ *nth element of* LP_m
                                $A_m := rotate(A_m, p_m)$
                                        if$(m < M)$ **then** $dim := search(dim+1)$
                                                else $dim := record(p_1, \ldots, p_M)$
                        end
                $p_m := 0$
                        $return(dim-1)$
        end

procedure $update(A_m, LD_m)$
        begin
                for $i = 1$ *to* m - 1
                begin *if* $(i \, \epsilon LD_m)$ *AND* $(q_{im} \neq p_i)$
                        then begin
                                        if $(q_{im} < 0)$ **then** $v_{im} := vector(i,m)$
                                        $A_m := rotate(A_m, p_i)$
                                        $q_{im} := p_i;$ $p_m := $ -1
                                end
                        end
                if $(p_m < 0)$ **then** $v_{mm} := vector(m,m)$
                for $k = m+1$ *to* $M;$ *if* $(q_{mk} \geq 0)$ **then** $v_{mm} := vector(m,m)$
                $p_m := 0$
                $return \, (A_m)$
        end

procedure $validate(LA_m, LP_m)$
        begin
                for $r = 1$ *to sizeof(*LA_m*)* **while***(*$LP_m \neq \phi$*)*
                begin
                for $s = 0$ *to* m-1 **while***(*$LP_m \neq \phi$*)*
                        begin
                                for $k = 1$ *to sizeof(*LA_s*)* **while***(*$LP_m \neq \phi$*)*
                                begin $i := $ *kth element of* LA_s
                                        $(f,d) := interatom(a_i, a_j, c_{ij})$
                                        if $(f < 0)$ **then** $LP_m := \phi$
                                        else *if* $(f > 0)$ **then** $LP_m := compress(d, LP_m)$
                                end
                        end
                end
                $return(LP_m)$
        end

Figure 3.

The coordinate set representing the position of atoms in substructure $j \geq 1$ rotated by p_i around axis i is denoted by $A_j(p_i)$. A rotation around the i-th axis by p_i is represented by $R(p_i)$ and, symbolically, $A_j(p_i) \leftarrow R(p_i)A_j$. Consequently, the rotation of a substructure whose position is dependent on multiple (ℓ) consecutive torsion angles is specified by:

$$A_j(p_1,p_2,\ldots,p_\ell) \leftarrow R(p_1,p_2,\ldots,p_\ell)A_j , \quad 1 \leq j \leq m \quad (19)$$

Further,

$$A_j(p_1,p_2,\ldots,p_\ell) \leftarrow R(p_2,\ldots,p_\ell)R(p_1)A_j \quad (20)$$

with $p_2 = \ldots = p_\ell = 2\pi$, specifies the conformations available to the substructure in the one dimensional space corresponding to the first axis; and, in general,

$$A_j(p_1,\ldots,p_j,p_{j+1},\ldots,p_\ell) \leftarrow R(p_{j+1},\ldots,p_\ell)R(p_i)\ldots R(p_1)A_j$$
$$1 \leq i \leq \ell \leq j \quad (21)$$

with $p_{i+1} = \ldots = p_\ell = 2\pi$, specifies the conformations available to the sustructures in the i-dimensional subspace corresponding to the 1-st,...,i-th axis. More generally, the position of substructure A_j is determined by an ordered set of torsional rotations corresponding to the edges on the path from E_o to E_j in the graph $E'(7)$, and we denote the effect of those rotations on A_j by $R_j(P)A_j$. Then for any flexible molecule, the coordinate set corresponding to a point P in discrete hyperspace can be described by the linear notation,

$$A(P) = A_0 + R_1(P)A_1 + \ldots + R_m(P)A_m ,$$

or by the recurrence relation shown below:

$$A(p_1,\ldots p_m) = A(p_1,\ldots p_{m-1}) + R_m(P)A_m$$
$$\text{where } A(p_1) = A_0 + R_1(P)A_1$$

Systematic search procedures have been programmed in a variety of languages, on multiple machines over a period of fifteen years in our laboratory [10b]. The most recent implementations are based on the recursive procedure shown below in Figure 3.

Initial data values are produced by a pre-processor [10c] which performs the topological analysis described in Sect. 2.1. These data are read by a main routine and are globally accessible to the recursive procedure named <u>search</u> and all of its subprocedures.

Data structures and variables are summarized below:

    N =   the number of atoms.
    M =   the number of variable torsion angles.
    A =   the coordinates of an initial sterically allowed
           conformation.
    B =   an M×2 array whose j-th row references the coor-
           dinates of the two atoms defining the axis of the
           j-th torsional rotation.
    C =   the van der Waals constraint array (N×N)

For each substructure A_m, $0 \leq m \leq M$,

LA_m = a list of the atoms contained in A_m.

LC_m = a list of the distance constraints which are applicable to A_m.

LD_m = a list of the torsional rotations which terminate in A_m.

LT_m = a list of the indices of trigonometric values corresponding to points in the m-th dimension of the discrete hyperspace.

Control structures and variables are:

dim = a scalar variable whose value determines the dimensionality of the discrete subspace containing the current sterically allowed conformation

P = a one dimensional array of size M whose entries are the coordinates of the point in discrete hyperspace corresponding to the current value of the coordinate set A.

Q = an M×M storage array called the control table. The super-diagonal entries of the m-th column indicate the status of the coordinates of A_m with respect to the first m-1 torsion angle variables.

T = a table of trigonometric values (cos, sin pairs).

Temporary storage arrays are:

LP_m = a temporary storage array for the set of indices of trigonometric values which when used to rotate A_m with respect to torsional axis m, will produce a set of sterically allowed conformations representable in the m dimensional subspace.

V_{im} = a storage array for the sets of vectors describing the motion of A_m with respect to the i-th torsional axis, $0 < i \leq m$.

The functions and subprocedures used in Systematic Search (Figure 3) are summarized below:

The function <u>update</u> is a generalized implementation of eq. (21) which uses information contained in LD_m to select only those torsional rotations, $R(p_i)$, $0 < i < m \leq M$, which are applicable to A_m. The values contained in the m-th column of the control table Q determine the status of the vector sets used to generate new coordinates of A_m; the first m-1 elements of the global variable P indicate the torsion angle values of the current conformation. The coordinates returned by update represent A_m updated by $(p_1, p_2, \ldots, p_{m-1})$.

The function _copy_ creates an initial set of indices of trigonometric values corresponding to points in the m-th dimension of discrete hyperspace. This set is used by the function _validate_ which eliminates from LP_m the indices of all trigonometric values do not produce sterically allowed conformations.

The function _interatom_ is used by _validate_ to generate the coefficients of the differential distance equations, $\delta_{ij}(\omega) = 0$ in (18), for all non-bonded atom pairs (a_s, a_r), $\forall\; a_s \in A_p$, $p < m$, $\forall\; a_r \in A_m$. The values returned by _interatom_ are d, the coefficients, and f, a variable reflecting the results of the discriminant analysis: $f < 0$ indicates that there are no values of ω_m for which $d^2_{sr}(\omega_m) \geq c^2_{sr}$; if $f > 0$, then some values of ω_m will cause a violation of the steric constraint between a_s and a_r.

The function _compress_ evaluates the differential distance equation using each cos and sin pair referenced by LP_m, and eliminates those entries for which $\delta_{sr}(\omega_m) \leq 0$. If $f = 0$, then all values of ω_m would, for the pair (a_s, a_r), produce sterically allowed conformations and it is not necessary to evaluate the differential distance equations.

The function _rotate_ implements the rotation operator defined in (21), $A_m \leftarrow R(p_m)A_m$.

The function _screen_ is invoked when there are additional constraints on the allowed distance between an atom in A_m and an atom in A_p, $p < m$. A constraint is specified by an atom pair (a_s, a_r) and the minimum and maximum allowed distances. The function _interatom_ is used by _screen_ to generate the coefficients of a differential distance equation for the atom pair using the minimum distance as the van der Waals constraint. A modified form of the _compress_ function is used to eliminate values in LP_m which would produce interatomic distances less than the minimum or greater than the maximum allowed values.

Significant reductions in the computational complexity of a Systematic Search can be achieved by first applying the search procedure to each of the substructures obtained by the composition of each pair of adjacent aggregates. There are M such substructures, each containing one torsional degree of freedom. Let LT_m, $0 < m \leq M$, be the set of indices of trigonometric values determined by Δ, $|LT_m| = r = 360/\Delta$. Then, the output produced by a search of each substructure, LT'_m, $|LT'_m| = r'_m$, includes only those values of LT_m which can produce sterically allowed conformations of the substructures. The excluded values are those which will always result in a violation of the steric constraints between two atoms in adjacent aggregates in the complete structure. If LT'_m, $0 < m \leq M$, are used in place of LT_m as inputs to a Systematic Search, then the computational complexity, measured in terms of the number of points (NP′) in discrete hyperspace is given by the formula:

$$NP' = \prod_{m=1}^{M} r'_m \ , \qquad r'_m \leq r \qquad\qquad (22)$$

In addition, the procedure validate may be modified to eliminate the calculation of differential distance equations for atom pairs contained in adjacent aggregates [10d].

THE MOLECULAR ENERGY CALCULATION

Computation of the energy complex molecular systems, i.e., systems composed of 500 atoms or more with very many local energy minima, is usually performed using molecular mechanics methods [11,12].

The molecular mechanics program (Maximin) developed in our laboratory uses a Simplex minimizer [15], White's force field [16], and energy functionals systematized in Table 1. The distinctive features which make Maximin extremely suitable for the specific requirements of computer-aided drug design are summarized below. Our approach requires the use of Systematic Search to identify representative and well-defined conformations for the system and the problem at hand, and their further refinement by Maximin, with molecular graphics [20] playing a secondary analysis role, see e.g., [17].

The symbols used in Table 1 have the following connotation: (1) N_B: no. of bonds in molecule; d_i: length of the i-th bond, Å; d_i^o: equilibrium length for the i-th bond, Å; k_i^d: bond stretching force constant, kcal mole^{-1} Å^{-2}; (2) N_A: no. of valence angles in molecule; θ_i: value of the i-th valence angle, degrees; θ_i^o: equilibrium value for the i-th valence angle, degrees; k_i^θ: angle bending force constant, kcal mole^{-1} degree^{-2}; N_{AOP}: no. of out-of-plane bending angles at trigonal atoms; δ_i: value of the i-th out-of-plane bending angle, k_i^δ: out-of-plane bending constant, kcal mole^{-1} degree^{-2}; (3) v_i: height of the i-th torsional barrier, kcal mole^{-1}; $|\eta_i|$: periodicity of the torsion angle, ω_i, $S_i = \text{sign}|\eta_i|$; N_T: no. of torsional angles; (4) q_i: net atomic charge at the i-th atom, $|e|$; r_{ij}: separation distance between charge carriers, Å; $\varepsilon(r_{ij})$: dielectric function. Due to the absence of obviously applicable techniques for dealing with the dependence of the dielectric function on r_{ij} [14], Maximin allows designation of three ranges for r_{ij}, and for each of these ranges it is given a choice of three functionals: $\varepsilon(r_{ij}) = ar_{ij}^b + c$, $\varepsilon(r_{ij}) = a[1-\exp(-br_{ij})] + c$, and $\varepsilon(r_{ij}) = a[1-\exp(-ar_{ij}^2)] + c$; (5) N_{At}: no. of atoms in molecule; $a_{ij} = r_{ij}/(R_i+R_j)$ where R_i is the van der Waals radius of the i-th atom.

Table 1. Energy functionals used in Maximin

#	Energy term	Functional form
1.	Bond stretching	$E_{str} = \frac{1}{2} \sum_{i=1}^{N_B} k_i^d (d_i - d_i^o)^2$
2.	Angle bending	$E_{ang} = \frac{1}{2} \sum_{i=1}^{N_A} k_i^\theta (\theta_i - \theta_i^o)^2 + \frac{1}{2} \sum_{i=1}^{N_{AOP}} k_i^\delta \delta_i^2$
3.	Torsional	$E_{tor} = \frac{1}{2} \sum_{i=1}^{N_T} v_i[1+S_i\cos(|n_i|/\omega_i)]$
4.	Electrostatic	$E_{ele} = 332.17 \sum_{\substack{i<j}}^{N_{At}} q_i q_j/\varepsilon(r_{ij})r_{ij}$ (i,j non-bonded)
5.	Van der Waals	$E_{vdw} = \sum_{\substack{i<j}}^{N_{At}} E_{ij}[-2.25/a_{ij}^6$ (i,j non-bonded) $+8.28\ 10^5 \exp(-13.586957\ a_{ij})]$ for $a_{ij} \geq 0.785555$, and $E_{vdw} = \sum_{\substack{i<j}}^{N_{At}} E_{ij}[1.0/a_{ij}^{12}-2.0/a_{ij}^6]$ (i,j non-bonded) for $a_{ij} < 0.785555$
6.	Hydrogen bond	The method described in [13].

The capability provided in Maximin to hold constant $N_{fix} \le 40$ parameters during a minimization greatly enhances its utility to drug designers. By setting a large value for the fixing force constant, k_I^{fix}, a small change, p_I, in the fixed parameter, p_I^o, is extremely unfavorable. The potential functions

$$E_{fix} = \sum_{I=1}^{N_{fix}} k_I^{fix} (p_I - p_I^o)^2 \qquad (23)$$

allow the absolute coordinates of an atom, the distance between two atoms, the angle formed by three atoms, the dihedral angle formed by four atoms, the angle between a vector defined by two atoms and a plane defined by three atoms, and the angle between two planes defined by two sets of three atoms each to be held constant. Energy minimization with subsidiary constraints provides a way to preserve the stereoelectronic requirements characterizing the receptor-substrate interaction.

Another feature of Maximin, which further increases its versatility, is the multi-molecule fitting option, i.e., a procedure to force atoms of different molecules to occupy approximately the same position in space, adjusting their geometry to relieve any strain while maintaining low energy. The procedure utilizes E_{multi} potentials

$$E_{multi} = \sum_{i=1}^{N_{ref}} k_{ij}^s \, d_{ij}^2 \, , \qquad (24)$$

where d_{ij} is the distance between the atom i of molecule j and r_i, the i-th reference point, k_{ij}^s is the spring constant connecting atom i to the reference point, and N_{ref} denotes the number of reference points.

Maximum provides alternative methods for comparing molecular geometries: The reference points may correspond to atom positions of a reference molecule which is treated as a rigid entity; or all molecules compared are treated as conformationally flexible entities and the coordinates of the reference points are the arithmetic means of the coordinates of the atoms to be superimposed. Further, the possibility to assign different weights, w_j, to the molecules considered, i.e., $E_{multi} = \sum w_j \, k_{ij}^s \, d_{ij}^2$, and the key-in-lock theory [19] allows one to explore and/or mimic to a certain extent, the conformational features of an unknown biological receptor. Note, also, that multi-molecule fitting offers an elegant and straightforward approach to the determination of the existence of a consistent pharmacophore hypothesis - the question of its uniqueness may be addressed by use of Systematic Search, as shown, e.g., in [18] for a series of angiotension converting enzyme inhibitors.

THE PHARMACOPHORE

Consider the biologically active analogs M_i, $i \le i \le n$, and let F_ℓ, $1 \le \ell \le m$, be the constituent functional groups responsible for analog recognition and subsequent activation of the receptor.

Let $\{P_{ij}\}$ collect the points j in the discrete conformational hyperspace available to M_i and corresponding to sterically allowed conformations, A_{ij} be the coordinate set which specifies the topography of P_{ij}, and d_{pq} be the 3m-6 pairwise distances between F_p and F_q, $1 \le p, q \le m$, $p < q$, in P_{ij}.

A <u>pharmacophore hypothesis</u> consists of the specification of $\{F_1\}_{\ell \le i \le m}$, the correspondence between $\{F_\ell\}$ and $\{M_i\}_{1 \le i \le n}$, and the specification* of d_{pq} as $d_{pq} \in D_{pq} = [d_{pq}^{min}, d_{pq}^{max}]$, with $d_{pq}^{min}, d_{pq}^{max} =$ given constants, $1 \le p, q \le m$, $p < q$.

If the set (25) is empty, there is no common three dimensional arrangement of $\{F_1\}$ in $\{M_i\}$ and, therefore, either the hypothesis is invalid, or some $M_i \in \{M_i\}$ act at different receptor sites.

$$\bigcap_{i=1}^{n} \{P_{ij}\} = \{P_{i1}|d_{pq} \in D_{pq}, \ 1 \le p, \ q \le m, \ p < q, \ \text{for any } i\} \quad (25)$$

If $\bigcap_i \{P_{ij}\} \neq \Phi$ but $\{d_{pq}\}$ cluster, e.g., as shown below (m=3):

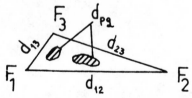

the hypothesis is not unique and additional information would be required for its resolution; if $\bigcap_i \{P_{ij}\} \neq \Phi$ and $\{d_{pq}\}$

---

*Evidently, if M_i's are conformationally rigid, or the topography of the receptor is known, $d_{pq}^{min} = d_{pq}^{max}$.

cluster, e.g., as below

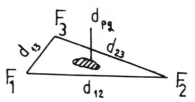

the hypothesis is unique, and $\{F_\ell\}$ together with $\{d_{pq}\}_{\substack{1 \leq p, q \leq m \\ p < q}}$
represent a <u>pharmacophore model</u>.

In the absence of any molecular information concerning the receptor, a pharmacophore model, once identified, should be regarded as a topographical and chemico-physical specification of the receptor and allows meaningful comparison of congeneric and/or non-congeneric molecules exhibiting the same biological action.

An illustrative example for identification and validation of pharmacophore hypothesis is provided [28] by a series of seven molecules, i.e., bamipine (1), clemastine (2), cyroheptadine (3), triprolidine (4), promethazine (5), chlorpheniramine (6), and carbinoxamine (7) - Figure 4 - which are only tenuously related structurally, possess relatively high conformational flexibility, and represent potent antogonists (commercially available drugs) of histamine H_1 receptor.

Figure 4.

F_ℓ, $a \leq \ell \leq 4$, correspond to the nitrogen of the cationic head ($NH_3^{\oplus}$) and N(1) atom of the imidazole ring of histamine, and, respectively, to the centroids of the two benzene rings of cyproheptadine. The two nitrogens were selected to ensure that histamine also fits the pharmacophore model, and to account for

the ionic (NH$_3^{\oplus}$) and hydrogen bond (N(1)) histamine H$_1$ receptor interactions. The two benzene rings were selected for
their probable implication in beneficial hydrophobic interactions.

Systematic conformational search calculations have proved
the uniqueness of the pharmacophore model shown in Figure 5.

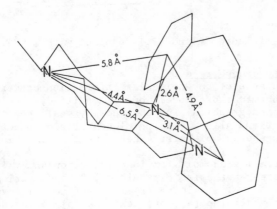

Figure 5.

Table 2.
Histamine H$_1$ antagonists: the energy of the most stable conformation and the best
fitted conformation

Antagonist	$E_1^{a)}$	$E_2^{b)}$
Bamipine	-5.8	1.9
Clameasitne	9.7	0.2
Cyproheptadine	9.7	1.7
Teriprolidine	8.1	4.9
Promethazine	-0.1	1.6
Chlorpheniramine	-0.1	2.3
Carbinoxamine	-3.5	3.3

a)   The lowest energy conformer (local mole^{-1}), minimized as a
free molecule.

b)   The difference (local mole^{-1}) between the energy of the
best fitted conformation (Figure 4) and E$_1$.

For further examples of pharmacophore model identification
one may consult [29a] (angiotensin converting enzyme inhibitors), [29b] (antiulcer drugs), [29c] (muscarinic agonists),
[29d] (histamine H$_2$ antagonists).

Inspection of Figure 6 indicates a good geometrical fit of the
pharmacophore model, and multi-molecule fitting calculations
clearly show the ability of the molecules considered to achieve
the pharmacophore topography at modest energy cost: an energy

well up to 5 local mole^{-1} (Table 2) is within the range that might be expected in drug-receptor interactions.

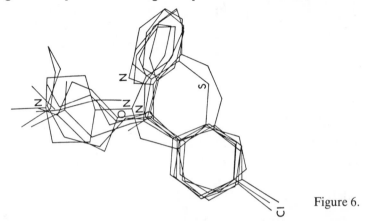

Figure 6.

THREE-DIMENSIONAL MOLECULAR SHAPE DESCRIPTORS

Efforts to extract three-dimensional information regarding drug-(unknown) receptor interaction from measurements of potency are generally centered on the QSAR paradigm [1], and appropriate characterization of the molecular shape is the major obstacle in the development of a topographical basis for QSAR.

The space occupied by a molecule can be defined conveniently in the framework of the hard-sphere approximation: each atom of the molecule is represented by an isotropic sphere centered at the equilibrium position (X_I, Y_I, Z_I) of the atomic nuclei $I = 1, 2, \ldots, N$, and having a radius equal to the van der Waals radius, r_I, of the atom; N is the number of atoms constituting the molecule. The locus of points (x,y,z) within the molecule satisfies the following inequalities:

$$(X_I-x)^2 + (Y_I-y)^2 + (Z_I-z)^2 \leq r_I^2 , \quad 1 \leq I \leq N \tag{25}$$

A molecular van der Waals (vdW) envelope may be uniquely defined as the surface of the intersection of the vdW spheres associated with the atoms in the molecule; consequently, the total volume inside the vdW envelope represents the vdW volume.

Evidently, for a given molecular geometry, the molecular shape and size depends to a major degree on the atomic vdW radii considered.

We have calculated [21] a set of effective atomic vdW radii, r_σ, which, unlike previous studies [22], reproduces adequately (see Figure 7) the accessible areas of conformational space in the protein. This indicates that r_σ's used with (25) offer physically significant estimates of molecular shape and size.

Figure 7. (a) Calculated map of NH–CαH (PHI,#1) and (b) High-resolution crystal structures of seven proteins CαH–CO (PSI,#3) torsional angles.

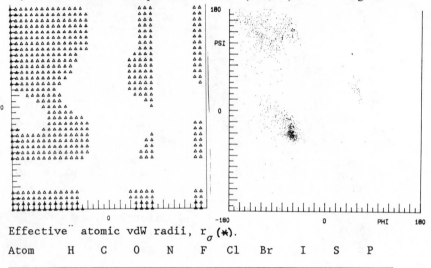

Effective atomic vdW radii, r_σ (*).

Atom	H	C	O	N	F	Cl	Br	I	S	P
r_σ	1.08	1.53	1.36	1.45	1.30	1.65	1.80	2.05	1.70	1.75

*r_σ implies a repulsive interaction of approximately 5 kcal mole^{-1} at closest approval.

Consider the molecules M_i, i = 1,2,...,n, which exhibit the biological activity via a common mode of action, and let M_r, 1 ≤ r ≤ n, be the most active compound in the data base. It follows [23-25], in agreement with the drug-receptor theory [19], that M_r is the best available "copy" of the receptor site, including steric features, and, therefore, it may be considered as a reference structure. Further, as the pharmacophore represents a logical frame of reference for the receptor, meaningful comparison of M_i, 1 ≤ i ≤ n, with M_r can be achieved by geometrical congruence of M_i over M_r, requiring that the pharmacophore groups occupy approximately the same position. These geometrical congruences can be obtained using methods described in Sect. 3.

As the receptor and substrate "feel" the shape and size of each other through van der Waals interactions, it follows that the number pair (26) represents the simplest three-dimensional (3D) molecular shape descriptor (MSD):

$$3D\text{-}MSD(r,i) = [OV(r,i); NOV(i,r)] , \qquad (26)$$

$OV(r,i)$ is the overlapping vdW volume of M_r and M_i, and $NOV(i,r)$ is the non-overlapping volume of M_i superimposed over

M_r according to the criteria described above. OV may be regarded as the polarizability volume available to the molecule M_i, and NOV may represent either regions occupied by the receptor and, therefore, not available for binding, and/or regions located away from the receptor and, therefore, sterically irrelevant.

We turn to numerical integration techniques [26] for obtaining accurate estimate of OV and NOV, respectively. Therefore, one determines the finite countable set $\underline{W}$ whose elements $W_j(W_{jx},W_{jy},W_{jz})$ are points in the E_3 space,

$$\underline{W} = \{W_j | P_1, P_2\} \tag{27}$$

and satisfy the following properties:

(P_1): W_j, for all j, lie within the parallelepiped

$d_1 \leq x \leq D_1$, $d_2 \leq y \leq D_2$, $d_3 \leq z \leq D_3$ which embeds the collection of spheres (25) representing the superimposed molecules M_r and M_i;

(P_2): W_j, for all j, are uniformly distributed independent random points.

Next, one determines the sets

$$\underline{V}^{\delta} = \{V_j^{\delta} | P_3^{(\delta)}, P_4^{(\delta)}\} \ , \ \delta = r \ or \ i \ , \tag{28}$$

whose elements $V_j^{\delta}(V_{jx}^{\delta}, V_{jy}^{\delta}, V_{jz}^{\delta})$ satisfy the properties

(P_3): $\underline{V}^{\delta} \subset \underline{W}$, $\delta = r \ or \ i$;

(P_4): V_j^{δ}, for all j, satisfy at least one of the inequalities (25) corresponding to the reference molecule M_r ($\delta = r$, $\underline{V}^r$) and, respectively, to the compared molecule M_i ($\delta = i$, $\underline{V}^i$).

As volume may be regarded as a non-negative continuous function in the closed bounded domain defined by the vdW envelopes, OV and NOV values are estimated by [27]:

$$OV(r,i) = g|\underline{V}^r \cap \underline{V}^i|/|\underline{W}| \ , \tag{29}$$

and, respectively,

$$NOV(i,r) = g|\underline{V}^i - \underline{V}^r|/|\underline{W}| \ , \ or \tag{30}$$

$$NOV(r,i) = g|\underline{V}^r - \underline{V}^i|/|\underline{W}| \ ;$$

here, $\underline{V}^i - \underline{V}^r = \{V_j | V_j \in \underline{V}^i, V_j \notin \underline{V}^r\}$ is the relative complement of $\underline{V}^i$ in $\underline{V}^r$, and $g = (D_1-d_1)(D_2-d_2)(D_3-d_3)$ represents the volume of the parallelepiped which embeds the vdW envelopes of M_i superimposed over M_r.

The $\underline{W}$ set is constructed using either Monte Carlo or strictly deterministic procedures. Within the first procedure $W_j \in \underline{W}$ is given by: $W_{jx} = d_1 + (D_1-d_1)\xi_1$; $W_{jy} = d_2 +$

$(D_2-d_2)\xi_2$; $W_{jz} = d_3 + (D_3-d_3)\xi_3$, where (ξ_1,ξ_2,ξ_3) are uniformly distributed independent random sequences on the unit interval. The deterministic procedure, which mimics the stochastic approach, divides the parallelipiped into $|\underline{W}|$ subspaces ("elementary" parallelipipeds) whose centers (W_{jx},W_{jy},W_{jz}) constitute $W_j \in \underline{W}$, $1 \le j \le |\underline{W}|$.

The accuracy of the estimates (29) and (30), for given g, is $\sim|\underline{W}|^{-1/2}$; this property implies relatively slow convergence of the procedure and a requirement for $\underline{W}$ sets with large cardinality (>100,000).

Note that the molecular shape analysis (MSA) descriptor [27a], V_o, is clearly the $OV(r,i)$ and, therefore, it is a particular case of the 3D-MSD. Further, the MSA procedure to evaluate V_o overestimates its value and introduces a sizeable error which increases with the branching of molecules considered. Also, MSA generally assigns higher shape similarity between molecules compared: In the framework defined by the reference structure and the criteria for geometrical congruence, the shape of each molecule M_i, $1 \le i \le n$, is characterized by the two dimensional vector $3D\text{-}MSD(r,i) = [OV(r,i); NOV(i,r)]$, and the n vectors are directly comparable.

A similarity measure [30], $S(M_p,M_q)$, of the shape of molecules M_p and M_q with respect to the same set of characteristics $\{OV,NOV\}$ may be defined using [27] the single-valued monotonically-decreasing function

$$S(M_p,M_q) = [1 + d(M_p,M_q)]^{-1}, \tag{31}$$

which has the following properties:

$0 \le S(M_p,M_q) \le 1$, $M_p \ne M_g$; $S(M_p,M_p) = 1$;
$S(M_p,M_q) = S(M_q,M_p)$; $S(d\to\infty) = 0$, and $S(d\to0) = 1$.

Here, $d(M_p,M_q)$ is the Euclidean metric (32),

$$d(M_p,M_q) = [|OV(r,p)-OV(r,q)|^2 + |NOV(p,r)-NOV(p,q)|^2]^{1/2}. \tag{32}$$

Note that from (32) and the triangle inequality and, respectively, (31), it follows that $S(M_p,M_q)$ with respect to the whole set of characteristics considered, i.e., $\{OV,NOV\}$, and $S^\delta(M_p,M_q)$ defined with respect to some of the characteristics, i.e., $\{OV = V_0\}$ satisfy the inequality

$$S(M_p,M_q) < S^\delta(M_p,M_q). \tag{33}$$

Further, it is easily seen that (30) allows accurate calculation of the three-dimensional version [33a] of the Minimal Steric Difference (MSD) descriptor [33b]:

$$MSD(r,i) = g|(\underline{V}^r-\underline{V}^i) \cup (\underline{V}^i-\underline{V}^r)|/|W| =$$
$$= g(|\underline{V}^r-\underline{V}^i| + |\underline{V}^i-\underline{V}^r|)/|W| = \tag{34}$$
$$= NOV(r,i) + NOV(i,r)$$

The MSD approach, arbitrarily prescribing [34] equal weights to the non-overlapping volumes $NOV(r,i)$ and $NOV(i,r)$ lacks the needed flexibility to adequately account for the effect of molecular shape on bioactivity.

The most active compound in the data base considered does not necessarily provide the best topographical fit to the receptor active site. One should consider the possibility of missing sterically relevant atoms and/or the presence of sterically superfluous atoms. The problem is to determine, from the bioactivity of the molecules available, a better topography of the active site.

A classification methodology, the SIBIS algorithm [35], has been developed to map the receptor space explored by the molecules under study, i.e., to identify areas which correspond to the active site, areas occupied by the receptor, and, respectively, areas which, pointing away from active site, offer little opportunity for interaction with the receptor and are, therefore, steric irrelevant. SIBIS is a self-consistent type procedure based on the least squares method with, or without subsidiary conditions; the convergence criterium is the best overall agreement between the observed (Y_i) and estimated $(\hat{Y}_i)$ bioactivities.

To obtain a framework on which to base a computational effort seeking the optimization of the reference structure, M_r, one must develop an appropriate basis from which to describe numerically the stereochemistry of the molecules considered.

One proceeds by superimposing the n molecules over M_r; the superposition procedure uses the multi-molecule fitting described in Sect. 2 supplemented with an allowance for contracting into a simple vertex all nearby atoms, i.e., the atoms $p \in M_i$, $1 \le i \le n$, and $q \in M_r$ occupy the same vertex if the distance $d(p,q) \le \delta$, where δ is given. The obtained pattern of vertices mimics the essential topographical features of the receptor space explored by the n molecules; it is called the investigated receptor space, abbreviated by IRS. One can use the IRS as coordinate system and ascribe to each M_i the m-dimensional row vector $\underline{X}_i = [X_{ij}]$, $1 \le j \le m$, with $X_{ij} = 1$ if the vertex j is occupied by an atom of M_i, and $X_{ij} = 0$ if it is empty; here, m denotes the number of IRS vertices.

Next, one derives the initial steric map, $<IRS>_{init}$, of the receptor as follows: (i) Consider the additional possible connectivities in the IRS by connecting vertices p and q if the resultant edge may represent a covalent bond; (ii) Partition the IRS vertices into three classes: vertices assigned to the active site (c), to the receptor backbone (w), and steric irrelevant areas (i). The c-type vertices correspond to the atoms of the M_r, and the other vertices are of w-type. For convenience, one introduces a dummy vertex of i-type and connects it with those IRS vertices for which one wishes to check the steric relevance.

The steric molecular descriptor SMD = [SMD(r,i); SMD(i,r)] is an approximate measure of the overlapping volume of M_r over M_i, (SMD(r,i)), and non-overlapping volume of M_i over M_r, (SMD(i,r)):

$$SMD(r,i) = \sum_{j\in\{c\text{-type}\}} S_{ij}X_{ij}, \quad SMD(i,r) = \sum_{j\in\{w\text{-type}\}} S_{ij}X_{ij}, \quad (35)$$

where S_{ij} is an additive measure of the size of the atom $j \in M_i$ and $X_{ij} \in \underline{X}_i$.

The optimization of the reference structure consists of the following steps:

1) Consider $<IRS>_{init}$ and compute the regression equation (36) and the corresponding correlation coefficient R_o:

$$\hat{Y}_i = \alpha_o + F(SMD(r,i), SMD(i,r)) + G(\sigma_1,\sigma_2,...) , \quad (36)$$

where $F = \beta_1 SMD(r,i) - \beta_2 SMD(i,r)$ or $F = \beta_1 SMD(r,i) - \beta_2 SMD(i,r) - \beta_3 [SMD(r,i)^2]^2$, if one wishes to obtain an estimate for the receptor bulk tolerance, i.e., $\partial\hat{Y}/\partial SMD(i,r) = 0$ and $SMD(r,i)_{max} = 2\beta_3\beta^{-1}$; $G(\sigma_1,\sigma_2,...)$ is (usually) a linear function of non-shape factors, quantified by σ_1, σ_2, ..., which may condition bioactivity, $G = \sum_k \gamma_k \sigma_{ik}$. G sorts out the non-shape component of $\hat{Y}$ in order to prevent the contamination of the SMD's due to their iterative correction.

2) Change the classification of $j \in <IRS>_{init}$ iff the following two conditions hold:

  i) the resultant eq. (36) has a better correlation coefficient $R_1 \geq R_0 + \Delta R$; and

  ii) the set of c- and i-type vertices, respectively, are all connected among themselves. Changes of vertex classification are performed until no further improvement of the correlation coefficient is observed.

The condition (ii) is designed to be consistent with the concept of reference structure which is represented by a connected graph (c-type vertices) and, respectively, with key-in-lock theory (i-type vertices).

3) The resultant <IRS> is then considered as $<IRS>_{init}$ and the step 2 is carried out for all j.

4) Repeat steps 2 and 3 until self-consistency is achieved, i.e., the vertex classification no longer changes within the given tolerance ΔR on repeated iteration.

The resultant <IRS> is optimal, $<IRS>_{opt}$, and the set of c-type vertices of $<IRS>_{opt}$ represents the "best" probable molecular shape complementary to the receptor active site.

The recently developed version [36a] of the MTD approach [36b,c] parallel closely the SIBIS algorithm.

Information energy method [32] can be used with the IRS concept to quantify the degree of relatedness of the molecular shape of M_i, $1 \le i \le n$: one associates with each M_i, $\underline{via}$ the $\underline{X}_i$ vectors, the finite probability scheme $\underline{P}_i$,

$$\underline{P}_i = (P_{ij})_{\substack{1 \le i \le n \\ 1 \le j \le m}} , \quad P_{ij} = S_{ij}X_{ij} / \sum_{j=1}^{m} S_{ij}X_{ij} ,$$

(37)

$$0 \le P_{ij} \le 1 , \quad \sum_{j=1}^{m} P_{ij} = 1, \ X_{ij} \in \underline{X}_i ,$$

where P_{ij} is the probability that M_i occupies the receptor space centered around the vertex j of IRS. Because $\underline{P}_i$ is related to the shape of the molecule M_i, the degree of relatedness of the probability schemes $\underline{P}_p$ and $\underline{P}_q$ will characterize the degree of shape relatedness of the molecules, M_p and M_q, provided that the vectors $\underline{X}_p$ and $\underline{X}_q$ were defined within the same IRS as coordinate system. The information correlation coefficient R(p,q), given by

$$R(p,q) = \sum_{j=1}^{m} P_{pj}P_{qj} / [E(p)E(q)]^{1/2} ,$$

(38)

expresses quantitatively the relationship between $\underline{P}_p$ and $\underline{P}_q$, and, accordingly, the relationship between the shapes of $\underline{M}_p$ and $\underline{M}_q$, i.e., R(p,q) = 1 if $\underline{P}_p$ and $\underline{P}_q$ are identical repartized (M_p and M_q have the same shape), and R(p,q) = 0 if $\underline{P}_p$ and $\underline{P}_q$ are indifferent (the shape of M_p and M_q are not related); the intermediate values $0 < R(p,q) < 1$ are judged using the criteria for the significance of the correlation coefficient [41]. Here, the quantity E(q) is called the information energy content of $\underline{P}_q$,

$$E(q) = \sum_{j=1}^{m} P_{qj}^2 ; \ 1/m \le E(q) \le 1 ,$$

(39)

and it is a measure of the uniformity of the system described by $\underline{P}_q$. Other estimates [38,40] of the degree of similarity between a pair of chemical structures are based on graph-theoretical and information-theoretical [37,39] concepts.

CONCLUSIONS

In our view, one can make a convincing argument that model building and energy calculations are techniques which will play an increasingly important role in drug design [3,12,42]. While each technique may be applied independently, their combined application has a sinergistic effect.

ACKNOWLEDGEMENTS

This work was supported by the National Institute of General Medical Sciences by a grant (GM 24483). The authors also wish to acknowledge the contributions of their collaborators, C.D. Barry, H.E. Bosshard, B.L. Kalman, S.F. Karasek, J. Labanowski, and C.E. Molnar.

REFERENCES

1. C. Hansch, in "Drug Design", ed. E.J. Ariens, vol. I, Academic Press. New York, 1971; C. Hansch, in "Structure Activity Relationships", ed. C.J. Cavallito, vol. I, Pergamon, Oxford, 1973; C. Hansch, in "Correlation Analysis in Chemistry-Recent Advances", eds. N.B. Chapman and F. Shorter, Plenum, New York, 1978; Y.C. Martin, "Quantitative Drug Design. A Critical Introduction", Dekker, New York, 1978; R.S. Osman, H. Weinstein and J.P. Green, in "Computer Assisted Drug Design", eds. E.C. Olsen and R.C. Christoffersen, ACS, vol. 112, Washington, D.C., 1979; A.T. Balaban, A. Chiriac, I. Motoc, Z. Simon, "Steric Fit in QSAR", Lecture Notes in Chemistry vol. 15, Springer, Berlin, 1980; M. Charton and I. Motoc, eds., "Steric Effects in Drug Design", Top. Curr. Chem. vol. 114, Springer, Berlin, 1983; J.G. Topliss, ed., "Quantitative Structure-Activity Relationships of Drugs", Academic Press, New York, 1983; R. Franke, "Theoretical Drug Design Methods", Elsevier, Amsterdam, 1984. M. Kuchar, ed., "QSAR in Design of Bioactive Compounds", Prous Science, Barcelona, 1985.

2. S.M. Free and J.W. Wilson, J. Med. Chem. $\underline{7}$, 395 (1964); D.R. Hudson, G.E. Bass, and W.P. Purcell, J. Med. Chem. $\underline{13}$, 1184 (1970); W.P. Purcell, G.E. Bass, and J.M. Clayton, "Strategy of Drug Design: A Guide to Biological Activity", Wiley, New York, 1973; T. Eujita and T. Ban, J. Med. Chem. $\underline{14}$, 148 (1971); H. Kubinyi, J. Med. Chem. $\underline{19}$, 587 (1976).

3. G.R. Marshall and I. Motoc, Top. Mol. Pharm. $\underline{3}$, in press.

4. S.C. Nyburg, Acta Cryst. $\underline{B30}$, 251 (1974); D.J. Duchamp, in "Computer Assisted Drug Design", eds. E.C. Olson and R.E. Christofferson, ACS, vol. 112, Washington, D.C., 1979.

5. P. Gund, Annu. Rep. Med. Chem. $\underline{14}$, 299 (1979); C. Humblet and G.R. Marshall, Annu. Rep. Med. Chem. $\underline{15}$, 267 (1980), G.R. Marshall, in "Quantitative Approaches to Drug Design", ed. J.C. Rearden, Elsevier, Amsterdam, 1983; D.H. Smith, J.G. Nourse, and C.W. Crendell, in "Struct. Correl. Predict. Tool Toxicol." (Pap. symp.), ed. L. Goldberg, Hemisphere, Washington, D.C., 1985, p. 171-191.

6. V. Prelog, J. Mol. Catalysis $\underline{1}$, 159 (1975/76); I. Motoc, J.N. Silverman, O.E. Polansky, and G. Olbrich, Theoret. Chim. Acta (Berl.) $\underline{67}$, 63 (1985).

7. F.P. Preparata and R.T. Yeh, "Introduction to Discrete Structures for Computer Science and Engineering", Addison-Wesley, Reading, 1973.

8a. L.G. Roberts, "Homogeneous Matrix Representation and Manipulation of N-Dimensional Constructs", Lincoln Laboratory, MIT, Document No. MS 1045, 1965. b) N. Go and H.A. Scheraga, Macromolecules 3, 178 (1970).

9. J.W. Gibbs, "Vector Analysis", Yale University Press, New Haven, Connecticutt, 12-th Ed., 1958.

10a. R.A. Dammkoehler and C.D. Barry, "Vector and Scalar Formulation of Coordinate Rotation and Variable Pairwise Distances in E-3 Space", Technical Memo No. 4, Department of Computer Science, Washington University, St. Louis, Mo., 1970; b) C. Humblet, "Systematic Conformational Search: A User Oriented Guide to VAUDEVILLE", Technical Memo No. 31, Department of Physiology and Biophysics, Washington University, St. Louis, Mo., 1981; B.L. Kalman, "User's Guide to the ENERGY-SEARCH System", Technical Memo No. 37, Department of Computer Science, Washington University, St. Louis, Mo., (1981); B.L. Kalman, "Modifications to SEARCH to Process Molecules with Rings", Technical Memo No. 48, Department of Computer Science, Washington University, St. Louis, Mo., 1982; B.L. Kalman, "An Experimental Time Comparison of Two Implementations of SEARCH", Technical Memo No. 59, Department of Computer Science, Washington University, St. Louis, Mo. 1983; c) S.F. Karasek, "A Procedure for Generating Aggregate Descriptors", Technical Memo No. 27, Department of Computer Science, Washington University, St. Louis, Mo., 1980; d) B.L. Kalman, "ONESTEP: A Description and Explanation", Technical Memo No. 23, Washington Universtiy, St. Louis, Mo., 1980.

11. A.J. Hopfinger, "Conformational Properties of Macromolecules", Academic Press, New York, 1973; S.R. Niketic and K. Rasmusen, "The Consistent Force Field", Lecture Notes in Chemistry vol. 3, Springer, Berlin, 1977; A.J. Stuper, W.E. Bruger, and P.C. Jurs, eds., "Computer Assisted Studies of Chemical and Biological Function", Wiley, New York, 1979; V. Burkert and N.L. Allinger, "Molecular Mechanics", ACS Monograph no. 177, ACS, Washington, D.C., 1982.

12. P. Kollman, Account. Chem. Res. 18, 105 (1985).

13. A.T. Hagler and S. Lifson, J. Am. Chem. Soc. 101, 5111 (1979).

14. A.J. Hopfinger, "Conformational Properties of Macromolecules", Academic Press, New York, 1973, ch. 2; D.A. Greenberg, C.D. Barry, and G.R. Marshall, J. Am. Chem. Soc. 100, 4020 (1978); A. Warshell, J. Chem. Phys. 83, 1640 (1979); O.L. Olatunji and S. Premilat, Biochem. Biophys. Res. Comm. 126, 247 (1985), J.B. Matthew, Annu. Rev. Biophys. Chem. 14, 387 (1985).

15. J.A. Nader and R. Mead, Computer J. 7, 308 (1965).

16. D.N.J. White, Computer and Chemistry 1, 225 (1977).

17. S. Naruto, I. Motoc, G.R. Marshall, S.B. Daniels, M.J. Sofia, and J.A. Katzenellenbogen, J. Am. Chem. Soc. 107, (1985), in press.

18. D. Mayer, G.R. Marshall, and I. Motoc, in preparation.

19. J.M. van Rossum, ed., "Kinetics of Drug Action", Springer, Berlin, 1977.

20. Sybyl software, Tripos Assoc., St. Louis, MO, USA.
21. I. Motoc and G.R. Marshall, Chem. Phys. Lett. 116, 415 (1985).
22. G.N. Ramachandran, C.M. Venkatachalan, and S. Krim, Biophys. J. 6, 849 (1966).
23. Z. Simon and Z. Szabadai, Stud. Biophys. (Berlin) 39, 123 (1973); Z. Simon, Angew. Chem. 86, 802 (1974); Z. Simon, A. Chiriac, I. Motoc, S. Hoban, D. Ciubotarin, and Z. Szabadai, Stud. Biophys. (Berlin) 55, 217 (1976).
24. C. Humblet and G.R. Marshall, Annu. Rep. Med. Chem. 15, 26 (1980); J.R. Sulfrin, D.A. Dunn, and G.R. Marshall, Mol. Pharmacol. 19, 307 (1981); I. Motoc and G.R. Marshall, Z. Naturforsch, in press.
25. I. Motoc, Arzneim.-Forsch. 31, 290 (1981); R. Valceane and I. Motoc, Rev. Roum. Chim. 27, 225 (1982); I. Motoc, Z. Naturforsch. 38a, 1342 (1983); I. Motoc, Quant. Struct.-Act. Relat. 3, 43 (1984).
26. J.M. Hammersley and D.C. Handscomb, "Monte Carlo Methods", Wiley, New York, 1964; D.E. Knuth, "The Art of Computer Programming. Seminumerical Algorithms", vol. 2, Addison-Wesley, Reading, Mass., 1981; B.P. Demidovici and I.I. Maron, "Computational Mathematics", Mir, Moscow, 1973.
27. I. Motoc, G.R. Marshall, R.A. Dammkoehler, and J. Labanowski, Z. Naturforsch, in press, and references cited herein.
27a. A.J. Hopfinger, J. Am. Chem. Soc. 102, 7196 (1980); A.J. Hopfinger, J. Med. Chem. 24, 818 (1981); C. Battershell, D. Malhotra, and A.J. Hopfinger, J. Med. Chem. 24, 812 (1981); A.J. Hopfinger, Arch. Biochem. Biophys. 206, 153 (1981); S. Noor, A.J. Hopfinger, and D.R. Bickers, J. Theor. Biol. 102, 323 (1983).
28. S. Naruto, I. Motoc, and G.R. Marshall, Eur. J. Med. Chem., in press.
29a. C.H. Hassall, A. Krohn, C.J. Moody, and W.A. Thomas, J. Chem. Soc. Perkin Trans. I, 1984, 155; P.R. Andrews, J.M. Carson, A. Caselli, M.J. Spark, and R. Wood, J. Med. Chem. 28, 393 (1985). b) G. Trummlitz, G. Schmidt, H.V. Wagner, and P. Luger, Arzneim.-Forsch./Drug Res. 34, 849 (1984); c) H. Weinstein, S. Maayani, B. Pazhenchevsky, C. Venanzi, and R. Osman, Int. J. Quantum Chem., OBS 10, 309 (1983); S. Takemura, J. Pharmacobio.-Dyn. 7, 436 (1984); d) A.M. Bianucci, A. Martinelli, and A. DaSettino, Farmaco, Ed. Sci. 39, 686 (1984); e) J.M. Schulman, M.L. Sabio, and R.L. Disch, J. Med. Chem. 26, 817 (1983).
30. J. Fabian, A. Melhorn, and F. Fratev, Int. J. Quantum Chem. 17, 235 (1980); A. Melhorn, F. Fratev, O.E. Polansky, and V. Monev, Math. Chem. 15, 3 (1984).
31. I. Motoc, Z. Naturforsch, 38a, 1342 (1983).
32. O. Onicescu, C.R. Acad. Sci. Paris A263, 841 (1966); St. Cerc. Mt. 18, 1419 (1966).
33a. I. Motoc, S. Holban, R. Vancea, and Z. Simon, Stud. Biophys. (Berlin) 66, 75 (1977); b) Z. Simon and Z. Szabadai, Stud. Biophys. (Berlin) 39, 123 (1973).
34. I. Motoc and G.R. Marshall, Z. Naturforsch., in press.

35. I. Motoc and O. Dragomir, Math. Chem., I. Motoc, Quant. Struc.-Act. Relat. 3, 43, 47 (1984); I. Motoc, G.R. Marshall, and J. Labanowski, Z. Naturforsch., in press; I. Motoc, J. Labanowski, G.R. Marshall, and R. Dammkoehler, Eur. J. Med. Chem., submitted.

36a. Z. Simon, A. Chiriac, S. Holban, D. Ciubotariu, and G.J. Mihalas, "Minimum Steric Difference. The MTD Method for QSAR Studies", Research Studies Press, Letchworth, 1984. b) Simon, A. Chiriac, I. Motoc, S. Holban, D. Ciubotariu, and Z. Szabadai, Stud. Biophys. (Berlin) 55, 217 (1976); c) I. Motoc, Math. Chem. 5, 275 (1979); I. Motoc, Top. Curr. Chem. 114, 93 (1983).

37. A. Renyi, "Probability Theory", North-Holland, Amsterdam, 1970.

38. M. Barysz, N. Trinajstić, and J.V. Knop, Int. J. Quantum Chem., QCS 17, 441, (1983).

39. C. Shannon and W. Weaver, "Mathematical Theory of Communication", Univ. of Illinois, Urbana, IL., 1949; L. Brillouin, "Science and Information Theory", Academic Press, New York, 1956.

40. D. Bonchev and N. Trinajstić, J. Chem. Phys. 67, 4517 (1977).

41. P.R. Wells, "Linear Free Energy Relationships", Academic Press, London, 1968; J.G. Topliss and R.J. Costello, J. Med. Chem. 15, 1066 (1972).

42. P. Gund, "Present and Future Computer Aids to Drug Design, in "X-Ray Crystallography and Drug Action", eds. A.S. Horn and C.J. DeRanter, Oxford Univ. Press, Oxford, 1984; G. Jolles and K.R.H. Wooldrige, eds., "Drug Design: Fact or Fantasy", Academic Press, London, 1984; R.A. Maxwell, Drug. Dev. Res. 4, 375 (1984); P.J. Goodford, J. Med. Chem. 27, 551 (1984); N.C. Cohen, Trends Pharmacol. Sci. 4, 503 (1983); Y.C. Martin, K.H. Kim, T. Koschmann, and T.J. O'Donnell, Anal. Chem. Symp. Ser. 15, 285 (1983). W. Bartman and G. Snatzke, eds., "Structure of Complexes of Biopolymers with Low Molecular Weight Molecules", Wiley-Heiden, Chichester, 1982; J.P. Tallenaere, Trends Pharmacol. Sci. 3, 138 (1982); C. Humblet and G.R. Marshall, Drug Dev. Res. 1, 409 (1981).

Chapter 21

HOW STRONG IS THE GAUCHE^P –GAUCHE^M INTERACTION?

Eiji Ōsawa
Department of Chemistry, Faculty of Science,
Hokkaido University, Sapporo 060, Japan

ABSTRACT

Traditional areas of experimental organic chemistry have been slow in adopting mathematical methods and concepts, although the relation is improving. As an example of obtaining benefits from application of computation, molecular mechanics simulation of the dependence of gauche^P-gauche^M interaction on environments is presented.

INTRODUCTION

Mathematical concepts are generally not considered as the requisite for the study and practice of experimental organic chemistry, especially in those traditional areas like synthesis and natural products. Concepts prevailing in these areas are usually expressed in terms of abstract words frequently ending with 'ty', such as affinity, reactivity or polarizability. In fact, chemists practicing in these areas can do very well even without recourse to mathematics at all. Despite this tradition, the entire history of chemistry can be viewed as that of progressively turning the vague ideas into concrete theories which can somehow be formulated with the aid

of numerals.   Those familiar terms like aromaticity
[1], steric strain [2] and congestion [3] which once
had only abstract meanings have recently been well
quantified.
     One of the most successful among these ap-
proaches is the empirical formulation of molecular
force field [4].   Molecular mechanics, as it is
usually called, actually does not involve any novel
mathematical ideas but is a form of representing intramolecular potential energy or force with a set of
simple potential functions.   Remarkably high reputa-
tion that this method has in recent years acquired as
a practical means of predicting molecular properties
like shape and vibrational behavior, rests entirely
upon good set of parameters for potential functions.
We are interested in molecular mechanics partly because it is a new way of doing chemistry while at the
same time giving us ample opportunity to improve and
update the art of simulating molecular force field.
Several problems existing in molecular mechanics [5]
need fresh concepts, hopefully with mathematically
attractive ideas, but we do not touch on these problems here.
     This article is intended to illustrate how the
molecular mechanics can be applied to practical organic chemical problems.*   The problem discussed
here started from one of the vague 'ty' terms currently appreciated among organic chemists, namely,
flexibility of organic molecules.   Whereas some
classes of molecules, for example saturated cyclic
and acyclic alkanes, are generally considered flexible, the flexibility is more or less restricted by
the shape of its energy hypersurface.   Geometry of
such a molecule changes through energetically the
lowest possible paths along valleys and saddle points
on the surface.   Because of high dimensionality of
ordinary organic molecules, no complete detail of
energy hypersurface is known for molecules of practical interests to experimental organic chemists.
However, several structural features that limit the
flexibility under certain well-defined circumstances
have been known for some time and they have been
serving as indispensable guiding principles in conformational analysis.   The particular feature that
we are going to discuss here is the 'forbidden'
conformation of n-pentane, namely two adjacent
gauches of opposite sign, i.e. gauche^P-gauche^M (1),
or so-called 1,3-diaxial interaction when it is on
the ring (2), wherein close approach of end carbon
atoms (C1 and C5) of n-pentane or corresponding par-

----

* See also ref. [4] and [6] for other applications.

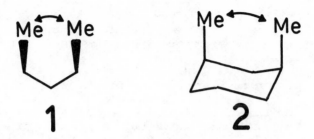

tial structure destabilizes the g^Pg^M half-ring conformation. However, not much was known about the way this notable steric interaction can be relieved by the available flexibility of molecule or, inversely, how the interaction can be strengthened. Energetics involved were also unknown. The situation was so, until molecular mechanics came to use. Now we can calculate this type of interaction under diverse circumstances and compare the computed results with experimental observations.

BACKGROUND

According to Dale [7], the g^Pg^M conformation of n-pentane is a shallow energy minimum, which is 14 kJ/mol more strained than the global minimum anti-anti conformer. Our MM2' [8] calculations confirm this point and reveals several other interesting features: (1) the g^Pg^M conformer has no symmetry element, namely the magnitudes of rotation about C2-C3 and C3-C4 bonds are not equal, (2) the C-C-C valence angles at C2, C3 and C4 are widened to 115 to 116°, and (3) the shortest H/H contact distance is 2.09 Å [9]. Clearly this conformation is unlikely to populate to a significant extent under normal conditions. Since the predicted strain is the result of complete relaxation, it can be anticipated that the interactions present in this type of conformation will readily increase if the relaxation is restricted by some means.

DISCOVERY OF STAGGERED ROTATIONAL BARRIER IN BICYCLO-ALKANES

Our entry into this project occurred by chance which at first appeared to have nothing to do with the g^Pg^M interaction, but seemed to be related with the molecular flexibility. In 1981, Ogawa et al. [10] reported on the dnmr determination of activation barriers for the rotation of N-N bond in substituted 1,1'-bipiperidines (3).   A point of interest in 3 is that the inversion of nitrogen atom is 'prohibited' by the presence of one or two equatorial alkyl substituents, which would give rise to 1,3-diaxial repulsions if the nitrogen atom (and hence the piperidine ring) is inverted.   In this way, it was expected that only the N-N bond rotation process should be visible in the appropriate temperature range. The observed barriers (74 to 79 kJ/mol) were assigned to a 'single passing barrier' (4) where two N-C bonds are eclipsed.

We were first intrigued by the seemingly too high barrier for a sym-tetraalkylhydrazine (reasonable guess is about 40 kJ/mol [11]) and tried to simulate the N-N bond rotation process by molecular mechanics [12].   Although MM2 force field [4] that we used has not been parameterized for N-N functionality, it contained parameters for aliphatic amines [13].   So we chose 1-(2'-methylcyclohexyl)-2-methylpiperidine (5) as a model of 3 and performed dihedral driver calculation on the N-C1' bond (Figure 1). Results were most surprising.   Starting from the global minimum conformation A and driving the 6-1-1'-2' dihedral angle toward larger positive values, we did reach the single-passing, eclipsed barrier (B). However, as shown, (B) is not at all remarkable compared to the pronounced barrier (C).   The calculated height of this barrier (about 60 kJ/mol) is the closest to what was observed, and in view of the recognized tendency of MM2 force field to grossly underestimate the rotational barrier height [6], this height can be considered reasonable.   Then the calculated height of (B) also appears as expected.

The 'unexpected' barrier (C) has perfectly staggered conformation about the rotating bond.   A close look at the calculated structure revealed that the strain arose from the long range, 1,5 type inter-

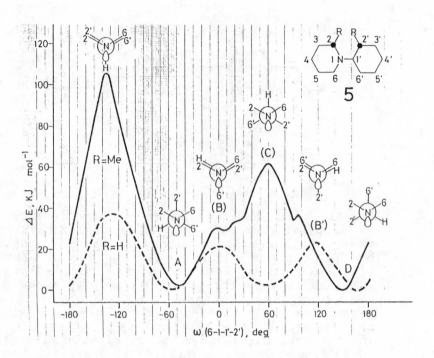

Figure 1. Torsional energy curve of N-cyclohexylpiperine (5, R=H, dotted line) and *meso*-2,2'-dimethyl derivative (5, R=Me, solid line) [12]

actions across the C-N bond, namely a pair of strongly repulsive $g^P g^M$ interactions, 6-1-1'-2'-Me and 6'-1'-1-2-Me. It should be emphasized here that only limited freedom is available at the top of the barrier, since any deformation in one of the $g^P g^M$ sequences to reduce the unfavorable interactions increases the strain in the other. Apparently the $g^P g^M$ pair is locked simultaneously and inescapably into the narrow saddle point and the strain per one $g^P g^M$ interaction has been intensified from 14 kJ/mol in n-pentane (vide supra) to 38 kJ/mol!

At this point, we still could not believe the results. First, we thought that the new barrier may be an artefact of single bond drive calculation. Then we performed two bond driver calculation, rotating 5-6-1-2 as well as 6-1-1'-2' dihedral angles of 5 [14]. This revealed a new pathway including twist form of piperidine ring and a new barrier of almost the same height as (C), but the essential feature remained unchanged; again a pair of strongly constrained and hardly relaxable $g^P g^M$ interactions as the source of barrier.

RESTRICTED ROTATION IN CANNABIDIOL

Incidentally, a dnmr study of cannabidiol (6) by Kane and Martin [15] was reported shortly after our work mentioned above was completed and their results came to our attention again by chance. They gave very high barrier of rotation about the pivot bond of **6** (61.5 kJ/mol) in contrast to literature value. Correspondence with the previous authors revealed their mistakes and the previous barrier height was corrected to 59.8 kJ/mol in good agreement with the Kane-Martin value. **6** is essentially a substituted phenylcyclohexene carrying 'ortho' substituents potentially capable of $g^P g^M$ and/or similar long-range nonbonded interactions across the pivot bond. Molecular mechanics driver calculations of **6** confirmed the above initial guess on the source of rotational barriers [15].

6

7

Here, the presence of sp^2-hybridized carbon atoms somewhat complicates the situation. Take phenylcyclohexane (**7**, R=H) as the simplest model: for the gauche-like interaction across the pivot bond in the well known equatorial-perpendicular conformation **8**, we proposed a name of 'progauche' interaction.

8

Progauche differs from gauche in at least two aspects. First, the dihedral angle is about 30° in the former whereas it is about 60° in the latter. Second, the two sp^2-hybridized carbon atoms in the former have different valence angles and bond lengths from those of the latter. A progauche interaction is probably slightly stronger than gauche.

PHENYLCYCLOHEXANES AND BICYCLOHEXYLS

It is now clear that the best model systems for systematic study of the staggered barrier are substituted phenylcyclohexanes (**7**) and bicyclohexyls (**9**).

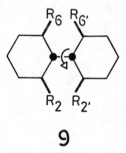

9

Although some of them have been subjected to preliminary calculations in the course of our works mentioned above [14,15], it was often difficult to exactly locate barriers by the dihedral drive technique of the MM2 program. Namely, the dihedral drive algorithm can be used only for simple cases. A better method of exploring a complex torsional energy hypersurface is to use full-matrix Newton-Raphson geometry optimization which converges at an energy maximum of any dimensionality. The dimensionality is equal to the number of negative eigenvalues of F-matrix (second derivatives of potential energy with respect to nuclear coordinates) which is obtained during the Newton-Raphson optimization [16].

Program BIGSTRN3 [17] is equipped with this capability and several other convenient options for our present purpose [15]. Hence this program was used in conjunction with MM2' force field [8] for an extensive study on the rotation about the pivot bond of variously methylated **7** and **9** [18]. Bicyclohexyls **9** followed expectedly complex dynamics when more and more 'ortho' positions are methylated. The highest calculated barrier was 109 kJ/mol for 1-(2',6'-dimethylphenyl)-2,6-dimethylcyclohexane (**7**, R's=Me).

At this barrier, two sets of long-range interactions (ggg/geg) occur simultaneously. During the calculation, one of the cyclohexyl ring was allowed to change its conformation, but the other ring was kept in chair form. This artificial constraint had to be imposed in order to reduce the amount of computational load to a manageable size. Hence the predicted pathway and barrier height must be regarded provisional. Even higher barrier (127 kJ/mol) was predicted for 1-(2',6'-dimethylphenyl)-2-methylcyclohexane (7, R_2=H, R_6=$R_{2'}$=$R_{6'}$=Me). Generally, the pathways calculated for the pivot rotation of 7 are less complex and have somewhat higher barriers than those of 9, reflecting the effect of more rigid phenyl compared to cyclohexane ring. Molecular mechanics predict that atropisomerism should be possible for appropriately substituted 7 and 9.

No experimental information is available, however, regarding the rotational barriers of 9. Bicyclohexyls 9 have anyhow been unpopular in organic chemistry and we wish to attract attention of experimental chemists for this unexplored class of hydrocarbon.[†]

A few experimental determinations of rotational barriers are reported for the derivatives of 8 and related molecules [18]. Only one example will be mentioned here. An indole derivative (10) showed

10

[†] See p. 270 of ref. [19]. In recent years, a large number of derivatives of 9 have been appearing in patents relating liquid crystals. We thank Mr. K. Yoshinaga of Canon Company for this information.

atropisomerism: isomers were separated by chromatography and identified by x-ray analysis [20].   Since
the amide nitrogen should be able to avoid congestion
by inversion, its substituent (COC_6H_4Cl) probably
does not contribute to the rotational barrier.
Then, the steric environments at the transition point
of this molecule during the rotation of pivot bond
correspond to those of trimethyl-**7**, the one which has
been predicted to have the highest barrier!

CONCLUSION

Upon reflection, we realize that conformational
analysis has long been dominated by the gauche effect
and biphenyl isomerism (**11**) [19].   The former is 1,4

11

type and the latter 1,5 and 1,6 types of nonbonded
interaction.   Hence the important classes of van der
Waals interaction had been covered, albeit in an
unsystematic way.   **7** and **9** can be regarded as the
extension of biphenyl isomerism to saturated and more
flexible analogues, which proved to offer rich material for the study of g^Pg^M, $g^Pg^Mg^P$ and similar long-
range interactions.
Our initial aim, to see if the g^Pg^M interaction
can be increased by imposing extra constraints, has
more or less been substantiated.   It would be interesting to seek other systems, than bicycloalkyls
and arylcycloalkyls, that provide straining circumstances to the long-range nonbonded interactions.

ACKNOWLEDGEMENTS

The author thanks Dr. Carlos Jaime for skillful-
ly carrying out all of the calculations referred in
this work.   We are indebted to Professor Y. Takeuchi
and Dr. K. Ogawa for valuable suggestions and plea-
sant discussions.   Partial financial supports were

provided by the Ministry of Eduction through Grants-in-Aid for Scientific Research.

REFERENCES

[1] S. Kuwajima, J. Am. Chem. Soc. 1984, 106, 6496 and references cited therein.

[2] (a) R. Fuchs, J. Chem. Educ. 1984, 61, 133. (b) A. Greenberg et al. J. Am. Chem. Soc. 1983, 105, 6855; Tetrahedron 1983, 39, 1533.

[3] M. L. Connolly, J. Am. Chem. Soc. 1985, 107, 1118; Science 1983, 221, 709.

[4] U. Burkert and N. L. Allinger, "Molecular Mechanics", American Chemical Society: Washington, D. C., 1982. A review of this book: P. Gund, J. Chem. Inf. Comput. Sci. 1983, 23, 88.

[5] T. Hirano and E. Ōsawa, Croat. Chem. Acta 1985, 58, 1573.

[6] E. Ōsawa and H. Musso, Angew. Chem. Int. Ed. Engl. 1983, 22, 1; Top. Stereochem. 1982 13, 117.

[7] J. Dale, "Stereochemistry and Conformational Analysis", Verlag Chemie, New York, 1978.

[8] C. Jaime and E. Ōsawa, Tetrahedron 1983, 39, 2769.

[9] Details to be published elsewhere.

[10] K. Ogawa, Y. Takeuchi, H. Suzuki and Y. Nomura, J. Chem. Soc., Chem. Commun. 1981, 1015; Chem. Lett. 1981, 607.

[11] S. F. Nelsen and G. R.Wisman, J. Am. Chem. Soc. 1976, 98, 3281.

[12] C. Jaime and E. Ōsawa, J. Chem. Soc., Chem. Commun. 1983, 708.

[13] S. Profeta, Jr. and N. L. Allinger, J. Am. Chem. Soc. 1985, 107, 1907.

[14] C. Jaime and E. Ōsawa, J. Chem. Soc., Perkin Trans. II 1984, 995.

[15] V. V. Kane, A. R. Martin, C. Jaime and E. Ōsawa, Tetrahedron 1984, 40, 2919.

[16] P. M. Ivanov and E. Ōsawa, J. Comput. Chem. 1984, 5, 307.

[17] Program BIGSTRN3 was kindly donated by Professor K. Mislow and Dr. R. B. Nachbar, Jr. See H.-B. Burgi, W. D. Hounshell, R. B. Nachbar, Jr. and K. Mislow, J. Am. Chem. Soc. 1983, 105, 1427.

[18] C. Jaime and E. Ōsawa, J. Mol. Struct. 1985, 126, 363.

[19] E. L. Eliel, "Stereochemistry of Carbon Compounds", McGraw-Hill Book Co.; New York, 1962.

[20] K. Harano et al. Cryst. Struct. Commun. 1981, 10, 165. See also, T. Kitamura et al. Heterocycles 1982, 19, 2015.

Chapter 22

TOPOLOGICAL EFFECTS ON MOLECULAR ORBITALS (TEMO)

O.E. Polansky
Max-Planck-Institut für Strahlenchemie D-4330 Mülheim a.d. Ruhr

ABSTRACT

Simple molecular topological considerations result in an interlacing theorem which produces TEMO. Its physical relevance is proved. The purely topological feature of TEMO is found to have a strong predictive power and, hence, dominates over several physical factors in determining MO pattern.

INTRODUCTION

The interest in the influence of topology on molecular properties has grown remarkably in the last few years. Much progress has been made in the various areas of chemical topology, for example the synthesis of sterically unusual compounds like catenanes, rotaxanes [1,2], and Möbius strips [3] as well as the topological analysis of the electron density function [4] and energy hypersurfaces [5].
In the present paper some progress in the field of molecular topology is reported [6,7]: the concept of *topologically related isomers*, termed *topomers*, as well as some *topological models* for their construction are presented; further a novel relation between the MO pattern of topomers is derived (*interlacing theorem*) and its physical significance is proved by both, quantum chemical calculations (at the *ab initio* SCF level) and experimental data.

MOLECULAR TOPOLOGY

Formally molecular topology is fully described by the simple graph $G = [V, E]$ associated with the chemical constitution of the molecule considered: The n vertices collected in the finite non-empty set V represent the atoms whilst the k unordered pairs of distinct vertices of V, forming the set E, correspond to the "chemical bonds" of the molecule in question. By means of the open set formalism it was shown [8] that a topological space is uniquely associated with any simple graph. A simpler alternative to this important work is offered by the neighbourhood formalism [9] which results in that topological space which is induced by the discrete metric defined upon a connected simple graph (note, by definition, a simple graph has no loops, no arcs, and no multiple edges). The alternative approach is briefly outlined as follows:

(i) A ball-neighbourhood $U_\varepsilon(x)$ of the vertex x of V is defined as the set of all elements y of V whose distance to x, $d(x,y)$, is smaller than an arbitrary positive nubmer ε:

$$U_\varepsilon(x) = \{y \mid y \in V, d(x,y) < \varepsilon\}. \tag{1}$$

(ii) A subset of V, $V' \subset V$, is said to be a neighbourhood U of the element x if and only if V' contains a ball-neighbourhood of x:

$$U_\varepsilon(x) \subset V'. \tag{2}$$

(iii) A topology $\underline{\underline{T}}$ is defined upon the set V by associating each element x of V with a system $U(x)$ of subsets of V, the so-called neighbourhoods U of x, obeying the axiomes:

[T 1] $x \in U$ for all $U \in U(x)$.
[T 2] If $U \in U(x)$ and $U' \subset U$, then $U' \in U(x)$.
[T 3] If U', $U'' \in U(x)$ then $U' \cap U'' \in U(x)$; $V \in U(x)$.
[T 4] For any $U \in U(x)$ there exists $U' \in U(x)$ such that $U \in U(y)$ for all elements $y \in U'$.

(iv) The set V together with its topology $\underline{\underline{T}}$ forms the *topological space* T.
In such a way each molecule M is unequivocally associated with a particular topological space $T(M)$ by its molecular graph G(M) which represents the constitution of the molecule under consideration. It may easily be verified that any fragment of the molecule, say $A \subset M$, corresponds with a distinct subspace of its topological space, $T(A) \subset T(M)$.

TOPOMERS AND TOPOLOGICAL MODELS

Two isomers are said to be *topologically related* if they are constituted from pairwise equal fragments, say A,B,C,... . Thus, the difference between the isomers of such a pair arises solely from the different mutual linkage of their fragments. Topologically related isomers are called *topomers*.
The topological spaces associated with a pair of topomers may be divided into subspaces $T(A)$, $T(B)$,... associated with the building fragments, A,B,...; evidently, these subspaces are pairwise isomorphic. Thus the topological spaces associated with a pair of topomers differ only with regard to the respective conjunctions of their subspaces.
A particular mode for the construction of topomers is termed a *topological model*. The number of topological models seems to be unlimited, but here we present only two such models of particular interest; more examples may be found elsewhere [10,11].
In *model 1* two fragments, A and B, are combined to the topomers S and T by $\ell \geq 2$ bonds. In *model 2* three fragments are used: the terminal fragments A and B are linked by $\ell \geq 2$ bonds with ℓ centers of the central moiety C. The two models are represented by the following schemes:

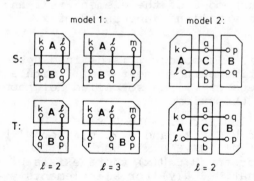

The topomers formed by means of one of these models are denoted by S and T according to

$$\lim_{x\to\infty} \phi(S,x) \quad < \quad \lim_{x\to\infty} \phi(T,x) \tag{3}$$

where $\phi(S,x)$ and $\phi(T,x)$ stand for the characteristic polynomials of the respective graphs.
It should be noted that in order to construct two different topomers at least two bonds are needed and, further, not all the centers where the linking takes place are allowed to be equivalent.
The use of the models is illustrated by Figure 1.

The topomeric pairs I-IV are constructed by means of model 1 while model 2 is used for V; as it can be seen from Figure 1 the pairs IV and V have the T isomer in common, IVT = VT.

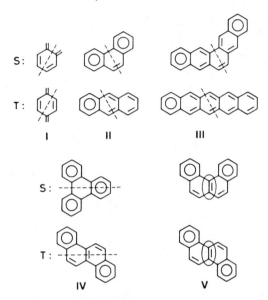

Figure 1. Examples of pairs of topomers.

INTERLACING THEOREM – TEMO

The characteristic polynomials of the topomers may be expressed in terms of the characteristic polynomials of their constituents. Thus for model 1, $\ell = 2$, one obtains

$$\phi(S,x) = \phi(A,x)\phi(B,x) - \phi(A-k,x)\phi(B-p,x) -$$
$$- \phi(A-\ell,x)\phi(B-q,x) + \phi(A-k-\ell,x)\phi(B-p-q,x) -$$
$$- 2[\Sigma\phi(A-P_{k\ell},x)][\Sigma\phi(B-P_{pq},x)],$$

$$\phi(T,x) = \phi(A,x)\phi(B,x) - \phi(A-k,x)\phi(B-q,x) -$$
$$- \phi(A-\ell,x)\phi(B-p,x) + \phi(A-k-\ell,x)\phi(B-p-q,x) -$$
$$- 2[\Sigma\phi(A-P_{k\ell},x)][\Sigma\phi(B-P_{pq},x)].$$

(4)

In these expressions A-k denotes the graph obtained from A by deleting the vertex k, etc.; the summations run over the complete sets of paths $\{P_{kl}\}$ connecting the vertices k,l $\in$ A and $\{P_{pq}\}$ connecting p,q,$\in$B, respectively.
Note that the characteristic polynomials as expressed by eq. (4) consist of a number of bilinear terms;

of the two factors forming such a term the one refers to subspace $T(A)$ and the other to $T(B)$. The difference of these polynomials defined by

$$\Delta(x) = \phi(T,x) - \phi(S,x) \tag{5}$$

is again a polynomial in x. In case of model 1, $\ell=2$, it follows from eq. (4) that

$$\Delta(x) = [\phi(A-k,x)-\phi(A-\ell,x)][\phi(B-p,x)-\phi(B-q,x)]. \tag{6}$$

Note that in eq. (6) all polynomials, but only those which are sensitive for the difference of the conjunction of $T(A)$ and $T(B)$ in $T(S)$ and $T(T)$, respectively appear. Thus $\Delta(x)$ incorporates these differences in polynomial form and, hence, it is considered as reflecting the *relative* topology of a pair of topomers.

The expression of eq. (6) takes the form of a perfect square if the moieties A and B are isomorphic, A $\simeq$ B, i.e. the centers k and ℓ of A may be mapped isomorphically onto the centers p and q of B, respectively. Under these conditions one obtains

$$\Delta(x) = [\phi(A-k,x) - \phi(A-\ell,x)]^2 \geq 0. \tag{7}$$

Obviously, in this specific case $\Delta(x)$ is non-negative within the complete range of its variable x; in view of eq. (7) $\Delta(x) = 0$ has either no real roots or only real roots of even degeneracy.
From eqs. (5) and) it follows that

$$\phi(S,x) \leq \phi(T,x), \qquad x \in (-\infty, +\infty). \tag{8}$$

This relationship between the characteristic polynomials of a pair of topomers is schematically depicted in Figure 2. Because both polynomials must

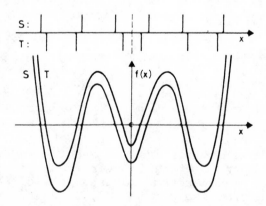

Figure 2. Schematic behaviour of $\phi(S,x)$ and $\phi(T,x)$ in case of model 1, $\ell = 2$, A $\simeq$ B.

have the same number of zeros, say N, in view of eq.
(8) the zeros of the polynomials mutually interlace
as follows: in the intervals given by two successive
zeros of $\phi(S,x)$ there are alternately two or no zeros
of $\phi(T,x)$:

$$x_1^S \overset{\leq}{=} x_1^T \overset{\leq}{=} x_2^T \overset{\leq}{=} x_2^S \overset{\leq}{=} \ldots \overset{\leq}{=} x_{2k-1}^S \overset{\leq}{=} x_{2k-1}^T \overset{\leq}{=} x_{2k}^T \overset{\leq}{=} x_{2k}^S \overset{\leq}{=} \ldots \tag{9}$$

Two independent and detailed proofs of eq. (9) are
found elsewhere [12,13]; in [12] it is shown further
that the interlacing theorem, eq. (9), is also valid
for a class of polynomials, not all of which necessarily characteristic polynomials.
When the interlacing theorem, eq. (9), is applied
to fully conjugated π-electron systems the zeros of
the respective characteristic polynomials may be
identified as the Hückel (HMO) eigenvalues of the
topomers. Thus, eq. (9) relates in a novel manner
the eigenvalue spectra of topologically related
compounds which have been considered to be independent hitherto. Since eq. (9) stems from $\Delta(x)$ which
incorporates the relative topology of the topomers,
the chemical application of the interlacing theorem
has been termed *topological effect on MO* (TEMO).
It should be mentioned that by means of a Hückel-
like method the validity of TEMO also for σ-MO has
been shown.
Finally the other two models given above should be
considered. In the case of model 2, $\ell = 2$, $A \underset{\sim}{} B$,
one obtains

$$\Delta(x) = [\phi(A-k,x) - \phi(A-\ell,x)]^2 \phi(C-a-b,x).$$

Under some particular conditions demanded for the
structure of the central moiety C, the polynomial
$\phi(C-a-b,x)$ is a perfect square, say $\phi(C-a-b,x) =$
$[\gamma(x)]^2$; then eqs. (8) and (9) are obtained again,
i.e. the MO spectra of the topomers in question exhibit the TEMO pattern as described above.
In the case of model 1, $\ell = 3$, the polynomial $\Delta(x)$
consists of four bilinear terms which have pairwise
opposite sign; thus no conclusion can be drawn about
the sign of $\Delta(x)$, even if $A \underset{\sim}{} B$. With increasing ℓ
the number of terms of $\Delta(x)$ also increases. The same
is true for model 2 as well as for additional models
not mentioned here.

PHYSICO-CHEMICAL CONSEQUENCES OF TEMO

Here we restrict the considerations to the π-electron
systems of those pairs of topomers which may be constructed by means of model 1 or 2, $\ell = 2$, $A \underset{\sim}{} B$,
$\phi(C-a-b,x) = [\gamma(x)]^2$. The mutual interlacing of the

MO of the S and the T topomer according to eq. (9) is schematically shown in Figure 3. Let N denote the number of π-electrons of S and T, respectively. It is easy to show that N is even in all cases considered here. If $N = 4\nu + 2$, an odd number of MO is doubly occupied in the ground state of the topomers and the HOMO (LUMO) of S lies below (above) that of T, i.e. the HOMO-LUMO separation is larger in S than in T (Figure 3a); if $N = 4\nu$, the opposite is true (Figure 3b).

This consequence of the TEMO pattern should be exhibited by pertinent UV absorption bands provided the state transitions in question are mainly determined by MO transitions. Since this requirement is well met by the p-bands of polycyclic aromatic hydrocarbons (PAH), they have been examined with the view of probing the predictions of TEMO theorem. These were found to be in excellent agreement with the experimental data (see Table 15 in [6]).

A similar consequence of the TEMO pattern should be shown by the first ionisation potentials, IP_1.

As seen from Figure 3 $IP_1^S \geq IP_1^T$ should hold for $N = 4\nu + 2$ but $IP_1^S \leq IP_1^T$, for $N = 4\nu$. This prediction also was found in accord with the experimental data [6].

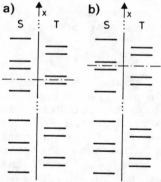

Figure 3. Schematic illustration of eq. (8). The broken line separates the doubly occupied and the empty MO. The number of π-electrons, N, is assumed to be $N = 4\nu + 2$ in a) but $N = 4\nu$ in b).

This first evidence for the physical relevance of TEMO theorem is supported by the results of some quantum chemical calculations as well as photoelectron (PE) spectra of topomers. These additional data will be discussed briefly in Sect. 7.

INVERSIONS

Let us return to eq. (6) valid for model 1, $\ell = 2$, $A \nleq B$. In the case of non-isomorphic moieties, in general, quadratic forms of the difference polynomial $\Delta(x)$ will not be obtained and consequently $\Delta(x)$ will change its sign when the variable x goes through one of the real roots of $\Delta(x) = 0$ provided the roots in question are not even-degenerated. Let us denote the real roots of odd degeneracy in decreasing order as follows:

$$x_1^I > x_2^I > \ldots > x_{2j-1}^I > x_{2j}^I > x_{2j+1}^I > \ldots \ . \tag{10}$$

They determine open intervals within which $\Delta(x)$ is either non-negative or non-positive; but $\Delta(x)$ becomes zero within the interval if x takes the value of a root of even degeneracy. As a consequence of eqs. (3) and (4) $\Delta(x)$ will be positive for sufficiently large values of x. Thus, one generally finds:

$$\Delta(x) \geq 0, \text{ iff } \quad x \in (x_{2j+1}^I, x_{2j}^I); \tag{11a}$$

$$\Delta(x) \leq 0, \text{ iff } \quad x \in (x_{2j}^I, x_{2j-1}^I), \tag{11b}$$

$$j = 0,1,2,\ldots \ ; \quad x_O^I \to +\infty.$$

For all the intervals where $\Delta(x) \leq 0$, one obtains from eq. (5) $\phi(S,x) \leq \phi(T,x)$. Using the same arguments as before, one derives from this inequality the sequence of eigenvalues as follows:

$$\ldots \leq x_{2j}^I \leq \cdots \leq x_{2k-1}^T \leq x_{2k-1}^S \leq x_{2k}^S \leq x_{2k}^T \leq \cdots \leq x_{2j-1}^I \leq \cdots \ . \tag{12}$$

As shown by eq. (12) the interlacing theorem holds even if $A \nleq B$, but within the intervals given by eq. (11b) the order of eigenvalues is inverted. Thus the real roots of odd degeneracy of $\Delta(x) = 0$ have been termed *inversion points*. Inversions as expressed by eq. (12) are said to be topologically induced. Attention should be paid to the following points:

(i) TEMO theorem predicts the mutual interlacing of the eigenvalues of a pair of topomers within defined intervals as alternatively expressed by eqs. (9) and (12).

(ii) Topologically induced inversions within the TEMO pattern as discussed above are a part of the generalized TEMO theorem.

(iii) The appearance of inversion points in a given TEMO pattern may be excluded when particular topological models are used; such an exclusion is due to an inherent property of the topological model

in question.
(iv) No general rule can be given for the actual
realization of inversions if they cannot be excluded; note: No inversion can be actualized if the
polynomials in question have not at least one zero
within the intervals given by eq. (11b).

PHYSICAL RELEVANCE OF TEMO

The quantum chemical calculation of the MO energies
of a pair of topomers may serve as a test of the
physical relevance of TEMO theorem. However, in such
an appraisal one has to bear in mind that in deriving eq. (9) only two interactions of centers of A
with those of B are considered, namely the interactions of k and ℓ of A with their neighbours in B.
In contrast to that in any non-empirical quantum
chemical calculation of MO energies, each center of
A is found to interact with each center in B and
vice versa; this leads to $a.b \gg \ell$ interactions where
a and b denote the number of centers of A and B,
respectively. Thus, the situation met in non-empirical
MO calculations resembles to model 1, $\ell > 2$ (where
inversions cannot be excluded!) more than to $\ell = 2$.
Consequently a large number of inversions might be
expected even if model 1, $\ell = 2$, $A \sim B$ is used. Surprisingly, the spectra of σ- and π-$\overline{MO}$ of o-(IS) and
p-benzoquinodimethane (IT) [6], shown in Table 1,
exhibit only a few inversions: In the range from -0,4
to -0,5 [au] where σ- and π-MO energies overlap the
sequence of the respective MO inverted; only within
the σ-MO pattern there are two additional inversion
intervals ranging from -0,53 to -0,56 and from
-0,62 to -0,73 [au], respectively. In some other
calculations of that kind [15,16] even less inversions have been found. As seen from Table 1 the MO
pattern of the topomeric pair I exhibits perfectly
the interlacing of the MO according to eqs. (9) and
(12). The number of inversion intervals is surprisingly low. Thus, the TEMO theorem stands very well
the test by non-empirical MO calculations.
Some examples [16-18] indicate that the physical
origin of the additional inversions observed is
nearly always the non-nearest neighbour interaction
which is not considered in the course of deriving
eqs. (9) and (12); an additional origin of physically induced inversions could be traced only in
one case of heterocyclic topomers [17].
Another rigorous proof of the physical relevance of
the TEMO theorem is offered by PE spectroscopy. Provided Koopmans theorem holds the vertical ionisation
potentials as exhibited in the PE spectra should re-

Table 1. The σ and π-MO of o-(IS) and p-benzoquinodimethane (IT) in [au] taken from [6] (STO 3G basis, standard geometry; inversion intervals are marked by dotted lines).

IS	IT	IS	IT
σ-MO:			
-1,096 109		-0,523 718	
	-1,095 072	-0,516 540	
	-1,012 555		-0,503 316
-1,004 848		-0,496 719	
-0,986 414			-0,472 228
	-0,973 930		-0,453 912
	-0,921 337	-0,432 840	
-0,900 402		-0,412 978	
-0,822 836			-0,411 872
	-0,784 236		
	-0,767 785		
-0,750 425			
	-0,727 923	π-MO:	
-0,714 896			-0,463 535
-0,637 657		-0,462 871	
	-0,621 709		-0,356 040
-0,620 554		-0,339 030	
	-0,611 044	-0,334 342	
	-0,601 409		-0,317 534
-0,594 379			-0,203 487
	-0,553 383	-0,203 378	
	-0,540 443		
-0,540 166			

present the upper MO levels of the compounds investigated. Thus, the PE spectra of a pair of topomers should render the TEMO pattern. This is verified very well by the PE spectra of the topomeric pairs II, III, and V [19] collected in Table 2; some additional data are found in [20]. The examples given in Table 2 are taken from the class of PAH because the PE spectra of these compounds consists of a very large number of well-resolved peaks. But it should be mentioned that the PE spectra of more than a hundred pairs of topomers, selected from literature and exhibiting a wide variation in their constitutional characteristics, satisfy the TEMO theorem with astonishing fidelity [21].

CONCLUSIONS

(i) The interlacing theorem is a novel relation in mathematics; it has been proved rigorously.

Table 2. PE spectra of the topomeric pairs II, III, and V shown in Figure 1 (all data in [eV], taken from [19])

S	II	T	S	III	T	S	V	T
		7,41			6,61			7,59
7,86			7,27			7,60		
8,15			7,39			8,02		
		8,54			7,92			8,10
		9,19			8,32			8,68
9,28			8,54			8,98		
9,89			8,90			9,18		
		10,18			9,01			9,43
		10,28			9,39			9,72
10,59			9,53			9,96		
			9,66			10,22		
					9,80			10,52
					10,23			
			10,3					
			10,5					

(ii) The interlacing theorem is derived using the relations between the molecular-topological spaces associated with a pair of topomers.
(iii) Hence, it is applicable to the σ- and π-MO of organic compounds, thus producing the TEMO, provided the topological features of the MO are considered.
(iv) The test of TEMO by non-empirical MO calculations, UV absorption, and PE spectra shows that TEMO strongly superceeds several physical factors.
(v) From this fact one may conclude that molecular topology is not so much only the end of rigorous abstractions rather than one of the first principles; it seems it determines some sort of frame within which physical reality may be actualized.
(vi) Hence, besides its use in organic chemistry TEMO has some cognitive value too.

ACKNOWLEDGEMENT

The technical assistance of Mrs. T. Kammann, Mrs. I. Schneider, and Mr. R. Gruen is appreciated.

REFERENCES

[1] Schill G., Catenanes, Rotaxanes and Knots, Academic Press, New York 1971
[2] Frisch H.L., E. Wasserman, J. Amer. Chem. Soc. 83, 3789 (1961)

[3] Walba D.M., R.M. Richards, R.C. Haltiwanger, J. Amer. Chem. Soc. 104, 3219 (1982)

[4] Bader R.F.W., T.T. Nguyen-Dang, Y. Tal, Rep. Progr. Phys. 44, 893 (1981)

[5] Mezey P.G.,in Chemical Applications of Topology and Graph Theory (ed.: R.B. King), p. 75, Elsevier, Amsterdam 1983

[6] Polansky O.E., M. Zander, J. Mol. Struct. 84, 361 (1982)

[7] Zander M., O.E. Polansky, Naturwiss. 71, 623 (1984)

[8] Merrifield R.E., H.E. Simmons, Theoret. Chim. Acta (Berl.), 55, 55 (1980)

[9] Hausdorff F., Grundzüge der Mengenlehre, Veit, Leipzig 1914

[10] Graovac A., O.E. Polansky, Croat. Chem. Acta 57, 1595 (1984)

[11] Polansky O.E., G. Mark, Match (Math. Chem.) 18, 249 (1985)

[12] Gutman I., A. Graovac, O.E. Polansky, Chem. Phys Lett. 116, 206 (1985)

[13] Graovac A., I. Gutman, O.E. Polansky, J. Chem. Soc., Faraday Trans. 2, 81 (1985) in press

[14] Ruedenberg, K., J. Chem. Phys. 22, 1878 (1954); 29, 1232 (1958); 34, 1884 (1961)

[15] Motoc I., O.E. Polansky, Z. Naturforsch. 39b, 1053 (1984)

[16] Motoc I., J.N. Silverman, O.E. Polansky, G. Olbrich, Theoret. Chim. Acta (Berl.) 67, 63 (1985)

[17] Motoc I., J.N. Silverman, O.E. Polansky, Phys. Rev. A 28, 3673 (1983)

[18] Motoc I., J.N. Silverman, O.E. Polansky, Chem. Phys. Lett. 103, 285 (1984)

[19] Schmidt W., J. Chem. Phys. 66, 828 (1977)

[20] Polansky O.E., J. Mol. Struct. 113, 281 (1984)

[21] Mark G., O.E. Polansky, in preparation

Chapter 23

KEKULE VALENCE STRUCTURES REVISITED. INNATE DEGREES OF FREEDOM OF PI-ELECTRON COUPLINGS

Milan Randić
Department of Mathematics and Computer Science, Drake University,
Des Moines, Iowa 50311, USA
Douglas J. Klein   :LEUIRUN CUUPLINGS
Department of Marine Sciences, Texas A&M University,
Galveston, Texas 77553, USA

ABSTRACT

For individual Kekule valence structures we consider the smallest possible number of placements of CC double bonds such that a Kekule structure is fully determined. The number may be viewed as special weighting scheme for individual Kekule valence structures. Alternatively its reciprocal indicates the degree of a long-range order in a Kekule structure. Contributions from individual Kekule valence structure add to a novel structural invariant F, the innate degree of freedom associated to a conjugated system. We find that F correlates well with the molecular resonance energy.

INTRODUCTION

The question of the relative importance of Kekule valence structures has been frequently overlooked, implying by default that all Kekule valence structures have the same weight. In order to explain the reduced aromatic stability of nonbenzenoid systems in comparison with benzenoid systems having the same number of Kekule valence structures Longuet-Higgins and Dewar introduced the concept of parity for valence structures (1). One can interpret parity formally by assigning to some valence structures a weight +1 and to other -1. Valence structures of opposite parity then cancel each others contributions to molecular stability. The concept of parity, however, suffers from inconsistencies that become apparent when one extends the application to polycyclic systems having three odd fused rings (e.g., aceazulene) (2). Similarly it shows deficiency when applied to alternant systems having two or more four membered rings, in which cases the most stabilizing and the least stabilizing valence structures may appear with the same parity (3). In contrast Clar, using chemical intuition and logic, argued that in polycyclic systems individual rings may differ considerably, some showing a great similarity to benzene ring, others bearing little resemblance (4). Formally the approach of Clar amounts to assigning to most Kekule valence structures weight 1 and to a few weight 0. While the work of Longuet-Higgins, Dewar, and Clar, clearly point to a need to differentiate the relative importance of individual structures, the way to resolve this problem remains open. Recently several authors suggested a classification of valence structures into more than two classes (5). The present work reports yet another approach to this problem.

INNATE DEGREES OF FREEDOM OF VALENCE STRUCTURES

Most chemists will agree that valence structures in Fig. 1 do not represent useful depiction of the molecules shown. Can we quantify the chemical intuition that guides us in rejecting the valence structures of Fig. 1 as unimportant? Can one estimate the importance of a valence structure a priori?

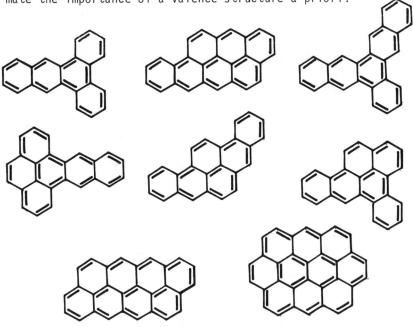

Figure 1. Selected Kekule valence structures for polycyclic benzenoid systems.

One clue is that the discriminated structures of Fig. 1 may
be identified with a type of long-range order. That is, there
are highly correlated couplings in the sense that the pairing
of vertices at large distances is dictated by arrangement of
CC double bonds in some local region. This suggests that one
can speak of an innate degree of freedom of individual Kekule
valence structures. The degree of freedom relates to the count
of steps at which in completion of the Kekule valence structure
one has the possibility to make a choice for CC bond type
(single or double). Consider two Kekule valence structures of
triphenylene:

For the left structure once we select any one C=C in the
central ring all other C=C are determined and the complete
Kekule valence structure can be written down. The valence
structure at left has a single inherent degree of freedom. In
the case of the valence structure at the right, however, by
assigning one C=C in the central ring only, two additional C=C
are determined (within the same side ring). In order to complete the Kekule valence structure two additional selections
have to be made, each to fix C=C bonds in the remaining peripheral rings. Hence, the structure at the right has three degrees of freedom. We can now present the definitions for the
innate degree of freedom of an individual valence structure
and the molecule as whole:
Definition 1: The innate degree of freedom of a Kekule valence structure is the smallest number f of choices of CC bond
types that fully defines the structure.
Definition 2: The innate degree of freedom of a molecule F
is given as the sum of inherent degrees of freedom for all
Kekule valence structures of the molecule.
The innate degrees of freedom f and F are structural invariants. For a given Kekule valence structure f is unique, even
though if one is to construct a Kekule valence structure the
number of choices may depend on the order in which CC bonds are
selected for considerations. That is the reason for emphasizing the smallest number of choices in the definition for f.

RESULTS

To find f and F is simple in some cases and more involved in
other. For linear acenes all Kekule valence structures have
f=1, and consequently F=K (K being the number of Kekule valence structures). This simply follows from the fact that in
such molecules each Kekule valence structure has a single

"vertical" C=C bond which determines the CC type of all other bonds. In such systems CC double bonds are strongly correlated, and as the size of the molecule increases, their RE/e values (the ratio of the resonance energy per pi-electron) decrease. The decreasing aromatic characteristics of such linear polycyclic systems is reflected in their low value for F. If we compare angularly fused structures with a single "kink":

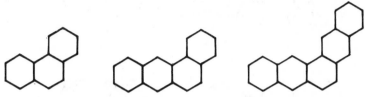

one finds that here there is for all such compounds a single valence structure which is fully determined by selecting one C=C bond in the central ring. The remaining Kekule structures may be viewed as composed from Kekule structures for the straight acenic fragments of rings on either side of the ring where the kink takes place. In these structures one needs to specify the location of two C=C double bonds transverse to the chain, one double bond in each straight acenic portion. Thus these structures have f=2, and F=2(K1)(K2) + 1, where K1 and K2 are the numbers of Kekule structures possible for the two straight acenic legs. It is not difficult to generalize this rule for other catacondensed acenes. Finding f for pericondensed systems is more involved. One essentially breaks down a larger structure into components of smaller size for which f values have already been found. A useful aid in determining f is to locate the smallest conjugated circuits R1 (6) (i.e., the rings with a benzene Kekule valence structure). The rationale for doing this is that f is at least as great as the (maximum) number of disjoint conjugated circuits that can be simultaneously identified to the Kekule structure. The conjugated 6-circuits are often the most important for benzenoids, and because they are the smallest circuits they lead to a greater number of such disjoint circuits to be found. Thence one sees that the number of circles in a Clar structure associated to the given Kekule structure is a lower bound for f.

The innate degrees of freedom provide a weighting for Kekule structure that though it parallels Clar's ideas is different. That is, we see that those Kekule structures associated to a Clar structure with many Clar's circles will have a high weight f. Those with no or few Clar's circles however are not given a weight of zero, as Clar initially suggested, but generally a lower weight. It is possible to have structures with arbitrary large values for f. This is seen for long chains where the number of disjoint Clar's circles can scale as the chain length (for some Kekule structures). Moreover these same type of chains yield examples with variety of different values of f.

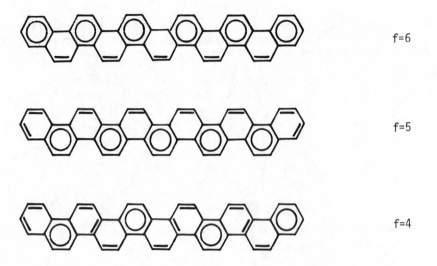

f=6

f=5

f=4

Figure 2. Long chain molecule with valence structures of variable f.

For instance, a chain molecule of Fig. 2 in addition to the f=6 valence structure also has f=5 and f=4 valence structures. In fact for such chains of M rings f takes all the integer values such that $(M+2)/3 \lesssim f \lesssim (M+1)/2$. In Table 1 we list the F values for numerous molecules.

RESONANCE ENERGIES

A logarithmic relationship between RE and structure count has been indicated in the past (7). There is qualitative reason to anticipate such a correlation when we interpret RE as the correction to an energy of a single Kekule structure. An "appreciable" RE/e should occur under the circumstances that a typical Kekule structure interacts "directly" with "many" others, the strongest interactions being between two structures that differ but just slightly. For an appreciable RE/e the number of such strong interactions should scale with system size, say as measured by the number N of pi-centers. Thence modifications to a typical Kekule structure should be possible in a number of local areas scaling proportionally to N, each modification being independent of the others. Though a structure differing by a single local modification admixes most strongly, the possibility of making different modifications in different local regions indicates their independence. The total number of Kekule structures then scales proportional to M**S with M a mean number of modifications possible per local region and S the number of different local regions. But we have already argued that the total RE scales with S (when RE/e is appreciable) and of course S is proportional to N. Thus log K is anticipated to scale as the RE, at least in terms of

Table 1. Polycyclic conjugated molecules considered, their F and K numbers (F is the number of innate degrees of freedom and K is the number of Kekule valence forms) shown as F/K.

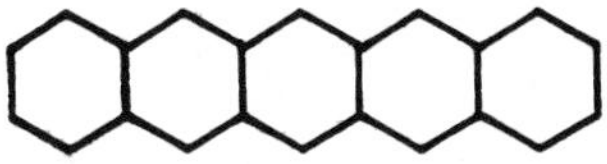

6/6

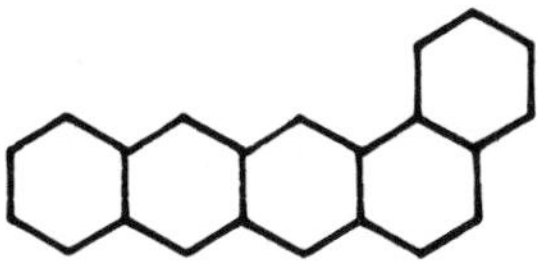

40/14

17/9

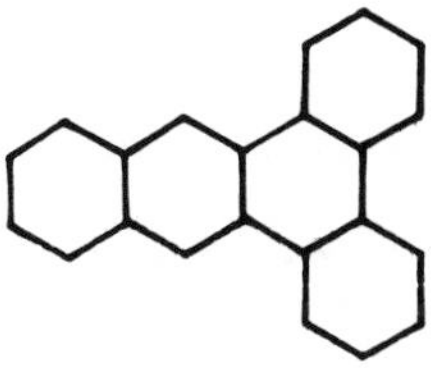

22/11

37/13

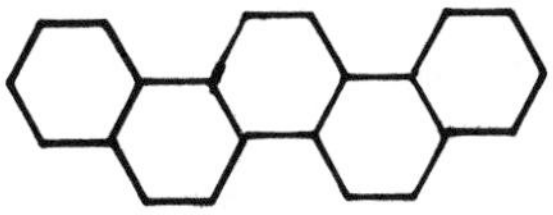

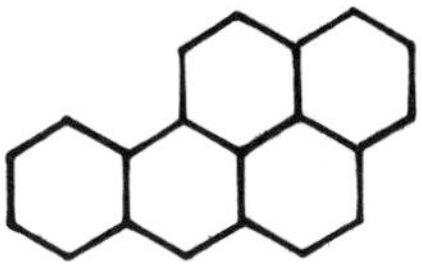

34/13

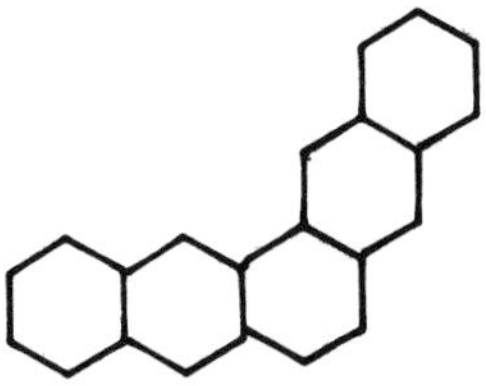

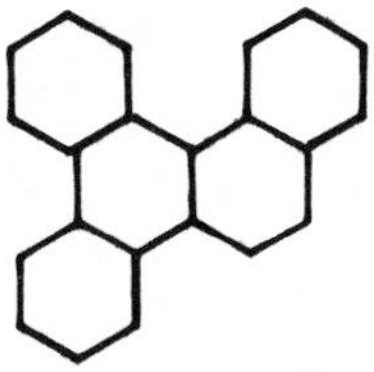

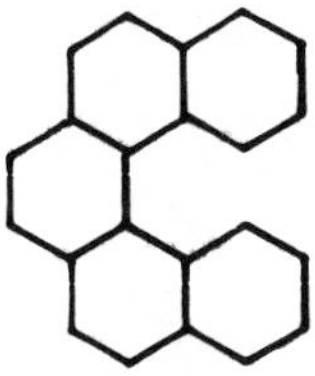

34/13

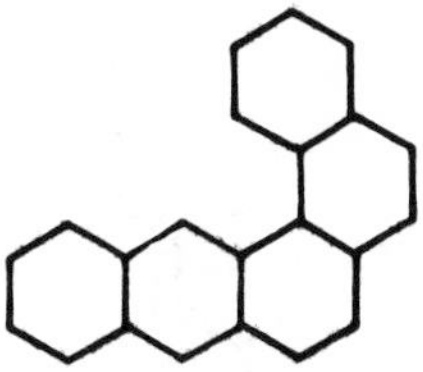

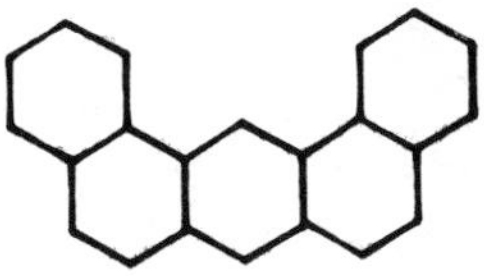

32/12

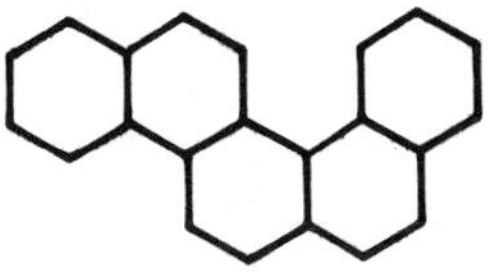

19/10

34/13

17/9

18/9

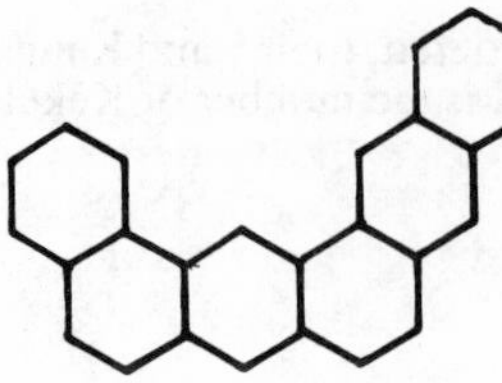

46/17

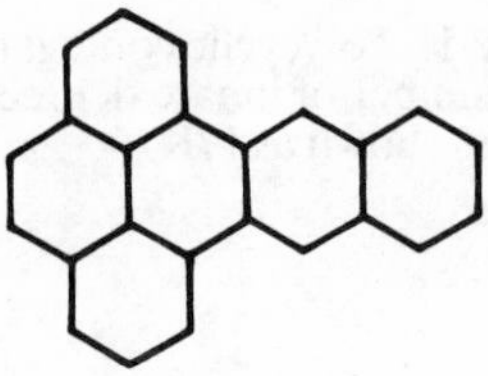

44/16

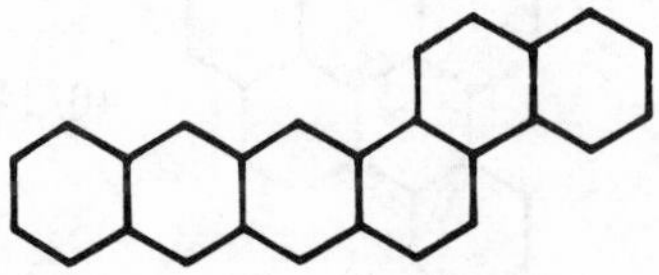

28/14

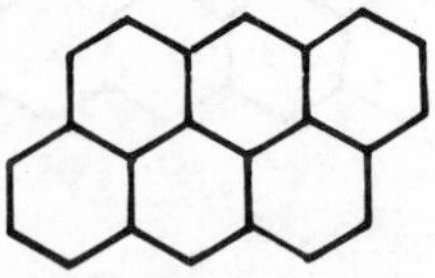

18/10

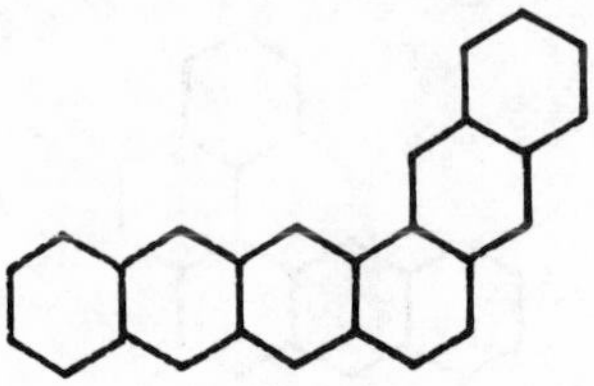

25/13

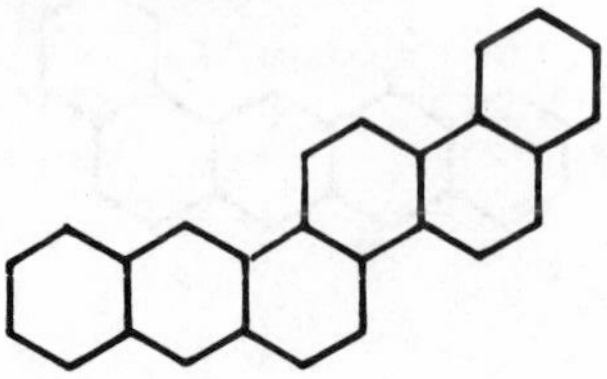

51/18

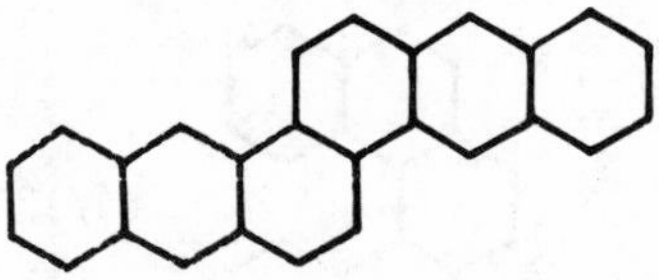

30/15

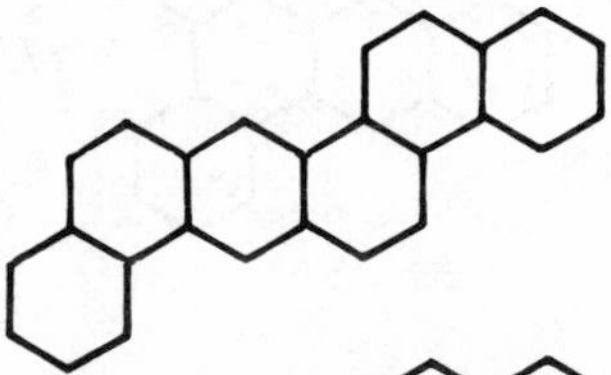

54/19

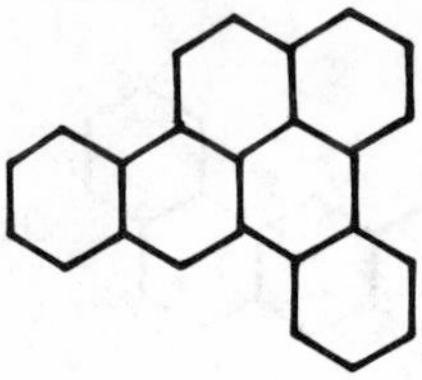

49/17

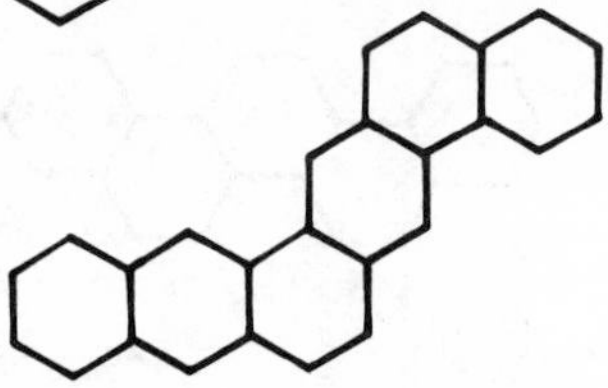

46/17

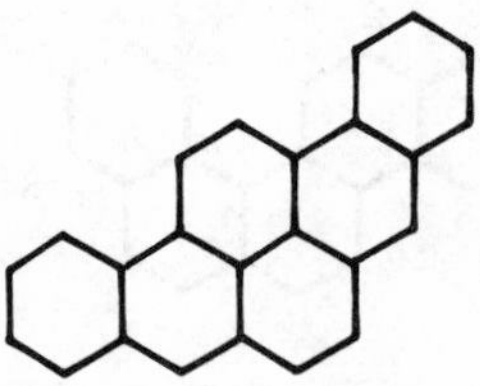

35/14

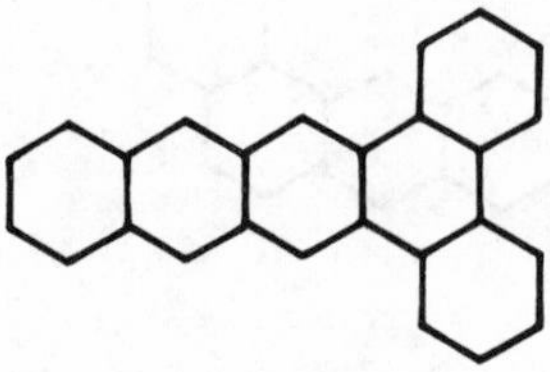

49/17

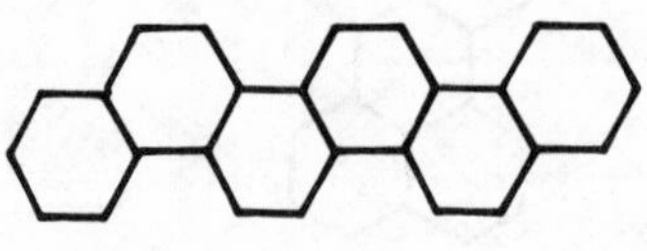

62/21

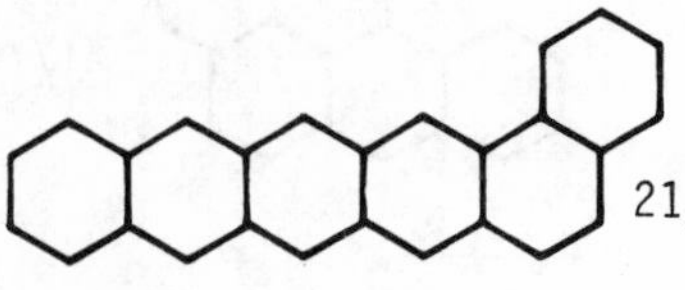

21/11

its "bulk" contributions. The rationale for a correlation between log K and RE can be summarized more briefly in a framework of multiplicative versus additive properties of structures. Kekule structure count K can be identified as "multiplicative" (when RE/e is appreciable) and RE as "additive" -- the log function being the well-known transformation from multiplicative to additive quantities. Quite similar comments apply to the identification of F as multiplicative quantity, so that here too we anticipate a linear correlation between RE and log F. Indeed in Fig. 3 this is seen to occur to some degree (the RE have been taken from Dewar and de Llano (8)). As one sees there is a satisfactory correlation, which differ in detail, but not in quality from the similar correlation based on K. Because K and F do not measure precisely the same structural features, F being a "refined" (weighted) K quantity, we see that the cause for the correlation is less apparent than first guessed. The dependence of F on K is not simple, as there are numerous cases of structures having the same K but different F values.

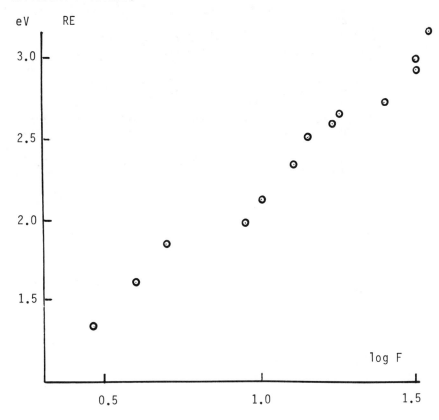

Figure 3. A plot of molecular resonance energy (RE) against log F.

In conclusion, the quest for elucidation of molecular structure of complex molecules may well critically depend on the generation of novel concepts, rather than on the next generation of computers! With this in mind we have presented one such novel concept, the innate degree of freedom of valence structures. Possibly it may be of direct use or it may stimulate other novel inquiries in molecular structure.

ACKNOWLEDGEMENTS

This work was supported in part by the office of the Director of Ames Laboratory. Ames Laboratory is operated for the U.S. Department of Energy by Iowa State University under contract no. W-7405-ENG-82.

REFERENCES

(1) M. J. S. Dewar and H. C. Longuet-Higgins, (1952) Proc. Roy. Soc., A 214, 482
(2) I. Gutman and N. Trinajstić, (1973) Croat. Chem. Acta, 45, 539
(3) M. Randić, (1977) Mol. Phys., 34, 849
(4) E. Clar, (1972) The Aromatic Sextet, Wiley, London
(5) P. Křivka and N. Trinajstić, (1985) Coll. Czech. Chem. Comm. 50, 291
(6) M. Randić, (1976) Chem. Phys. Lett. 38, 68
    M. Randić, (1977) Tetrahedron, 33, 190
    M. Randić, (1977) J. Amer. Chem. Soc., 99, 444
(7) C. F. Wilcox, Jr., (1968) Tetrahedron Lett., 795
    R. Swinborne-Sheldrake, W. C. Herndon and I. Gutman, (1975) Tetrahedron Lett., 755
(8) M. J. S. Dewar and C. de Llano, (1969) J. Amer. Chem. Soc. 91, 789

HAMILTONIAN vs. NON-HAMILTONIAN DYNAMICS IN MOLECULAR PROCESSES: AN ADIABATIC SUBSYSTEM APPROACH

William Rhodes
Department of Chemistry and Institute of Molecular Biophysics, Florida State University
Alka Velenik
Faculty of Pharmacy and Biochemistry, University of Zagreb

INTRODUCTION

Dynamic processes in molecules are usually regarded in terms of a few select degrees of freedom (modes). For example, the description of chemical reactions makes use of the reaction coordinate concept and molecular electronic spectroscopy makes use of electronic energy surfaces as functions of vibrational coordinates. It is both customary and reasonable to consider the smallest possible number of degrees of freedom of a molecular system required to understand its properties. On the other hand, it is generally recognized that every real system is composite in nature, consisting of very many modes that are at least weakly coupled to one another. In fact, since no part of the universe is truly isolated, every observed system is in principle coupled to the rest of the universe.

Our approach is to regard any given molecular system as consisting of a Principal Subsystem (referred to as the PS) and a Background set of modes (referred to as the BG, or in some contexts as the Bath). The PS represents those modes that are directly involved in the process under observation, for example, a chromophore which absorbs light within a certain frequency range. The BG consists of all modes that are directly or

indirectly coupled to the PS, for example, the substituent
groups on a chromophore, solvent or other medium, radiation
field modes, and phonon modes. Thus, the BG may be (in part)
structurally an integral part of the PS and it may also include
quite literally the surroundings.

In the theory of chemical reactions and of molecular spectroscopy, the role of the BG as a source and sink for energy is
generally taken for granted. Most processes involve at least
part of the BG being in thermal equilibrium and the BG thus
tends to direct the PS toward thermal equilibrium. However,
the BG also plays a vital role in coherence relaxation in the
PS; i.e., in the loss of a definite phase relation among its
quantum states. Modulation of the PS by BG modes leads to
dissipation of coherence as well as redistribution of energy.
This is equivalent to a memory loss in the PS (see below) and
can greatly modify its dynamics. In summary, the BG plays the
important dual role of energy and coherence dissipation in the
PS. The latter is of fundamental importance in the interplay
of (coherent) dynamics and (incoherent) kinetics, the concept
of irreversibility, and the approach to equilibrium of the PS.

In contrast to the dissipative forces of the BG there can be
coherent forces acting on the PS due to potentials V^P which are
either extrinsic or intrinsic in nature. Such potentials tend
to create or maintain definite phase relations among the PS
states. Examples are applied coherent fields and intramolecular potentials producing nonradiative transitions. The
theory developed here deals with the interplay of coherent and
dissipative potentials.

There are numerous other theoretical approaches to macroscopic
system dynamics and we mention a few that are relevant to ours.
Grigolini (1985; 1981) has made extensive studies in terms of a
reduced model theory, in which projection operator methods
formulated by Zwanzig (1961) and by Mori (1965) are used.
Kubo (1969) has developed a stochastic Liouville equation
method and similar stochastic equations have been used by
Oxtoby (Bagchi and Oxtoby, 1982), Silbey (Jackson and Silbey,
1981), Lin (Boeglin et al., 1983) and many others. These agree
in general with the classic work of Redfield (1965) on spin
relaxation. Finally, the subdynamics formalism of the
Prigogine school (Prigogine, 1981) should be noted.

We define a subsystem basis $\{|i>\}$ which spans the Hilbert
Space of the PS. The matrix elements $<i|\rho|j>$ of the entire
system's density operator ρ are operators with respect to the
BG subsystem. A key feature of our method is to make the
identification

$$\rho_{ij}(t) = <i|\rho|j> \equiv \rho^B_{ij}(t)\ \sigma_{ij}(t) \qquad (1)$$

$$\sigma_{ij}(t) \equiv tr_B\ \rho_{ij}(t),$$

which is used provided σ_{ij} is nonzero. Otherwise, the notation ρ_{ij} is retained. This defines the elements σ_{ij} of the reduced density operator for the PS and the associated operator for the BG. Our purpose is to formulate and evaluate the equations of motion for the σ_{ij} for various kinds of molecular processes.

The diagonal element $\sigma_{ii}(t)$ represents the probability for state $|i>$ of the PS, while σ_{ij} $(i \neq j)$ represents the coherence components (elements) of the PS. The latter is a measure of the coexistence of (probability) amplitudes for the state $|i>$ and $|j>$ in association (correlation) with the same wavefunction component of the BG. The coherence components play an important role in the dynamics of probabilities and other physical properties (e.g., average values of observables) of the PS.

By using eq. 1, it can be shown that the Liouville equation for the full system,

$$\dot{\rho}(t) = -\, i\, L\, \rho(t) \, , \tag{2}$$

may be reduced to an effective equation of motion for the PS having the form

$$\dot{\sigma}_{ij}(t) = -i \sum_{k\ell} \mathcal{L}_{ijk\ell}(t)\sigma_{k\ell}(t) \, . \tag{3}$$

The effective Liouvillian depends on the bath operators $\rho^B_{k\ell}(t)$ in a manner implied by eq. 1. In general, its components depend on time t and the history of the system, including earlier values of the various σ_{mn}. Eq. 3 permits the most complicated kind of dynamics in which the PS can have a major perturbation effect on the background, resulting in nonlinear equations for the σ_{mn}. We are interested here, however, in the dissipative limit, whereby the BG is very large and "spongy" and its properties do not change with time. Examples are non-stationary processes in molecules imbedded in a condensed medium at thermal equilibrium or molecular dynamics in a thermally equilibrated radiation field. For this limit $\mathcal{L}_{ijk\ell}$ does not depend on t.

Even for the dissipative limit the dynamics of the PS can be quite complicated. The PS is generally composite and may consist of many degrees of freedom, some of which may be strongly coupled. We may wish to focus on one or a few modes, whereby further reduction of the PS is needed.

The discussion of background modulation effects on PS dynamics will develop in several stages. (1) The motion of the BG will be treated adiabatically. This means that the Hamiltonian of the BG depends on the state of the PS and the state wavefunctions of the BG tend to evolve on the energy surfaces of the PS in a manner which modulates the dynamics of the latter. (2) The Liouville equation for the entire system is projected onto the diagonal components for the BG, the dissipative limit is used for the BG and reduction is made with respect to all background modes. The resulting reduced Liouville equation (RLE) describes coherent dynamics of the PS as modified by the dissipative background. (3) The non-Hamiltonian character of the RLE is examined and the criteria for an effective Hamiltonian component of the RLE is considered. (4) The special case of a two-level PS is discussed and results of computer calculations are presented for different coupling strengths and initial conditions. (5) Several important phenomenological consequences are considered for the effects of memory relaxation in the dissipative limit. These include structure stabilization, kinetic stabilization, and kinetic enhancement within the PS.

THE ADIABATIC HAMILTONIAN

The entire system consists of molecules and fields and their interactions. We tacitly assume that the Primary Subsystem (PS) contains a set of modes which can be clearly defined (e.g., a chromophore composed of bound electrons and nuclei). The system Hamiltonian is then

$$H = H^P + H^F + H^A + V^P + V^{FP} + V^{AP} \, , \qquad (4)$$

where the first three terms represent the Hamiltonia for the free PS, nonadiabatic fields (e.g., radiation or applied fields), and the adiabatic part of the background (adiabatic-BG), respectively, and the last three terms represent intra-PS interactions, field-PS interactions and interactions between the adiabatic-BG and the PS, respectively.

The adiabatic formulation (Rhodes, 1981, 1982, 1983) involves the PS Schrödinger equation

$$(H^P + V^{AP}) \, |i> = \omega_i(q) \, |i> \, , \qquad (5)$$

where q represents the set of BG coordinates that are to be treated adiabatically and $\omega_i(q)$ is the energy surface for the PS corresponding to PS state $|i>$. Strictly, $|i>$ depends at least weakly on q, but this is neglected here. Each $\omega_i(q)$ then serves as a potential for the motion of the BG modes. Thus,

$$[H^A + <i|H^A|i> + \omega_i(q)] \, |\lambda_\alpha^i> \qquad (6)$$

$$\equiv h_i|\lambda_\alpha^i> = \lambda_\alpha^i|\lambda_\alpha^i>$$

is the BG equation corresponding to PS state $|i>$. The adiabatic part of the system Hamiltonian may then be written

$$h = \sum_i |i>h_i<i| \; ; \tag{7}$$

and the full system Hamiltonian becomes

$$H = H^O + V^P + V^{FP} \tag{8}$$

where

$$H^O \equiv H^F + h$$

It is understood that residual nonadiabatic terms that are tacitly omitted in eq. (7) can be included by redefinition of the potentials V^P and V^{FP}.

REDUCED LIOUVILLE EQUATION: DISSIPATIVE LIMIT

We begin with a Zwanzig-type projection (Zwanzig, 1961) in which diagonal elements of the background subsystem are projected by an operator P. This gives the standard structure

$$P \dot{\rho}(t) = -i\, P\,L\,P\, \rho(t)$$
$$- \int_0^t dt_1 P\,L\,Q\, e^{-iQLQ(t-t_1)} QLP\, \rho(t_1) \tag{9}$$
$$-i P\,L\,Q\, e^{-iQLQt} Q\, \rho(o) \; ,$$

Where $Q = 1-P$. The last term depends on those elements of the initial density operator that are nondiagonal in BG states (for a given reference basis). For convenience, we assume that this term vanishes. Note that eq. 9 does not project diagonal components of the PS, so is fundamentally different from the usual Zwanzig projection. Next, we use the dissipative limit for eq. 9, whereby eq. 1 is used, reduction is made with respect to background modes, and the BG is assumed to be unperturbed by the PS (except for possible transient local perturbations). The resulting reduced Liouville equation (RLE) takes the form

$$\dot{\sigma}_{ij} = -i\, \bar{\omega}_{ij}\, \sigma_{ij} - i \sum_k [V^P_{ik}\sigma_{kj} - \sigma_{ik}V^P_{kj}]$$
$$- \gamma^A_{ijij}\, \sigma_{ij}$$
$$-\tfrac{1}{2} \sum_{km} [\gamma^F_{imkj}\, \sigma_{kj} + \sigma_{ik}\, \gamma^F_{ikmj}]$$
$$+ \sum_{k\ell} \gamma^F_{ijk\ell}\, \sigma_{k\ell} \; . \tag{10}$$

This form of eq. 3 is the fundamental equation for the dissipative limit. The components of $\mathcal{L}$ are

$$\bar{\omega}_{ij} = tr_B[H_i^O \rho_{ij}^B - \rho_{ij}^B H_j^O] \ , \tag{10a}$$

$$\gamma_{ijij}^A = tr_B \pi (h_i - h_j)^2 \rho_{ij}^B \tag{10b}$$

$$\equiv tr_B \pi \mathcal{V}_{ij}^2 \rho_{ij}^B$$

$$\gamma_{imkj}^F = tr_B 2\pi V_{im}^F V_{mk}^F \rho_{kj}^B \tag{10c}$$

$$\gamma_{ikmj}^F = tr_B 2\pi \rho_{ik}^B V_{km}^F V_{mj}^F \tag{10d}$$

$$\gamma_{ijk\ell}^F = tr_B 2\pi V_{ik}^F P \rho_{k\ell}^B V_{\ell j}^F \tag{10e}$$

Eq. 10a is the average frequency difference for PS states $|i>$ and $|j>$. Note the two-sided character of the commutator-like structure. Eq. 10b is the very important memory relaxation (pure dephasing) constant resulting from adiabatic motion of the BG modes on the two PS surfaces. It is the average value of the squared difference potential between surfaces ω_i and ω_j (denoted by $\mathcal{V}_{ij}$). The conditions of validity for eq. 10b are that the diagonal BG components of ρ_{ij}^B be real and slowly varying with BG energy levels. The γ^F terms in eqs. 10 c-e are coefficients for coherence transfer, probability and probability amplitude relaxation, and probability transfer, depending on the PS indices.

The RLE of eq. 10 describes the interplay of coherent and dissipative dynamics. It represents the play of nonequilibrium dynamics of the PS on a background of thermal equilibrium. This implies that ρ^B and consequently the γ's depends on temperature. The coherent potentials are given by V^P and the coherence transfer components of γ^F, while the dissipative potentials are given by γ^A and the probability (amplitude) relaxation and probability transfer components of γ^F.

The importance of the adiabatic formulation of the BG lies in the resulting memory relaxation term γ^A of eq. 10b and the associated simple conceptual picture. According to this picture coherence loss (pure dephasing) between $|i>$ and $|j>$ is caused by a difference in the force acting on the BG coordinates in the two states (Rhodes, 1981,1982,1983). Other theories of dephasing have used collision models in which collision of a BG molecule with a PS molecule causes a phase disruption (Harris and Stodolsky, 1981). The adiabatic formulation shows clearly that dephasing depends on the difference in properties of the two PS states relative to the BG molecules.

EFFECTIVE HAMILTONIAN FORMULATION

In the most general situation, eq. 10 is a nonHamiltonian equation of motion for the PS. This means that $\mathcal{L}$ cannot be cast in commutator-like form involving a Hamiltonian. The presence of γ^A alone is sufficient to destroy the Hamiltonian character. In addition, however, the existence of γ^F terms, such as γ_{iikk} in eq. 10e describing probability transfer from state $|k>$ to state $|i>$, also destroys the Hamiltonian structure.

On the other hand, it can easily be seen, that there exists a set of γ^F components for which eq. 10 maintains (nonHermitian) Hamiltonian character. These consist of the elements in eqs. 10c and 10d corresponding to relaxation of amplitude of the PS states. For example, if the only surviving γ elements in eq. 10 are γ_{ikij} in eq. 10c and γ_{ijmj} in eq. 10d and if the former are independent of state $|j>$ and the latter are independent of state $|i>$, then eq. 10 takes the form

$$\dot{\sigma} = -i[\mathcal{H}\sigma - \sigma\mathcal{H}^{\dagger}] , \tag{11}$$

with the effective (nonHermitian) Hamiltonian

$$\mathcal{H} = h + V^P - i\tfrac{1}{2}\gamma .$$

It is tacitly assumed that the matrix elements of V^P are real Thus, $\mathcal{H}$ is a complex symmetric operator which can be diagonalized by a (complex) symmetric transformation. The discrete eigenvalues of $\mathcal{H}$ lie in the lower half of the complex plane.

The dynamics described by eq. 11 is one in which the amplitudes and probabilities for states having a nonzero γ component tend to decay exponentially. On the other hand, states coupled by V^P tend to have oscillating probabilities. The resulting pattern is one of decays and oscillatory decays. Consequently, $\mathcal{H}$ is not norm conserving, the reason being that use of an effective, nonHermitian Hamiltonian implies that a projection onto a subspace of the Hilbert space of the PS has tacitly been made. The subspace removed by projection is effectively a probability sink for the PS. Effective Hamiltonians can be of great practical value in describing intramolecular dynamics (Heller, Elert, and Gelbart, 1978).

TWO-LEVEL SYSTEMS

The preceding sections have shown how modulation of the Principal Subsystem by Background modes in the dissipative limit leads to coherence (memory) relaxation within the PS. Some of these dissipative (γ) terms contribute a nonHermitian Hamiltonian component while others destroy the Hamiltonian character of the effective Liouvillian, $\mathcal{L}$. In order to understand the phenomenological consequences of these dissipative terms, including the introduction of irreversibility and kinetics (vs. coherent dynamics), we now consider a prototype PS consisting of two states $|a>$ and $|b>$.

At first we let $\mathcal{L}$ contain only an effective Hamiltonian part, eq. 11, and a pure memory relaxation term, $\gamma^A_{abab} \equiv \gamma^A_{ab}$. Note that symmetry requires that $\gamma^A_{ab} = \gamma^A_{ba}$. (cf. eq. 10b). We assume that V has no diagonal elements and that $V_{ab} \equiv V$ is real. The equation of motion then becomes

$$
\begin{bmatrix} \dot{\sigma}_{aa} \\ \dot{\sigma}_{ab} \\ \dot{\sigma}_{ba} \\ \dot{\sigma}_{bb} \end{bmatrix} = \begin{bmatrix} -\gamma_{aa} & iV & -iV & 0 \\ iV & (-i\omega_{ab}-\gamma_{ab}) & 0 & -iV \\ -iV & 0 & (-i\omega_{ba}-\gamma_{ab}) & iV \\ 0 & -iV & iV & -\gamma_{bb} \end{bmatrix} \begin{bmatrix} \sigma_{aa} \\ \sigma_{ab} \\ \sigma_{ba} \\ \sigma_{bb} \end{bmatrix} \tag{12}
$$

where $\gamma_{aa} \equiv \gamma_{aaaa}$, $\gamma_{bb} \equiv \gamma_{bbbb}$, and $\gamma_{ab} \equiv \gamma^A_{ab} + \frac{1}{2}(\gamma_{aa}+\gamma_{bb})$.

The pure imaginary components in eq. 12 tend to produce coherent oscillation of σ components, while the pure real (γ) components tend to give monotonic relaxation. For simplicity, we now let $\gamma_{aa} = \gamma_{bb} = 0$, whereby the only dissipation is due to pure dephasing, γ^A_{ab}. We wish to compare the weak and strong coupling cases for the magnitude of V vs. γ^A_{ab}. Of course, for $\gamma^A_{ab} = 0$, we have simple, coherent, oscillatory Hamiltonian dynamics in which σ_{aa} and σ_{bb} oscillate (for nonstationary initial conditions) about mean values which depend on ω_{ab}. However, for $\gamma^A_{ab} > 0$ the system always approaches a limit point $\sigma_{aa} = \sigma_{bb} = \frac{1}{2}$, provided $V \neq 0$. This is shown in Figs. 1a and 1b for the strong coupling $(V > \gamma^A_{ab})$ and weak coupling

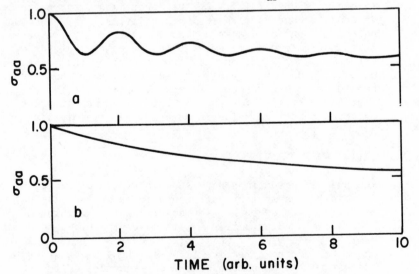

Figure 1. Two-state relaxation dynamics. (a) Strong coupling: $V = 1.0$; $\omega_{ab} = \sqrt{6}$; $\gamma_{ab} = 0.4$; $\gamma_{aa} = \gamma_{bb} = 0$. (b) Weak coupling: $V = 1.0$; $\omega_{ab} = 10.0$; $\gamma_{ab} = 10.0$; $\gamma_{aa} = \gamma_{bb} = 0$.

$(V < \gamma^A_{ab})$ cases, respectively.

Physical processes in which the strong coupling case is observed include optical nutation in laser driven excitation and quantum beats in molecules excited by light pulses. The weak coupling case is prototype for simple kinetic relaxation to equilibrium. For this limit the dynamics of σ_{aa} and σ_{bb} are accurately described by kinetically reversible equations in which the forward and reverse rate constants are equal and have the value

$$k = 2V^2/(\omega^2_{ab} + \gamma^2_{ab})^{\frac{1}{2}}. \tag{13}$$

It can be shown by projecting eq. 12 onto σ_{aa} and σ_{bb} that the resulting equation is a Markovian master equation (Zwanzig, 1961) in the limit $\gamma^A_{ab} \rightarrow \infty$ and V/γ^A_{ab} = constant.

Excitation energy transfer between molecules provides another good example of the dichotamy presented by strong and weak coupling. For strong coupling excitation transfer is a coherent delocalization of the excitation in which σ_{ab} (coherence component) maintains a viable role in the dynamics, while for weak coupling σ_{ab} is damped and maintained at a low level so that the excitation is transferred by an incoherent probability migration (hopping) mechanism (Förster limit).

Another possible role of γ^A_{ab} in eq. 12 is molecular structure stabilization. In principle, many molecular conformations can undergo isomerization through quantum mechanical tunneling. If γ^A_{ab} is larger than the tunneling matrix elements, there will be a kinetic stabilization of conformational isomers through modulation by the medium (Harris and Stodolsky, 1981). In this regard, medium effects via γ^A may play a role in the symmetry-breaking formulation of stable molecular structures. Such symmetry-breaking has been presented as a puzzle in recent papers (Wooley, 1980; Trindle, 1980).

Perhaps the simplest possible example of nonHamiltonian dynamics is provided by the case in which the only nonzero element of $\mathcal{L}$ in eq. 12 is γ^A_{ab}. We then have

$$-i\mathcal{L} = -\gamma \begin{bmatrix} 0 & 0 & 0 & 0 \\ 0 & 1 & 0 & 0 \\ 0 & 0 & 1 & 0 \\ 0 & 0 & 0 & 0 \end{bmatrix} \tag{14}$$

For components of σ in the (a,b) basis the only dynamics generated by $\mathcal{L}$ is the relaxation of σ_{ab} and σ_{ba}. A system initially in pure state $|a>$ or $|b>$ is stationary under $\mathcal{L}$, while a coherent superposition of $|a>$ and $|b>$ relaxes to a statistical mixture. This looks like rather dull dynamics until we consider the equation of motion in a different basis. For example, if we make the unitary transformation to the basis $|1>$ and $|2>$,

$$|1,2> = 2^{-\frac{1}{2}}(|a> \pm |b>) , \qquad (15)$$

we obtain

$$-i\mathcal{L} = -\tfrac{1}{2}\gamma \begin{bmatrix} 1 & 0 & 0 & -1 \\ 0 & 1 & -1 & 0 \\ 0 & -1 & 1 & 0 \\ -1 & 0 & 0 & 1 \end{bmatrix} . \qquad (16)$$

The dynamics in this new basis now looks very interesting. The equation of motion, eq. 3, is two uncoupled sets of equations, one for the probabilities σ_{11} and σ_{22} and one for the coherence components σ_{12} and σ_{21}. The coupled probability equations describe Markovian kinetics. A system initially in state $|1>$, for example, undergoes a kinetic transformation to a statistical mixture of $|1>$ and $|2>$. The coherence components are zero for all t.

In applying the results of eq. 16 to a model calculation we have included the parameters $\gamma_{aa} = 0.2$ and $\gamma_{bb} = 0.1$ in order to make the dynamics more interesting. The results for the value $\gamma_{ab}^{A} = 0.5$ are given in Fig. 2.

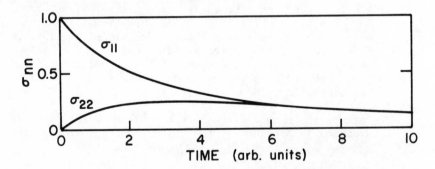

Figure 2. Two-state relaxation dynamics due solely to dissipative relaxation in the basis $|a>$ and $|b>$. Parameters $\gamma_{ab}^{A} = 0.5$; $\gamma_{aa} = 0.2$; $\gamma_{bb} = 0.1$; $V = \omega_{ab} = 0$.

Analysis shows σ_{11} to have biexponential decay resulting from competitive kinetics for the transition $|1> \rightarrow |2>$ and dissipative probability loss due to γ_{aa} and γ_{bb}.

The important feature exhibited by eq. 16 and Fig. 2 is that purely dissipative dynamics (in particular, pure coherence relaxation) in one basis produces a kinetic enhancement in another basis. Such processes may occur in molecular excited state relaxation, where coherent excitation prepares the molecule in a nonstationary excited state while coherence relaxation occurs among the set of superposed states comprising the prepared state.

DISCUSSION

We have shown how the modes of the Principal Subsystem (PS) are modulated by the Background (BG) modes in the dissipative limit. Eq. 10 is the most general result, but eq. 12 shows familiar features more clearly for a special case of a two-level system. The dissipative components γ_{aa} and γ_{bb} are often referred to as longitudinal relaxation (T_1) and γ_{ab} as the transverse relaxation (T_2). These equations contain the interplay of coherent and dissipative dynamics. Of particular importance is the role of the nondiagonal (coherence) components σ_{ab} for the PS. The coherent potential, V^P, couples σ_{ab} to the diagonal elements (probabilities). Thus, damping of σ_{ab} affects probability rates. Furthermore, σ_{ab} contributes directly to average values of PS observables.

Our key result is the adiabatic formulation of memory relaxation (pure dephasing), γ_{ab}^A, given in eq. 10b. The idea is that transitions within the PS due to V^P produce a motion of the BG modes ("shaking" of the bath), which motion in turn modulates the dynamics of transitions in the PS (adiabatic modulation). In the dissipative limit a strong correlation evolves between the PS and the BG. In terms of wavefunctions the BG wave-packets associated with different PS states tend toward orthogonality with rate γ_{ab}.

The damping of σ_{ab} by γ_{ab} can greatly modulate the dynamics of σ_{aa} and σ_{bb}, producing a Markovian master (kinetic) equation in the strong dissipative limit. Kinetics replaces coherent dynamics in the sense that σ_{ab} is maintained at a low level (due to large γ_{ab}). Consequently the rate of transitions due to V_{ab}^p is greatly lowered, resulting in kinetic stabilization. Eq. 16 shows how γ_{ab} produces kinetic enhancement in another basis. For transitions among more than two states, memory relaxation can produce sequential transitions, in which quantum interference terms are eliminated. This provides the distinction, for example, between coherent light scattering and resonance fluorescence (Rhodes, 1981).

Memory relaxation also gives a basis for understanding entropy production and the approach to equilibrium. A PS initially in a pure quantum state evolves into a statistical state with an increase in entropy from an initial value zero.

ACKNOWLEDGEMENTS

This work was supported in part by Contract No. DE-ASO5-78EVO5784 between the Division of Biomedical and Environmental Research of the Department of Energy and Florida State University and National Institutes of Health Grant No. 1RO1GM23942.

REFERENCES

Bagchi, B. and Oxtoby, D., J. Phys. Chem. 86, 2197 (1982).

Boeglin A., Villaeys, A., Voltz, R. and Lin, S., J. Chem. Phys. 79, 3819 (1983).

Grigolini, P., Adv. Chem. Phys. (in press, 1985); Nuovo Cimento 63B, 174 (1981).

Harris, R. and Stodolsky, L., J. Chem. Phys. 74, 2145 (1981).

Heller, D. F., Elert, M. L., Gelbart, W. M., J. Chem. Phys. 69, 4061 (1978).

Jackson, B. and Silbey, R., J. Chem. Phys. 75, 3293 (1981).

Kubo, R., Adv. Chem. Phys. 16, 101 (1969).

Mori, H., Prog. Theor. Phys. 33, 423 (1965); 34, 399 (1965).

Prigogine, I., in Order and Fluctuations in Equilibrium and Nonequilibrium Statistical Mechanics, G. Nicolis, G. Dewel, and J. Turner, eds., John Wiley, New York, 1981.

Redfield, A. G., Adv. Magn. Reson. 1, 1 (1965).

Rhodes, W., J. Chem. Phys. 75, 2588 (1981); J. Phys. Chem. 86, 2657 (1982); J. Phys. Chem. 87, 30 (1983).

Trindle, C., Israel Jour. of Chem. 19, 47 (1980).

Wooley, R., Israel Jour. of Chem. 19, 30 (1980).

Zwanzig, R., Lectures in Theoretical Physics, Vol. 3, W. Brittin, B. Downs, and J. Downs, eds., Interscience, New York, 1961.

Chapter 25

THE ROLE OF THE TOPOLOGICAL DISTANCE MATRIX IN CHEMISTRY

D.H. Rouvray

Department of Chemistry, University of Georgia, Athens, Georgia 30602, USA

ABSTRACT

The mathematical properties of the topological distance matrix are briefly surveyed and the numerous applications of this matrix to various branches of chemistry are then described. A detailed discussion is devoted to the Wiener index, defined as one half the sum of the entries in the distance matrix. This index has proved to be one of the most valuable topological indices in several different chemical contexts.

PRELIMINARY MATHEMATICAL SURVEY

Although in its origins the distance matrix can be traced back to the work of Cayley [1], it was not until 1895 that the matrix was first introduced in embryonic form by Brunel [2]. The distance matrix, $D(G)$, for a graph G is defined as a real, square, symmetrical matrix of order n, with entries, d_{ij}, representing the distance traversed in moving from vertex i to vertex j in G. The d_{ij} entries must always satisfy the following criteria:

(i) $d_{ii} = 0$ (identity relation)
(ii) $d_{ij} = d_{ji}$ (symmetry relation)
(iii) $d_{ij} \leq d_{ik} + d_{kj}$ (triangle inequality)

where i, j, k = 1,2,...., n. In the present context, the d_{ij} will refer only to topological distances and not to geometrical distances, though $D(G)$ is defined in the same way for both. Only in the case of a tree graph, T, will the d_{ij} be uniquely defined. To accommodate cyclic graphs, however, Hosoya [3] generalized the definition of the distance d_{ij} by taking it to be the <u>minimum</u> number of edges traversed in moving from i to j.

Mathematicians have studied $D(G)$ very extensively with the major emphasis focusing on tree graphs. Thus, Graham and Pollak [4] proved that for any tree, T, on n vertices, the determinant of $D(G)$ will assume the form $(-1)^{n-1} (n - 1)2^{n-2}$. It follows that $D(T)$ will always have one positive eigenvalue and $(n-1)$ negative eigenvalues, irrespective of the nature of the tree. The properties of the distance matrix polynomials obtained by expanding the determinants have been investigated by several authors [5-8], and it has been found that such polynomials are highly unsuitable for the unique characterization of trees [9], even though every finite tree is uniquely determined up to isomorphism by $D(T)$ [10]. The conditions under which a given $D(G)$ can be realized by a graph, and especially the conditions under which unique realization is possible, have been widely studied [11-15]. The determinants for distance matrices arising from weighted, directed graphs have also been explored [16]. Since distance matrices are closely related to finite metric spaces, it is not surprising that $D(G)$ has been employed in the study of isometric embeddings of graphs into the cartesian products of metric spaces [17-18].

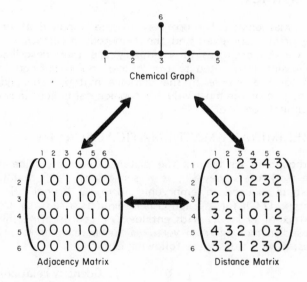

Figure 1. The adjacency and distance matrices for 3-methylpentane

The redundancy of much of the information contained in $D(G)$ has been commented upon by several authors [19-22]. Thus, for any graph G, $D(G)$ contains no more information than that contained in the adjacency matrix, $\mathbf{A(G)}$, a matrix similar to $\mathbf{D(G)}$ but with all $d_{ij} > 1$ set to zero. Both of these matrices are illustrated in Figure 1 for the graph of the molecule of 3-methylpentane. Once $\mathbf{A(G)}$ is known, the d_{ij} entries for $\mathbf{D(G)}$ may be readily determined iteratively by raising $\mathbf{A(G)}$ to the powers $2,3,\ldots,n$. The a_{ij} entry in $[\mathbf{A(G)}]^n$ yields the number of walks of length n in moving from vertex i to vertex j of $\mathbf{G}$ [23-24]. Algorithms for calculating $\mathbf{D(G)}$ nowadays, however, make use of faster procedures than matrix multiplication of powers of $\mathbf{A(G)}$. Bersohn [25], Peredunova et al. [26], and Herndon [27] have all used very efficient algorithms for calculating $\mathbf{D(G)}$ for any graph. Recently, algorithms have also been published for determining the shortest path between two given vertices in $\mathbf{D(G)}$ [28,29]; the shortest spanning tree in $\mathbf{D(G)}$ [30]; the next to shortest and longest paths between a pair of vertices [29,31]; the maximal degree of a tree [32]; and the number of branch points and their respective degrees in trees [33].

USES OF THE DISTANCE MATRIX

The distance matrix has found widespread application in a variety of different guises in the physical, biological and social sciences. Among disciplines relatively remote from chemistry, $\mathbf{D(G)}$ has been employed, for instance, in anthropology for the study of tribal grouping patterns [34]; in archaeology for the classification of ancient artifacts [35]; in electrical engineering for the modelling of loop switching [4]; in geography for the planning of transportation networks [36]; in geology for hierarchical cluster analysis [37]; in ornithology for the classification of bird song [38]; in philology for the study of semantics [39]; in psychology for the definition of psychological distance [40]; and in sociology for the modelling of social structures [41].

Turning now to disciplines more closely allied to chemistry, $\mathbf{D(G)}$ has been adopted in biochemistry for the comparison of DNA restriction maps [42], and for nucleic acid and protein sequencing in macromolecules [43,44]; in biology for the investigation of evolutionary distances in DNA sequences [45,46]; in genetics for the interpretation of phylogenetic relationships in macromolecules [47,48]; and in physics for the study of microclusters [49].

In terms of the number of applications made to date, chemistry has proved to be the most important client science for the services of the distance matrix. $\mathbf{D(G)}$ has been employed in chemistry in both explicit and implicit forms, and the d_{ij} entries in the matrix have been represented by geometric as well as topological distances. We shall not pursue here the uses of $\mathbf{D(G)}$ based on geometrical distances; interested readers are referred to reviews on this topic [50,51]. The first explicit use of $\mathbf{D(G)}$ in chemistry with the d_{ij} representing topological distances was made in 1975 by Clark and Kettle [52]. In a general study of stereochemically nonrigid molecules, these workers distinguished between the various interconversion mechanisms for pairs of permutational isomers by determining the shortest path sequence necessary to effect the rearrangement. These shortest paths were then used as the d_{ij} entries in the construction of an appropriate distance matrix. Although this made the first explicit use of $D(G)$, numerous other workers had previously used such an approach in implicit form. Examples are to be seen in the stereochemical matrices of Muetterties [53], used in the investigation of polytopal rearrangements; in the adjacency matrices and their powers used by Balaban [54] to represent intramolecular isomerizations of octahedral complexes; and in the definition of chemical metrics involving the distance between matrices representing chemical transformations [55].

INTRODUCTION TO THE WIENER INDEX

In spite of the fact that the earliest implicit usage of $D(G)$ in chemistry dates back some four decades, very important ramifications of this early work have manifested themselves in recent years. In 1947 Wiener started publishing several papers [56] in which he introduced for alkane species the concepts of the path number and the polarity number. The path number he defined as the sum of the chemical bonds existing between all the pairs of carbon atoms in the molecule; and the polarity number as the number of pairs of carbon atoms separated by three C-C bonds. At about the same time, Platt [57] added a third graph invariant which he called the first neighbor sum and defined as the sum of the first C-C neighbors for every C-C bond in the species. Clearly, all three of these invariants are closely related to the distance matrix.

It was shown by Hosoya [3], for instance, that the Wiener path number is equal to one half the sum of the d_{ij} entries in $D(G)$ for the graph G of the molecule in question, namely that

$$W(G) = \tfrac{1}{2} \sum_i \sum_j d_{ij} \,,$$

where $W(G)$ is the symbol for the Wiener path number, nowadays usually referred to as the Wiener topological index [58]. The polarity number, $P(G)$, and the first neighbor sum, $F(G)$, were also shown [58] to be expressible in terms of simple formulas:

$$P(G) = \tfrac{1}{2} \sum W_3(G),$$

where $W_3(G)$ is an off-diagonal entry in $D(G)$ having the distance three, and

$$F(G) = \sum_{i=1}^{n_e} \deg(e_i),$$

where n_e is the number of edges in G and $\deg e_i$ is the degree of the ith edge in G.

The Wiener index has been employed extensively for the correlation of the physicochemical properties of hydrocarbon species. Wiener himself [56] correlated first the boiling points of alkanes, t_B, using a biparametric relationship of the form:

$$t_B = a\,\mathbf{W(G)} + b\,\mathbf{P(G)} + c,$$

where a, b and c were constants for a given group of isomers, and then went on to consider heats of isomerization and vaporization, the Antoine vapor pressure constants, the specific dispersion, the surface tension, and the critical solution temperature in aniline. The good correlations he obtained encouraged him to speculate on the significance of the invariants he used. By setting $a = 98/n^2$, he concluded that the boiling point of a molecule varies inversely with its degree of compactness. He interpreted $\mathbf{P(G)}$ as a measure of the intramolecular attraction forces transmitted through the carbon chain.

Platt [57] made Wiener-type correlations with molecular volumes and molar refractivities, and also suggested that $\mathbf{W(G)}$ might well be applicable to hydrocarbons other than alkanes. The good correlations he obtained were interpreted by hypothesizing that $\mathbf{W(G)}$ provides a measure of the mean external contact area of the molecule. Use was also made of $\mathbf{W(G)}$ by Stiel and Thodos [58] to predict the various critical constants of the alkanes. Rouvray [59] first extended such studies to the alkenes, alkynes and arenes, in addition to the alkanes, using an index, $\mathbf{R(G)}$, equal to the sum of the d_{ij} in $\mathbf{D(G)}$. The Rouvray index, $\mathbf{R(G)}$, is clearly equal to $2\mathbf{W(G)}$. Good correlations were again obtained with six parameters ranging from melting point to viscosity, though now a Walker-type relationship [60] was employed. This assumes that a bulk property, X, is related to the index $\mathbf{W(G)}$ in the following way:

$$\mathbf{W(G)} = \alpha\,[X]^\beta,$$

where α and β are constants which were determined by plotting ln $\mathbf{W(G)}$ against ln X. It was then suggested by Randić [61] that $\mathbf{W(G)}$ might be useful in simplifying the prediction of chromatographic retention times by greatly reducing the number of parameters needed for this purpose. This idea was taken up by Papazova et al. [62] who used $\mathbf{W(G)}$ with other indices in their correlations with the isoalkanes, and by Bonchev et al. [63] who used $\mathbf{W(G)}$ alone in a Walker-type relationship [60] for correlation with the alkylbenzenes. Both groups achieved very good correlations; in the latter case the correlation coefficient exceeded 0.999.

EVALUATION OF THE WIENER INDEX

Because of the great usefulness of $\mathbf{W(G)}$ in various branches of chemistry, analytic and recursive formulas have been developed for its evaluation for many graphs of chemical interest. The first analytic expression was obtained by Wiener himself [56], who showed that $\mathbf{W(G)}$ for an unbranched path on n vertices is equal to $(n^3 - n)/6$. In 1976 Entringer et al. [64] published several basic formulas and established, inter alia, that for cyclic graphs on n vertices

$W(G) = (n^3 - n)/8$ for n odd whereas $W(G) = n^3/8$ for n even; and that the complete graph, K_n, has $W(G) = (n^2 - n)/2$. Bonchev and Trinajstić [65] obtained results for many different types of tree graphs, including the star graph, which has $W(G) = (n - 1)^2$, but were unable to treat the case of trees with branched branches. Closed formulas were also obtained by Bonchev et al. [66] for molecular species existing in the form of strips, such as linear ribbons of fused cycles and the edge-fused propellanes. A general, recursive procedure for determining $W(G)$ for any tree, regardless of the amount of branching present, was recently presented by Canfield et al. [67].

A large number of rules have been devised to characterize trends in the numerical values assumed by $W(G)$ within specific classes of molecules. Rules have been put forward for species having graphs in the form of trees [65], monocycles [68], monocycles with acyclic branches [69], polycycles which are fusion- [66], bridge- [70], and spiro-linked [71], and polycycles with acyclic branches [72]. $W(G)$ appears to be a very convenient device for expressing quantitatively regularities and variations in the molecular topology of both acyclic and cyclic compounds. Some of the rules are intuitively obvious whereas others are by no means so. Thus, it is evident, in going from a tree graph having the form of an unbranched path through a tree with branches to a star graph for a constant n, that the amount of branching is increasing and that $W(G)$ will decrease [65]. One's intuition is not so helpful, however, in determining that the value of $W(G)$ passes through a minimum in isomeric graphs comprised of a monocycle with a nonbranched side chain [69].

Our preceeding remarks have pointed up the fact that $W(G)$ provides a very good measure of molecular compactness: the more compact the graph of a molecule is the smaller its $W(G)$ value will be. This renders $W(G)$ a very useful index, for it is well-known that a vast number of properties - including physicochemical, thermodynamic, and quantum-chemical ones - are determined primarily by molecular size and shape. Although $W(G)$ does not provide a totally reliable means of discriminating between different isomers, it does reflect better than most topological indices the degree of branching present in species. For comparison of the amount of branching in different types of graphs, a mean branching index has been proposed. This index, $\overline{W}(G)$, has been defined [65] as $W(G)$ divided by the number of distances in the graph, i.e. $\overline{W}(G) = R(G)/n(n - 1)$; analytic expressions for $\overline{W}(G)$ have been published for certain simple graphs [65,68]. However, since even $\overline{W}(G)$ has not proved to be completely adequate for isomer discrimination, a highly sensitive new index, called the mean information on distance equality, has been introduced for this purpose [65,73,74].

The total information on distance equality is defined as follows:

$$I_D^E(G) = \frac{n(n-1)}{2} \log_2 \frac{n(n-1)}{2} - \sum_{i=1}^{m} k_l \log_2 k_l$$

and its mean is given as:

$$\bar{I}^{E}_{D}(G) = -\sum_{l=1}^{m} \frac{2k_l}{n(n-1)} \log_2 \frac{2k_l}{n(n-1)} \;,$$

where the distance l appears $2k_l$ times in $D(G)$ and m is the highest value of l. The mean index has been employed for the prediction of chromatographic retention times in the alkylbenzenes; it not only discriminated between all the differing isomeric structures but also yielded very high correlation coefficients [75]. Moreover, a closely related information-theoretical index, known as the information on distance magnitudes and defined [65] as:

$$I^{W}_{D}(G) = W(G) \log_2 W(G) - \sum_{l=1}^{m} k_l \; l \log_2 l \;,$$

where the distance l appears k_l times in the $W(G)$ partition, correlated extremely well with several different parameters, including heats of reaction and boiling points of the alkylbenzenes [76]. As limitations of space preclude further discussion here of informationtheoretical indices based on $D(G)$, the reader is referred instead to a review [77] and a book [78] dealing with the subject.

RECENT NOVEL USES OF THE WIENER INDEX

In addition to the correlational studies described above, a number of workers have begun to explore several exciting new directions in which applications of $W(G)$ are now leading. The first of these concerns the use of $W(G)$ to study the energies in different types of molecules. It has been demonstrated [71], for instance, that there is an excellent inverse correlation of $W(G)$ with the σ-electron energies in spiro compounds: the smaller $W(G)$ within a group of isomers the larger the extended Hückel energy will be. Similarly, a good inverse correlation exists [66] for the π-electron energies in arene systems: when $W(G)$ decreases, i.e. cyclicity increases, the HOMO energy decreases whereas the LUMO energy increases. In the case of cyclic molecules with acyclic branches, both the branching and the cyclicity affect the π-electron energy. If either of these factors is held constant, however, it is possible to obtain a reliable dependence of the other factor on $W(G)$ [69]. Bicyclic systems exhibit an inverse proportionality with $W(G)$ for $(4n + 2)$ systems and a direct proportionaltiy for $4n$ systems [70]. Such results are very important, for it is well-known that the frontier electron orbitals largely govern the molecular behavior in these systems. On the basis of cyclicity rules, it becomes possible to determine how the first ionization potential, the electron affinity, the maximum amount of light absorption, and the electrical conductivity will depend upon structural changes in the systems. Prediction of the relative stabilities and properties of unknown compounds is thus feasible. Moreover, it appears that the same rules can also be employed to classify molecular rearrangements [69,70].

By extending strings of such molecular systems to infinity, i.e. by treating them as monodimensional crystals, one is led to the prediction of the electronic energy as well as the energy gap in polyenes which are potential conductors or semiconductors. The method which has been used [79] involves essentially five steps, viz. (i) derivation by mathematical induction of a polynomial expression for $W(G)$ for the monomer of a given polymer homologous series of compounds; (ii) normalization of the derived $W(G)$; (iii) calculation of the normalized $W(G)$ for an infinite polymer chain; (iv) correlation between the normalized $W(G)$ and some property for several oligomers in the polymer series; and (v) prediction of the value of this property for the infinite polymer chain. The method has been utilized on a variety of different polymer series, including alternant and non-alternant benzenoid systems, polymers, and radialenes [79,82]. In general, high correlation coefficients were obtained for π-electron energies, and the higher the degree of the polynomial expression for $W(G)$ the better the results. The method has been tried out using both Hückel [79] and Pariser-Parr-Pople [80,81] formalisms. Physical properties which have been studied include the melting point, the refractive index, and the specific rotation; these have been predicted with an accuracy rivalling that of other wellknown methods such as the Padé method [82].

Further extension of the above notions to three-dimensional systems has made it possible to use $W(G)$ to model the behavior of solids. In particular, $W(G)$ has been used in the study of crystal vacancies; these have important industrial applications in areas such as corrosion control, catalysis and chemisorption. The fundamental idea is that a system may be regarded as being in its minimum energy state when $W(G)$ for the system has its minimum value. Using this approach [83], structures having vacancies in the most favored positions energetically can be recognized by minimizing their $W(G)$ value or by maximizing $\Delta W(G) = W(G) - W_0(G)$, where $W_0(G)$ represents the Wiener index for the ideal crystal without any lattice vacancies. By use of a procedure similar to that outlined above for polymer chains, it was possible to predict the most favored vacancy positions in the lattice. ΔW was first expressed in polynomial form in terms of the number of atoms present and the various positions of the defect and then partially differentiated with respect to the positions and set to zero. The minima thereby obtained yield the positions of vacancies in the crystal corresponding to the $W(G)$ minima. The method has also been applied to the study of double and triple vacancies [84], the migration of vacancies along preferred diffusion paths [84], the determination of the optimal positions of defect atoms in crystal lattices [85], and in the modelling of the crystal growth process [86]. More recently, a similar approach involving the study of the ordered structures of adsorbed gases in host lattices [87], was correctly able to predict the structure of γ-$PdD_{0.5}$.

It has been suggested by Mekenyan et al. [69] that the manifold rules giving trends in $W(G)$ values for a variety of different systems could well provide the starting point for a novel approach to both structure-property and structure-activity correlations. One major advantage of this approach is that it offers an optimal selection of isomer samples for usage in a given correlation [69]. To some extent this approach is already being tried out. Basak et al. [88-90] used $W(G)$, $I_D^W(G)$ and $\bar{I}_D^W(G)$ for the investigation of structure -activity relationships in bioactive molecules ranging from alcohols to barbiturates. To make $W(G)$ appropriate for molecules containing heteroatoms, it has been proposed that weighted graphs be used to determine $D(G)$. Lall and Srivastava [91] suggested using an edge-weighting factor based on Hückel parameters, whereas Barysz et al. [92] thought the nuclear charage on the atom should be used. A discussion on discriminating isomers by means of three new information-theoretical indices based on $D(G)$ has recently been presented [93]. The prediction of carcinogenicity in arene systems using indices based on $D(G)$ has also been proposed; Lall [94] used a first neighbor degree sum, and Seybold [95,96] used the atomic path code of Randić [97] which is equal to the sum of the entries in one row of $D(G)$.

ACKNOWLEDGEMENTS

Thanks are tendered to the U.S. Office of Naval Research for partial support of this work, and to Professor N. Trinajstić for the provision of a helpful list of his publications in this field.

REFERENCES

[1] Cayley, A., Cambridge Math. J. 2, 267 (1841).
[2] Brunel, M.G., Mémoires Soc. Sci. Bordeaux [4], 5, 165 (1895).
[3] Hosoya, H., Bull. Chem. Soc. Japan 44, 2332 (1971).
[4] Graham, R. and Pollak, H.O.; Bell Syst. Tech. J. 50, 2495 (1971).
[5] Edelberg, M. Garey, M.R., and Graham, R.L., Discrete Math. 14, 23 (1976).
[6] Graham, R.L. and Lovász, L., Lect. Notes in Math. 642, 186 (1978).
[7] Graham, R.L. and Lovász, L., Adv. in Math. 29, 60 (1978).
[8] Křivka, P. and Trinajstić, N., Aplikace Mat. 28, 357 (1983).
[9] McKay, B.D., Ars Combinatorica 3, 219 (1977).
[10] Zykov, A.A., Mat. Sbornik 24, 163 (1949).
[11] Hakimi, S.L. and Yau, S.S., Quart. Appl. Math. 22, 305 (1964).
[12] Simões-Pereira, J.M.S., J. Comb. Theory 6, 303 (1969).
[13] Boesch, F.T., Quart. Appl. Math. 26, 607 (1969).
[14] Patrinos, A.N. and Hakimi, S.L., Quart. Appl. Math. 30, 255 (1972).
[15] Simões-Pereira, J.M.S. and Zamfirescu, C.M., Lin. Alg. and its Appl. 44, 1 (1982).
[16] Graham, R.L. Hoffman, A.J., and Hosoya, H., J. Graph Theory 1, 85 (1977).
[17] Neumaier, A., Ned. Akad. Wetensch. Indag. Math. 43, 385 (1981).
[18] Graham, R.L. and Winkler, P.M., Trans. Am. Math. Soc. 288, 527 (1985).

[19] Smolenskii, E.A., U.S.S.R. Comput. Math. and Math. Phys. 2, 396 (1963).
[20] Dewdney, A.K., Inform. and Control 40, 234 (1979).
[21] Yushmanov, S.V., Math. Notes Acad. Sci. U.S.S.R. 31, 326 (1982).
[22] Yushmanov, S.V., Math. Notes Acad. Sci. U.S.S.R. 35, 460 (1984).
[23] Anderson, S.S., Graph Theory and Finite Combinatorics, Markham, Chicago, 1970, p.7-8.
[24] Biggs, N., Algebraic Graph Theory, Cambridge University Press, London, 1974, p.11-12.
[25] Bersohn, M.J. Comput. Chem. 4, 110 (1983).
[26] Peredunova, I.V., Kuzmin, V.E., and Konovorotskii, Y.P., Russ. J. Struct. Chem. 24, 645 (1983).
[27] Herndon, W.C., Unpublished work; the program is available from the author.
[28] Even, S., Graph Algorithms, Computer Science Press, Rockville, Maryland, 1979, p.11-18.
[29] Deo, N. and Pang, C., Networks 14, 275 (1984).
[30] Chachra, V., Ghare, P.M. and Moore, J.M., Applications of Graph Theory Algorithms, North Holland, New York, 1979, p.67-77.
[31] Katoh, N. Ibaraki, T., and Mine, H., Networks 12, 411 (1982).
[32] Zelinka, B., Kybernetika 9, 361 (1973).
[33] Iwahori, N., Sci. Papers Coll. Arts Sci. Univ. Tokyo 33, 63 (1983).
[34] Hage, P., Anthropolog. Forum 3, 280 (1973).
[35] Hodson, F.R., Sneath, P.H.A. and Doran, J.E., Biometrika 53, 311 (1966).
[36] Tinkler, K.J., Trans. Inst. Brit. Geog. 55, 17 (1972).
[37] Warshauer, S.M. and Smosna, R., Math. Geol. 13, 225 (1981).
[38] Bradley, D.W. and Bradley, R.A. in Time Warps, String Edits, and Macromolecules, ed. Sankoff, D. and Kruskal, J.B., Addison-Wesley, Reading, Massachusetts, 1983, chap. 6.
[39] Boyd, J.P. and Wexler, K.N., J. Math. Psychol. 10, 115 (1973).
[40] Cunningham, J.P., J. Math. Psychol. 17, 165 (1978).
[41] Klaua, D., Zeit. Psychol. 183, 141 (1975).
[42] Waterman, M.S., Smith, T.F. and Katcher, H.I., Nucleic Acids Res. 12, 237 (1984).
[43] Waterman, M.S., Smith, T.F. and Beyer, W.A., Adv. Math. 20, 367 (1976).
[44] Goad, W.B. and Kanehisa, M.I., Nucleic Acids Res. 10, 247 (1982).
[45] Sellers, P.H., S.I.A.M. J. Appl. Math. 26, 787 (1974).
[46] Mirkin, B.G. and Rodin, S.N., Graphs and Genes, Springer, Berlin, 1984, chap. 3.
[47] Li, W.-H., Proc. Natl. Acad. Sci. U.S.A., 78, 1085 (1981).
[48] Hogeweg, P. and Hesper, B., J. Mol. Evol., 20, 175 (1984).
[49] Hoare, M.R., Adv. Chem. Phys. 40, 49 (1979).
[50] Crippen, G.M., J. Comput. Phys. 24, 96 (1977).
[51] Havel, T.F., Kuntz, I.D., and Crippen, G.M., Bull. Math. Biol. 45, 665 (1983).
[52] Clark, M.J. and Kettle, S.F.A., Inorg. Chim. Acta 14, 201 (1975).
[53] Muetterties, E.L., J. Am. Chem. Soc. 91, 4115 (1969).
[54] Balaban, A.T., Rev. Roum. Chimie 18, 841 (1973).
[55] Dugundji, J. and Ugi, I., Topics Curr. Chem. 39, 19 (1973).
[56] Wiener, H., J. Am. Chem. Soc. 69, 17, 2636 (1947); J. Chem.

Phys. 15, 766 (1947); J. Phys. Chem. 52, 425, 1082 (1948); J. Phys. Colloid Chem. 52, 425, 1082 (1948).

[57] Platt, J.R., J. Chem. Phys. 15, 419 (1947); J. Phys. Chem. 56, 328 (1952).

[58] Stiel, L.I. and Thodos, G., A.I.Ch.E.J. 8, 527 (1962).

[59] Rouvray, D.H., Math. Chem. 1, 125 (1975); Rouvray, D.H. and Crafford, B.C., S. Afr. J. Sci. 72, 74 (1976).

[60] Walker, J., J. Chem. Soc. 65, 725 (1894).

[61] Randić, M., J. Chromatogr. 161, 1 (1978).

[62] Papazova, D., Dimov, N. and Bonchev, D., J. Chromatogr. 188, 297 (1980).

[63] Bonchev, D., Mekenyan, O., Protić, G. and Trinajstić, N., J. Chromatogr. 176, 149 (1979).

[64] Entringer, R.C., Jackson, D.E. and Synder, D.A., Czech. Math. J. 26, 283 (1976).

[65] Bonchev, D. and Trinajstić, N., J. Chem. Phys. 67, 4517 (1977).

[66] Bonchev, D., Mekenyan, O. and Trinajstić, N., Int. J. Quant. Chem. 17, 845 (1980).

[67] Canfield, E.R., Robinson, R.W. and Rouvray, D.H., J. Comput. Chem. 6, 000 (1985).

[68] Bonchev, D., Mekenyan, O., Knop, J.V. and Trinajstić, N., Croat. Chem. Acta 52, 361 (1979).

[69] Mekenyan, O., Bonchev, D. and Trinajstić, N. Croat. Chem. Acta 56, 237 (1983).

[70] Mekenyan, O., Bonchev, D. and Trinajstić, N., Int. J. Quant Chem. 19, 929 (1981).

[71] Mekenyan, O., Bonchev, D. and Trinajstic, N. Math. Chem. 6, 93 (1979).

[72] Mekenyan, O., Bonchev, D. and Trinajstić, N., Math. Chem. 11, 145 (1981).

[73] Bonchev, D., Knop, J.V. and Trinajstić, N., Math. Chem. 6, 21 (1979).

[74] Bonchev, D. and Trinajstić, N., Int. J. Quant. Chem., Quant. Chem. Symp. 16, 463 (1982).

[75] Bonchev, D. and Trinajstić, N., Int. J. Quant. Chem., Quant. Chem. Symp. 12, 293 (1978).

[76] Mekenyan, O., Bonchev, D. and Trinajstić, N., Int. J. Quant. Chem. 18, 369 (1980).

[77] Bonchev, D., Mekenyan, O. and Trinajstić, N., J. Comput. Chem. 2, 127 (1981).

[78] Bonchev, D., Information-Theoretic Indices for Characterization of Chemical Structures, Research Studies Press, Chichester, U.K., 1983, p.127.

[79] Bonchev, D. and Mekenyan, O., Z. Naturforsch. 35A, 739 (1980).

[80] Bonchev, D., Mekenayn, O. and Polansky, O.E., Z. Naturforsch. 36A, 643 (1981).

[81] Bonchev, D., Mekenyan, O. and Polansky, O.E., Z. Naturforsch. 36A, 647 (1981).

[82] Mekenyan, O., Dimitrov, S. and Bonchev, D., Eur. Polym. J. 19, 1185 (1983).

[83] Bonchev, D., Mekenyan, O. and Fritsche, H.-G., Phys. Stat. Sol. 55A 181 (1979).

[84] Mekenyan, O., Bonchev, D. and Fritsche, H.-G., Phys. Stat. Sol. 56A, 607 (1979).

[85] Mekenyan, O., Bonchev, D. and Fritsche, H.-G., Z. Phys. Chem. 265 959 (1984).

[86] Bonchev, D., Mekenyan, O. and Fritsche, H.-G., J. Cryst. Growth, 49, 90 (1980).

[87] Fritsche, H.-G., Bonchev, D. and Mekenyan, O., Cryst. Res. Technol. 18, 1075 (1983).

[88] Basak, S.C., Gieschen, D.P., Magnuson, V.R. and Harriss, D.K., I.R.C.S. Med. Sci. 10, 619 (1982).

[89] Basak, S.C., Gieschen, D.P. and Magnuson, V.R., Env. Toxicol. Chem. 3, 191 (1984).

[90] Basak, S.C., Harriss, D.K. and Magnuson, V.R., J. Pharm. Sci. 73, 429 (1984).

[91] Lall, R.S. and Srivastava, V.K., Math. Chem. 13, 325 (1982).

[92] Barysz, M., Jashari, G., Lall, R.S., Srivastava, V.K. and Trinajstić, N., Studies Phys. Theor. Chem. 28, 222 (1983).

[93] Raychaudhury, C., Ray, S.K., Ghosh, J.J., Roy, A.B. and Basak, S.C., J. Comput. Chem. 5, 581 (1984).

[94] Lall, R.S., Math. Chem. 15, 251 (1984).

[95] Seybold, P.G., Int. J. Quant. Chem., Quant. Biol. Symp. 10, 95 (1983).

[96] Seybold, P.G., Int. J. Quant. Chem., Quant. Biol. Symp. 10, 103 (1983).

[97] Randić, M., Math. Chem. 7, 5 (1979).

Chapter 26

COVARIANT AND CONTRA-VARIANT TRANSFORMATIONS IN CHEMISTRY

L.J. Schaad, B.A. Hess, Jr. and P.L. Polavarapu
Department of Chemistry, Vanderbilt University, Nashville, Tennessee 37235, USA

ABSTRACT

Let $\mathbf{C}$ be the transformation matrix from atomic orbitals to molecular orbitals in an LCAO calculation, and let the contribution of AO j to MO i be the element c_{ij}. In the similar molecular vibration problem one describes the importance of internal motion j to normal mode i not by the corresponding element of the apparently analogous transformation matrix, but by this element of the transpose inverse matrix. The resolution of this seeming discrepancy is understood most clearly by comparing the covariant transformation of a basis in vector space and the contravariant transformation of the coordinates of a point in such a space.

INTRODUCTION

In the molecular orbital (MO) problem one writes [1]

$$\psi = \phi\, \mathbf{C} \tag{1}$$

where ψ is a row matrix of MO's

$$\psi = (\psi_1, \psi_2 \cdots \psi_n), \tag{2}$$

and ϕ is a row matrix of atomic orbitals (AO's)

$$\phi = (\phi_1, \phi_2 \cdots \phi_n). \tag{3}$$

The columns of the square matrix $\mathbf{C}$ are the eigenvectors of the LCAO SCF problem

$$\mathbf{FC} = \mathbf{SC}\boldsymbol{\epsilon} \tag{4}$$

where $\mathbf{F}$ is the Fock matrix, $\mathbf{S}$ the overlap matrix and $\boldsymbol{\epsilon}$ a diagonal matrix of orbital energies.

Molecular vibrational problems are similar in structure with the force constant matrix taking the place of the Fock matrix $\mathbf{F}$ in (4) and the kinetic energy matrix taking the place of $\mathbf{S}$. Solutions of the vibrational eqns. are usually given in the form [2]

$$\mathbf{s} = \mathbf{LQ} \tag{5}$$

where $\mathbf{s}$ is a column matrix of internal coordinates and $\mathbf{Q}$ a column matrix of normal modes. There are two differences between (1) and (5). The more trivial is that the vibrational results are in terms of column matrices rather than row matrices. The less trivial is that (1) transforms from the starting AO's to the solution MO's while (5) transforms from the solution normal modes back to the starting internal coordinates. To put (5) into a form apparently analogous to (1), multiply on the left by $\mathbf{L}^{-1}$ and then take the transpose to give

$$\tilde{\mathbf{Q}} = \tilde{\mathbf{s}}\tilde{\mathbf{L}}^{-1} \tag{6}$$

where $\tilde{\mathbf{Q}}$ and $\tilde{\mathbf{s}}$ are row vectors, and $\tilde{\mathbf{L}}^{-1}$ of (6) takes the place of $\mathbf{C}$ of (1).

The question to be examined here is this: If (1) and (6) are analogous, why is it that if a quantum chemist wants to know the importance of the 5th AO in the 3rd MO (say), he looks at the element in the 5th row of the 3rd column of $\mathbf{C}$, while the vibrational spectroscopist who wants the importance of the 5th internal coordinate in the 3rd normal mode does not look at the element in the 5th row of the 3rd column of $\tilde{\mathbf{L}}^{-1}$, but rather at this element in $\mathbf{L}$? It is true that the MO and vibrational problems are stated in somewhat different form, but as will be seen this does not provide the answer to our question. Instead the key to the problem is that if one makes a transformation in a vector space, the base vectors transform one way (covariantly) while the coordinates of a point in that space transform a different way (contravariantly). If the matrix $\mathbf{A}$ describes the first transformation, $\mathbf{A}^{-1}$ describes the second. In the SCF problem one works directly with the base vectors, but with the coordinates of a point in the vibrational problem.

COVARIANT AND CONTRAVARIANT TRANSFORMATIONS

Consider a vector space with the base vectors $\underline{e}_1$, $\underline{e}_2$, ... $\underline{e}_N$. A point in this space can be specified by its position vector

$$\underline{v} = \sum_{i=1}^{N} x_i \underline{e}_i \; . \tag{7}$$

Two vectors are added by adding corresponding components. A vector can be multiplied by a scalar by multiplying all components by that scalar. Scalar products of two vectors, norms of vectors, the angle between two vectors, and the distance between points in the vector space need not be defined.

Now define a linear non-singular transformation to new base vectors

$$\underline{E}_j = \sum_{i=1}^{N} a_{ij} \underline{e}_i \qquad j=1,2...N \tag{8}$$

or in matrix notation

$$\mathbf{E} = \mathbf{e}\,\mathbf{A} \tag{9}$$

where $\mathbf{E}$ and $\mathbf{e}$ are row matrices and $\mathbf{A}$ is square. Alternatively one could choose $\mathbf{E}$ and $\mathbf{e}$ to be column matrices and write

$$\mathbf{E} = \mathbf{A}\,\mathbf{e} \; . \tag{10}$$

The transformation would then use the rows rather than the columns of $\mathbf{A}$. This choice is not key; let us choose the first alternative. Note that the $\underline{E}_j$ are not normalized since no norm is defined in the vector space. Normalization could be defined, but it is not necessary here.

The position vector $\underline{v}$ can be written in terms of the new base vectors

$$\underline{v} = \sum_{j=1}^{N} X_j \underline{E}_j \; . \tag{11}$$

To find the new coefficients X_j, substitute (8) into (11) and equate to (4).

$$\underline{v} = \sum_j x_j \underline{E}_j = \sum_j x_j \left(\sum_i a_{ij} \underline{e}_i \right)$$

$$= \sum_i \left(\sum_j a_{ij} x_j \right) \underline{e}_i = \sum_i x_i \underline{e}_i. \tag{12}$$

Therefore

$$x_i = \sum_j a_{ij} x_j = \sum_j \tilde{a}_{ji} x_j \tag{13}$$

or

$$\mathbf{X} = \mathbf{X} \tilde{\mathbf{A}} \tag{14}$$

where $\mathbf{X}$ and $\mathbf{X}$ are row matrices. Multiplying (14) on the right by $(\tilde{\mathbf{A}})^{-1} = (\widetilde{\mathbf{A}^{-1}})$ gives

$$\mathbf{X} = \mathbf{X} \tilde{\mathbf{A}}^{-1}. \tag{15}$$

Thus if the base vectors transform as the columns of $\mathbf{A}$ by (9), the coordinates of a point in this vector space transform as the columns of $\tilde{\mathbf{A}}^{-1}$ by (15). The base vectors are said to tranform covariantly and the coordinates contravariantly. Had $\mathbf{E}, \mathbf{e}, \mathbf{X}$ and $\mathbf{X}$ been written as column rather than row vectors, the covariant and contravariant transformations would use rows (rather than columns) of $\mathbf{A}$ and $\tilde{\mathbf{A}}^{-1}$ respectively.

In the special case of an orthogonal transformation $\tilde{\mathbf{A}}^{-1} = \mathbf{A}$ and there is no distinction between covariant and contravariant behavior.

Consider a 3-dimensional example using $\underline{i}, \underline{j}, \underline{k}$ in place of $\underline{e}_1$ ... and x, y, z in place of x_1, ...

Base Vectors (covariant)

$\mathbf{E} = \mathbf{e} \mathbf{A}$	$\mathbf{e} = \mathbf{E} \mathbf{A}^{-1}$
$\underline{I} = 2\underline{i} + \underline{j} + \underline{k}$	$\underline{i} = \underline{I} - \underline{J} + 0\underline{K}$
$\underline{J} = \underline{i} + \underline{j} + \underline{k}$	$\underline{j} = \underline{I} - 2\underline{J} + \underline{K}$
$\underline{K} = 0\underline{i} + 2\underline{j} + \underline{k}$	$\underline{k} = -2\underline{I} + 4\underline{J} - \underline{K}$

$$\tag{16}$$

Coordinates of a point (contravariant)

$$\mathbf{X} = \mathbf{x}\,\tilde{\mathbf{A}}^{-1} \qquad\qquad \mathbf{x} = \mathbf{X}\,\tilde{\mathbf{A}}$$

$$
\begin{array}{lll}
X = & x + & y - 2z \\
Y = & -x - & 2y + 4z \\
Z = & 0x + & y - z
\end{array}
\qquad
\begin{array}{l}
x = 2X + Y + 0Z \\
y = X + Y + 2Z \\
z = X + Y + Z
\end{array}
\qquad (17)
$$

For example, the transformation of base vectors might be thought of as a set of recipes saying that 1 cake (I) is made by combining 2 times 1 cup of sugar ($\underline{i}$), 1 cup of flour ($\underline{j}$) and 1 cup of water ($\underline{k}$); similarly for 1 pie (J) and 1 loaf of bread (K). The contravariant transformations tell how the number of cups of sugar (x), number of cups of flour (y) and number of cups of water (z) are related to the number of cakes (X), number of pies (Y) and number of loaves of bread (Z) in some mixture (caused perhaps by the wreck of the bakery truck).

COMPARISON OF FOUR MATRIX PROBLEMS

The MO problem and the molecular vibration problem are both versions of the same abstract problem of diagonalizing two symmetric (or unitary) matrices simultaneously with the restriction that one of the two matrices be positive definite. Both problems could be worked in exactly parallel ways though it is traditional not to do so. To see that the methods actually used are equivalent we let $\mathbf{A}$ and $\mathbf{B}$ be square symmetric matrices, where $\mathbf{A}$ is positive definite, and examine four formulations of the same problem.

Find a Transformation that Diagonalizes A and B.

Since $\mathbf{A}$ is symmetric there exists a square matrix $\mathbf{U}$ such that

a. $\mathbf{U}$ is orthogonal ($\tilde{\mathbf{U}} = \mathbf{U}^{-1}$)

b. $\tilde{\mathbf{U}}\mathbf{A}\mathbf{U} = \mathbf{D} = [d_i \delta_{ij}]$ (a diagonal matrix).

Define

$$
\mathbf{D}^{\frac{1}{2}} =
\begin{bmatrix}
d_1^{\frac{1}{2}} & & & 0 \\
& d_2^{\frac{1}{2}} & \cdots & \\
& & & d_n^{\frac{1}{2}} \\
0 & & &
\end{bmatrix}
\qquad (18)
$$

and

$$\mathbf{D}^{-\frac{1}{2}} = \begin{bmatrix} d_1^{-\frac{1}{2}} & & & \\ & d_2^{-\frac{1}{2}} & \cdot & \mathbf{0} \\ & & \cdot & \cdot & \\ \mathbf{0} & & & \cdot & d_n^{-\frac{1}{2}} \end{bmatrix} . \tag{19}$$

These have the properties

$$\mathbf{D}^{1/2}\mathbf{D}^{1/2} = \mathbf{D} \quad ; \quad \mathbf{D}^{1/2}\mathbf{D}^{-1/2} = \mathbf{I} \quad ; \quad \mathbf{D}^{-1/2}\mathbf{D}^{-1/2} = \mathbf{D}^{-1} . \tag{20}$$

Let

$$\mathbf{V} = \mathbf{U}\mathbf{D}^{-1/2} \tag{21}$$

so that

$$\tilde{\mathbf{V}}\mathbf{A}\mathbf{V} = \mathbf{D}^{-1/2}\tilde{\mathbf{U}}\mathbf{A}\mathbf{U}\mathbf{D}^{-1/2} = \mathbf{D}^{-1/2}\mathbf{D}\mathbf{D}^{-1/2} = \mathbf{I} . \tag{22}$$

Note that a similarity transformation instead of the congruence transformation of (22) would not work since

$$\mathbf{V}^{-1}\mathbf{A}\mathbf{V} = \mathbf{D}^{1/2}\mathbf{U}^{-1}\mathbf{A}\mathbf{U}\mathbf{D}^{-1/2} = \mathbf{D}^{1/2}\mathbf{D}\mathbf{D}^{-1/2} = \mathbf{D} . \tag{23}$$

Let

$$\mathbf{B}' = \tilde{\mathbf{V}}\mathbf{B}\mathbf{V} . \tag{24}$$

$\mathbf{B}'$ is symmetric since

$$\tilde{\mathbf{B}}' = \tilde{\mathbf{V}}\tilde{\mathbf{B}}\tilde{\tilde{\mathbf{V}}} = \tilde{\mathbf{V}}\mathbf{B}\mathbf{V} = \mathbf{B}' . \tag{25}$$

Therefore there exists a $\mathbf{C}$ such that

 a. $\mathbf{C}$ is orthogonal

 b. $\tilde{\mathbf{C}}\mathbf{B}'\mathbf{C} = \mathbf{\Lambda} = [\lambda_i \delta_{ij}]$ (diagonal).

Further

$$\tilde{\mathbf{C}}\mathbf{B}'\mathbf{C} = \mathbf{\Lambda} = \tilde{\mathbf{C}}\tilde{\mathbf{V}}\mathbf{B}\mathbf{V}\mathbf{C} = \tilde{\mathbf{P}}\mathbf{B}\mathbf{P} \tag{26}$$

where

$$\mathbf{P} = \mathbf{V}\mathbf{C} = \mathbf{U}\mathbf{D}^{-1/2}\mathbf{C} \tag{27}$$

and

$$\tilde{\mathbf{P}}\mathbf{A}\mathbf{P} = \tilde{\mathbf{C}}\tilde{\mathbf{V}}\mathbf{A}\mathbf{V}\mathbf{C} = \tilde{\mathbf{C}}\mathbf{C} = \mathbf{I} . \tag{28}$$

Thus by (26) and (28), a congruence transformation with $\mathbf{P}$ diagonalizes both $\mathbf{A}$ and $\mathbf{B}$. $\mathbf{P}$ is not orthogonal since

$$\mathbf{P}\tilde{\mathbf{P}} = \mathbf{VC}\tilde{\mathbf{C}}\tilde{\mathbf{V}} = \mathbf{UD}^{-1/2}\mathbf{D}^{-1/2}\tilde{\mathbf{U}} = \mathbf{UD}^{-1}\tilde{\mathbf{U}} = \mathbf{A}^{-1}. \tag{29}$$

Hence (26) and (28) are not orthogonal transformations.

Find the Eigenvectors of $A^{-1}B$

Consider the similarity transformation of $\mathbf{A}^{-1}\mathbf{B}$ with $\mathbf{P}$

$$\mathbf{P}^{-1}\mathbf{A}^{-1}\mathbf{B}\mathbf{P} = \mathbf{P}^{-1}\mathbf{A}^{-1}\tilde{\mathbf{P}}^{-1}\tilde{\mathbf{P}}\mathbf{B}\mathbf{P} = (\tilde{\mathbf{P}}\mathbf{A}\mathbf{P})^{-1}(\tilde{\mathbf{P}}\mathbf{B}\mathbf{P})$$
$$= \mathbf{I}^{-1}\mathbf{\Lambda} = \mathbf{\Lambda}. \tag{30}$$

Note that here a congruence transformation with $\mathbf{P}$ would not work

$$\tilde{\mathbf{P}}\mathbf{A}^{-1}\mathbf{B}\mathbf{P} = \tilde{\mathbf{P}}\mathbf{A}^{-1}\tilde{\mathbf{P}}^{-1}\tilde{\mathbf{P}}\mathbf{B}\mathbf{P} = (\tilde{\mathbf{P}}\mathbf{A}\tilde{\mathbf{P}}^{-1})^{-1}\mathbf{\Lambda}. \tag{31}$$

Multiply (30) on the left by $\mathbf{P}$

$$(\mathbf{A}^{-1}\mathbf{B})\mathbf{P} = \mathbf{P}\mathbf{\Lambda} \tag{32}$$

so that the columns of $\mathbf{P}$ are eigenvectors of $\mathbf{A}^{-1}\mathbf{B}$, and problems 1. and 2. are equivalent.

Solve the Generalized Eigenvalue Problem $BP = AP\Lambda$.

Multiply (32) on the left by $\mathbf{A}$ to give the generalized eigenvalue equation

$$\mathbf{B}\mathbf{P} = \mathbf{A}\mathbf{P}\mathbf{\Lambda} \tag{33}$$

so that problems 1., 2. and 3. are equivalent.

Diagonalize Two Quadratic Forms Simultaneously.

Start with the quadratic forms

$$Q_1 = \sum_{i,j} b_{ij} x_i x_j = \mathbf{X}\mathbf{B}\tilde{\mathbf{X}} \tag{34}$$

and

$$Q_2 = \sum_{i,j} a_{ij} x_i x_j = \mathbf{X}\mathbf{A}\tilde{\mathbf{X}} \tag{35}$$

where $\mathbf{X}$ is a row matrix. From (34)

$$Q_1 = \mathbf{x} \mathbf{B} \tilde{\mathbf{x}} = \mathbf{x} \tilde{\mathbf{P}}^{-1} \tilde{\mathbf{P}} \mathbf{B} \mathbf{P} \mathbf{P}^{-1} \tilde{\mathbf{x}}. \tag{36}$$

Let

$$\mathbf{x}' = \mathbf{x} \tilde{\mathbf{P}}^{-1} \tag{37}$$

so that

$$Q_1 = \mathbf{x}' \tilde{\mathbf{P}} \mathbf{B} \mathbf{P} \tilde{\mathbf{x}}' = \mathbf{x}' \mathbf{\Lambda} \tilde{\mathbf{x}}' = \sum_i \lambda_i {x'_i}^2. \tag{38}$$

Similarly

$$Q_2 = \mathbf{x} \mathbf{A} \tilde{\mathbf{x}} = \mathbf{x} \tilde{\mathbf{P}}^{-1} \tilde{\mathbf{P}} \mathbf{A} \mathbf{P} \mathbf{P}^{-1} \tilde{\mathbf{x}}$$

$$= \mathbf{x}' \mathbf{I} \tilde{\mathbf{x}}' = \sum_i {x'_i}^2. \tag{39}$$

Hence transformation with $\tilde{\mathbf{P}}^{-1}$ takes both quadratic forms to canonical form, and problems 1., 2., 3. and 4. are all equivalent.

The x_i's in these quadratic forms are the coordinates of a point in space and transform contravariantly. According to (37) they transform like the columns of $\tilde{\mathbf{P}}^{-1}$. Therefore the columns of $\mathbf{P}$ (the eigenvectors) transform covariantly (like base vectors).

APPLICATION TO THE MOLECULAR ORBITAL AND VIBRATIONAL PROBLEMS

The MO and vibrational problems are of essentially the same form. The matrix $\mathbf{B}$ above becomes the Fock matrix $\mathbf{F}$ in the MO problem and the force constant matrix $\mathbf{F}$ (in internal coordinates) in the vibrational problem. $\mathbf{A}$ becomes the overlap matrix $\mathbf{S}$ in the MO problem and the kinetic energy matrix $\mathbf{G}^{-1}$ in the vibrational case.

In discussions of the MO problem it is usual to use the generalized eigenvalue form (4) and (33), but solutions are usually found by going to the equivalent problem of diagonalizing $\mathbf{F}$ and $\mathbf{S}$ simultaneously.

The vibrational problem is completely analogous, but it is usual to write it in form (32), where $\mathbf{A}^{-1}\mathbf{B}$ is $\mathbf{GF}$ of the Wilson method [3]. This is usually solved by the Miyazawa [4] method which diagonalizes $\mathbf{G}$ (rather than $\mathbf{G}^{-1}$) and $\mathbf{F}$ simultaneously.

In both cases one wants the transformation from old base vectors (AO's or internal coordinates) to new base vectors

(MO's or normal modes). As seen above, these are given by the columns of the modal matrix of eigenvectors. In the MO problem one works directly with the covariant base vectors, the AO's and the MO's, when one writes an expression for the molecular energy. In the vibrational problem one works with the quadratic forms for kinetic and potential energy and, as seen in (34)-(39), with the contravariant coordinates of a point. Eq. (6) therefore describes the transformation of these contravariant coordinates and is <u>not</u> analogous to (1) which gives a transformation between covariant base vectors. The transformation between base vectors in the vibrational problem is given by the transpose inverse of $\mathbf{L}^{-1}$, i.e. by $\mathbf{L}$ itself.

In summary, in the MO problem one works with a transformation between the covariant AO and MO base vectors. In the vibrational problem one works instead with the transformation from the contravariant coordinates of a point in normal mode space to those in internal coordinate space. Then in reporting the results one takes the inverse transpose of this transformation to give the proper relation between the internal coordinate and normal mode base vectors.

REFERENCES

[1] C. C. J. Roothaan, Revs. Mod. Phys., <u>23</u>, 69 (1951).
[2] S. Califano, "Vibrational States," John Wiley and Sons, 1976, Chapters 2 and 4.
[3] E. B. Wilson, Jr., J. C. Decius and P. C. Cross, "Molecular Vibrations," McGraw-Hill, 1955, Chapt. 4.
[4] T. Miyazawa, J. Chem. Phys., <u>29</u>, 246 (1958).

Chapter 27

ON ALGEBRAIC AND COMPUTATIONAL ASPECTS OF ISOMERIC-CHEMISTRY

Z. Slanina
The J. Heyrovský Institute of Physical Chemistry and Electro-chemistry,
Czechoslovak Academy of Sciences, CS-121 38 Prague 2, Czechoslovakia

ABSTRACT

Algebraic chemistry as well as more conventional computational methods are at present frequently applied to problems of isomeric chemistry. For example, with evaluation of reaction graphs a reliable estimation of equilibrium and, particularly, rate constants is required. The contemporary possibilities of the joint quantum-chemical and statistical-thermodynamical treatment for evaluation of equilibrium and rate constants and its combination with algebraic approach to synthesis design are surveyed.

INTRODUCTION

The concept of isomerism, introduced into chemistry under this name by Berzelius /1/, is continually being tested and broadened. New types of isomerism are being recognized and even very simple compounds (e.g. /2-4/, ClO_2, N_2O_4, H_2SO_4, or the smallest amino acid, glycine /5/) can be observed (or expected) to exhibit isomerism. This procedure is assisted considerably by the theoretical approaches that at present enable the discovery of new isomers (e.g. the closed form of ozone /6/ or the prediction of hitherto unforseen reactions within an isomeric system (e.g., rearrangements /7,8/). A number of chemically bizarre isomers of small organic molecules has recently been characterized in the study of interstellar species (for a review, see /9/).

The oldest theoretical means for studying the phenomenon of isomerism were algebraic methods, used for isomer enumeration (for
reviews, see /10-15/); they are at present once again the subject
of considerable interest, e.g. /16-75/. In addition to enumeration, algebraic techniques (set, group, and graph as well as information theory) are used especially for classification purposes - e.g. /76-80/. The algebraic generalization of the notion
of isomerism was formulated for computer-assisted design of syntheses (e.g., /81-85/). Later the theoretical techniques were
enriched by addition of quantum- and statistical-mechanical,
quantum-chemical and molecular mechanics methods. In the two
latter cases, the individual isomers and the relations between
them are treated as minima and their interrelations on the corresponding potential energy hypersurface(s), e.g., /86-90/.
Description of an isomeric system in terms of local energy minima and saddle points usually represents the present state-of-
-the-art limit in theoretical studies. Hundreds of systems (for
a review, see /91/) have been computationally characterized in
this way.

GENERALIZED ISOMERISM, REACTION GRAPHS, AND NON-EMPIRICAL DESIGN OF SYNTHESES

Ugi et al. /76, 81-83/ suggested to treat chemical isomerism as
an equivalence relation. The equivalence relation of isomerism
can further be generalized by transition from the individual molecules to ensembles of molecules. Consider a particular set of
atoms and form individual compounds or ensembles of compounds so
that all the atoms are employed. Each of these atomic arrangements thus represents a single isomeric ensemble of molecules
and all these ensembles form a family of isomeric ensembles of
molecules. This concept represents a generalization of the equivalence relation from molecules to ensembles of molecules.
The equality introduced by the equivalence relation of the isomerism of the ensembles of molecules between the individual members of the family of isomeric ensembles of molecules has a clear
quantum-mechanical justification. All the members of a given family of isomeric ensembles of molecules correspond to the same
total Hamiltonian. In the framework of the Born-Oppenheimer approximation, multi-membered ensembles of molecules are included
in the potential energy hypersurface as regions of dissociation
products at infinity. The use of the equivalence relation in
this connection is related /91/ to the uncertainty principle.
The possibility of obtaining detailed information concerning
systems is limited by this principle regardless of further improvements in instrument precision. Thus it was suggested /91/
the generation of models of chemical systems that represents
certain equivalence classes of states rather than the states
themselves.
Generalized isomerism provides a useful tool for computer-assisted synthesis design as every chemical reaction can be interpreted as an interconversion between two isomeric ensembles of mo-

lecules /81-83/. Planning of syntheses consists of two phases: first, pathways are sought that generally lead from the readily available starting materials to the target molecule T; then, from among these conceivable pathways, the optimal one is selected on the basis of certain selection criteria. In terms of generalized isomerism this implies the establishment of a family of isomeric ensembles of molecules that contain the target molecule T in at least one ensemble and the initial materials of the synthesis in another ensemble. The synthesis design then implies finding a set of pathways connecting the ensemble of molecules including T with the ensemble of molecules containing the starting material. In the solution of the problem of synthesis design, three levels can be distinguished /81/: non-empirical, semi-empirical, and empirical.

The non-empirical approach to synthesis design assumes the construction of a family of isomeric ensembles of molecules that includes all the chemical compounds that must be considered in the solution of the given problem. Then all of the pathways connecting the ensemble of the starting material with the target molecule ensemble are sought. A non-empirical (topological) program based on these concepts generates a complete set of synthetic pathways for a given target molecule and indiscriminately incorporates known and unknown chemical reactions into the synthetic pathway generated. The non-empirical approach is, of course, limited /81/ to purely topological relationships and does not include the selection of in some sense optimal synthetic pathways from the topologically posible ones. Within the framework of the purely non-empirical approach the selection can be carried out /83/ using the concept of the shortest reaction pathway introduced in terms of chemical distance /83/. However, it is apparent that all the proposed topologically possible pathways, or rection graphs, should be classified in terms of a reasonable yield and reasonable rate. From the most general point of view, the rate constant for each elementary step in every particular reaction graph (even in both directions of the step) should be available for such a classification. To keep the non--empirical character of the synthesis design, the sets of rate constants should also be derived non-empirically, i.e. on the basis of quantum-chemical and statistical-thermodynamical methods.

QUANTUM-CHEMICAL AND STATISTICAL-THERMODYNAMICAL EVALUATION OF EQUILIBRIUM AND RATE CONSTANTS

The present theory of chemical reactivity is entirely based /92, 93/ on representation of energy hypersurfaces by means of their stationary points. Location and identification of these points is however only the first step with calculations of the characteristics of equlibrium and rate processes. For this purpose it is necessary to link effectively both quantum chemistry and statistical thermodynamics, i.e. two fields of science traditionally somewhat disconnected /94,95/. Recent comprehensive studies /96,

97/ testing the applicability of quantum-chemical methods as
sources of molecular data for evaluation of partition functions
have been successful. The used RRHO approximation /93/ of partition functions does not seem to depreciate the quality of the
calculated reactivity characteristics. Thus, it has become possible in the case of the calculation of thermodynamic functions to
replace the molecular parameters which are conventionally derived from (usually spectroscopic) experiment or merely estimated
by those obtained from theoretical calculations.
The same holds for the evaluation of rate characteristics /93/
by means of the activated-complex theory. In fact the linking
up of the activated-complex theory with quantum-chemical methods has brought the theory to its renaissance. Besides the generation of reliable characteristics of activated complexes
which follows from this symbiosis, one more thing is contribu-
ting to its boom: test studies comparing in case of very simple
rate systems the values of characteristics of the rate processes
obtained on the basis of the activated-complex theory and on the
basis of exact quantum-mechanical calculations have shown surprisingly good agreement /93/. Thus, a period of evaluating reaction characteristics (standard as well as activation ones)
completely independently of experimental information (with the
exception of masses, fundamental physical constants, and the
form of Coulomb's law) has begun. Characteristics of a large
number of equilibrium and/or rate processes in the gas phase have been calculated in this manner /93,94/. Hence, it is also
meaningful to consider, at least in principle, a possibility to
generate all the necessary elementary rate constants for the
classification of all the topologically possible synthesis paths
(and, thus, for a selection of an optimal pathway) exclusively
from the joint quantum-chemical and statistical-thermodynamical
treatment.

ISOMERISM OF REACTION COMPONENTS OF EQUILIBRIUM AND RATE PROCESSES

Systematic investigation of a potential energy hypersurface often reveals several different local energy minima all represented by one species in an experiment and/or several different
saddle points corresponding to activated complexes in a single
rate process. Potential energy criteria can sometimes prove only
one structure to play an important role. However, it may also
happen that two or more isomeric structures of comparable stability coexist and are indistinguishable under given experimental conditions. Then any structure-dependent observable can be
considered as an average value resulting from contributions of
all the isomers in question. In view of this reaction components
isomerism, a new class of generalized chemical equilibria (viz.
equilibria of which each component is a mixture of isomers /98/)
as well as generalized rate processes with parallel /99,100/
or sequential /101,102/ isomerism of activated complexes have
been introduced. Several examples of considerable differences

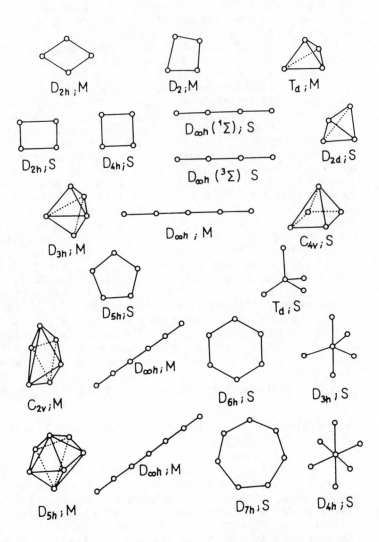

Figure 1. Schemes of stationary points found /103/ on the MINDO/2 potential energy hypersurfaces of C_n (n = 4-7); M is a minimum, S a saddle point (courtesy of American Chemical Society).

Table 1. Partial and Total MINDO/2 Standard Enthalpy ΔH_T^o and Entropy ΔS_T^o Terms of 4C(s) $\rightleftharpoons$ C$_4$(g) Equilibrium at T = 2400 K^a

Process	ΔH_T^o (kJ/mol) ^{b}	ΔS_T^o (J/K/mol) ^{b}
4C(s) $\rightleftharpoons$ C$_4$(g; D$_{2h}$)	904.1	210.2
4C(s) $\rightleftharpoons$ C$_4$(g; D$_2$)	952.2	213.6
4C(s) $\rightleftharpoons$ C$_4$(g; T$_d$)	1301.5	205.5
4C(s) $\rightleftharpoons$ C$_4$(g; total)	909.9	213.6

^{a}According to Ref. /103/.
^{b}Standard state: ideal gas at 101325 Pa pressure.

between one-isomer and multiple-configuration equilibrium or rate characteristics have been reported, e.g. /103-108/.
In contemporary quantum-chemical practice, a special case of the general reaction components isomerism is frequently met, viz. processes where theory has demonstrated that only a single component is a mixture of more (n) isomers. Then weighting treatment for a quantity ΔX can be simply expressed by:

$$\Delta X = f(\Delta X_i, w_i, T); \qquad (i=1, 2, \ldots, n), \tag{1}$$

where ΔX_i denotes the quantity corresponding to the process considered, however, realized through the i-th isomer, and w_i denotes the weighting factor of the isomer related at temperature T by:

$$w_i = \frac{q_i \exp(-e_o^{(i)}/kT)}{\sum_{j=1}^{n} q_j \exp(-e_o^{(j)}/kT)} \tag{2}$$

to the partition function, q_i, of the i-th isomer and to its ground-state energy, $e_o^{(i)}$. The most interesting quantities in the weighting treatment are the enthalpy and entropy changes. As an instructive example, the formation /103/ of C$_n$(g) aggregates can be presented. For each of the C$_n$ aggregates for n=4-7 the MINDO/2 calculations /103/ demonstrated the existence of at least two isomeric structures (Figure 1). The formation of C$_n$(g) aggregates was studied experimentally by mass spectrometry, i.e. the technique not distinguishing among isomers. While the experiment thus yielded the overall thermodynamic characteristics of the formation, the theory led primarily to partial values characterizing the formation of the individual isomers. Application of the weighting treatment for correct comparison of the theory and experiment was justified here as the high temperatures at which these experiments were carried out formed favourable con-

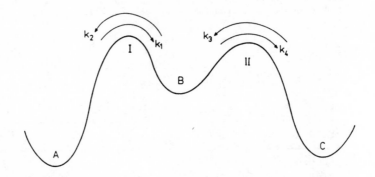

Figure 2. Reaction potential profile for a rate process A→C realized through one intermediate B, i.e. the process exhibiting double sequential isomerism of activated complexes (I,II).

ditions for attaining inter-isomeric equilibria. The weighting is illustrated in Table 1 on the C_4 system, for which the MINDO/2 approach predicted the existence of three minimum energy structures all with different point groups of symmetry (D_{2h}, D_2, T_d). It is evident that representation of the overall process merely by partial terms belonging to the most stable structure (D_{2h}) could be misleading. This is especially true for the entropy term: the change resulting from isomerism is of the same order as possible errors introduced by the use of the MINDO/2 molecular parameters instead of exact ones.

The isomerism of reaction components in chemical equilibria is particularly important for the correct comparison of theory and experiment and for prediction of equilibrium behaviour. For the synthesis design, however, the sequential isomerism of activated complexes is of a primary importance.

THE OVERALL, EFFECTIVE RATE CONSTANT FOR A SYNTHESIS PATHWAY

Let us start with a simple situation: a single local minimum, i.e. an intermediate, separating two activated complexes lying on a common pathway from the reactant to the product - see Figure 2. At an elementary rate constant level, the kinetics of the system is essentially described by the elementary rate constants k_i for the four partial rate processes involved. However, we can also be interested in an overall kinetics of the process A→C, and we can evaluate an effective rate constant of this complex process. In Refs. /101,102/ several such effective, overall rate constants were derived, for example:

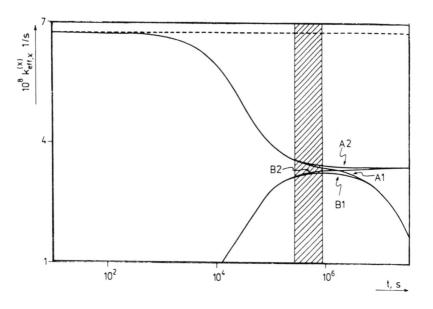

Figure 3. Time dependences of the effective rate constants for the boat pathway of the Cope rearrangement of 1, 5-hexadiene at 532.15 K treated as a process with double sequential activated-complex isomerism; for detailed description of the individual curves – see Ref. /102/ (courtesy of John Wiley & Sons, Inc.)

$$k_{eff} = \frac{k_1 k_4}{u_1 u_2 t} \ln \frac{u_2 - u_1}{u_2 \exp(-u_1 t) - u_1 \exp(-u_2 t)} , \qquad (3)$$

where:

$$u_{1,2} = \frac{1}{2}(k_1 + k_2 + k_3 + k_4 \mp ((k_1 + k_2 + k_3 + k_4)^2 - 4(k_1 k_3 + k_1 k_4 + k_2 k_3))^{\frac{1}{2}}). \qquad (4)$$

Originally, we considered four (time-independent) rate constants k_i, now we have one, however, time-dependent effective rate constant k_{eff}. Its time dependence can be very pronounced - cf. Figure 3. Instead of this effective rate constant in a time t we can alternatively use the actual concentration of the product in that time:

$$c_C(t) = c_A(0) \frac{k_1 k_4}{u_1 u_2 (u_1 - u_2)} (u_2 \exp(-u_1 t) - u_1 \exp(-u_2 t) + u_1 - u_2), \qquad (5)$$

where $c_A(0)$ designates the initial concentration of reactant A.

Let us discuss now a more complex picture: there is altogether n-1 intermediates between the reactant and the product, i.e. the sequential isomerism of activated complexes of order n. The system is now kinetically described by two n-membered sets of elementary rate constants. Let us have a set of such kinetic systems between the same reactant and product. To decide which one of these systems is the most convenient from the kinetic point of view, a one-dimensional representation of the system would be desirable. For the purpose we can again derive an effective, overall rate constant as a generalization of Eq. (3), i.e. a complicated function of these 2n elementary rate constants and of time. The system with the highest value of the effective rate constant in a chosen time may be considered as the optimal at that time. Alternatively, we can work on the level of the time dependences of the product concentrations and to consider the system with the highest value of the concentration in a chosen time as the optimal in that time. Clearly enough, in either of these criteria as well as in different time regions different systems can be found as the most convenient. The above effective rate constant may represent a proposal how to classify individual synthesis pathways in a one-dimensional representation in spite of many elementary rate constants involved. Of course, a real synthetic pathway is composed not from isomerizations only and, moreover, there are time discontinuities in the synthesis pathways (isolation of a synthesis intermediate). Thus, the one-dimensional representations of synthetic pathways in kinetic terms still represent an open problem. Summarizing, a really non-empirical synthesis design consists of three (non-trivial) steps: (i) generation of all topologically possible pathways, (ii) evaluation of all the elementary rate constants involved using the joint quantum-chemical and statistical-thermodynamical treatment, and finally (iii) classification of all the topologically possible pathways according to a (one-dimensional) kinetic criterion.

CONCLUDING REMARKS

The recent development of algebraic (or mathematical) chemistry continuosly demonstrates that the use of algebraic methods is gradually becoming a useful, powerful complement to more conventional computational methods of theory of chemical reactivity. Isomeric chemistry represents a pregnant example of the field where combined application of these theoretical approaches is relevant and fruitful. The algebraic study of isomerism can, e.g., improve the procedures for determining the numbers of stationary points on potential hypersurfaces, their classification and mapping relationships between them. However, the usefulness of algebraic predictions most certainly depends on the degree how the physical reality is included in the mathematical model used. In every application a delicate balance between generality of the mathematical model and a sufficient retention of physical or chemical reality should be respected.

REFERENCES

1. J. J. Berzelius, Ann. Phys. Chem. 19 (der ganzen Folge 95), 305 (1830).
2. S. W. Benson and J. H. Buss, J. Chem. Phys. 27, 1382 (1957).
3. F. Bolduan and H. J. Jodl, Chem. Phys. Lett. 85, 283 (1982).
4. R. L. Kuczkowski, R. D. Suenram, and F. J. Lovas, J. Am. Chem. Soc. 103, 2561 (1981).
5. R. D. Suenram and F. J. Lovas, J. Am. Chem. Soc. 102, 7180 (1980).
6. J. S. Wright, Can. J. Chem. 51, 139 (1973).
7. A. T. Balaban, D. Fărcaşiu, and R. Bănică, Rev. Roum. Chim. 11, 1205 (1968).
8. A. T. Balaban, Rev. Roum. Chim. 12, 875 (1967).
9. S. Green and E. Herbst, Astrophys. J. 229, 121 (1979).
10. D. H. Rouvray, Chem. Soc. Rev. 3, 355 (1974).
11. D. H. Rouvray, Endeavour 34, 28 (1975).
12. R. C. Read, in "Chemical Applications of Graph Theory", Ed. A. T. Balaban, Academic Press, London 1976.
13. A. T. Balaban, in "Chemical Applications of Graph Theory", Ed. A. T. Balaban, Academic Press, London 1976.
14. H. Frei, A. Bauder, and H. Hs. Günthard, Top. Curr. Chem. 81, 1 (1979).
15. J. V. Knop, W. R. Müller, K. Szymanski, and N. Trinajstić, "Computer Generation of Certain Classes of Molecules", SKTH, Zagreb 1985.
16. "Symmetries and Properties of Non-rigid Molecules", Eds. J. Maruani and J. Serre, Elsevier, Amsterdam 1983.
17. "Chemical Applications of Topology and Graph Theory", Ed. R. B. King, Elsevier, Amsterdam 1983.
18. N. Trinajstić, "Chemical Graph Theory", Vols. I, II, CRC Press, Boca Raton 1983.
19. Z. Slanina, "Contemporary Theory of Chemical Isomerism", D. Reidel, Dordrecht, 1985.
20. A. T. Balaban, Match 1, 33 (1975).
21. D. H. Rouvray and A. T. Balaban, in "Applications of Graph Theory", Eds. R. J. Wilson and L. W. Beineke, Academic Press, New York 1979.
22. A. T. Balaban, Theor. Chim. Acta 53, 355 (1979).
23. A. T. Balaban, Chem. Phys. Lett. 89, 399 (1982).
24. A. T. Balaban, Pure Appl. Chem. 54, 1075 (1982).
25. K. Balasubramanian, Theor. Chim. Acta 53, 129 (1979).
26. K. Balasubramanian, Int. J. Quantum Chem. 22, 385 (1982).
27. K. Balasubramanian, Theor. Chim. Acta 61, 307 (1982).
28. K. Balasubramanian, Theor. Chim. Acta 65, 49 (1984).
29. K. Balasubramanian, J. Comput. Chem. 5, 387 (1984).
30. J. R. Dias, J. Chem. Info. Comput. Sci. 22, 15 (1982).
31. J. R. Dias, J. Chem. Info. Comput. Sci. 22, 139 (1982).
32. J. R. Dias, Match 14, 83 (1983).
33. J. R. Dias, J. Chem. Info. Comput. Sci. 24, 124 (1984).
34. J. R. Dias, Carbon 22, 107 (1984).
35. R. B. King, Theor. Chim. Acta 59, 25 (1981).

36. R. B. King, Theor. Chim. Acta 63, 103 (1983).
37. R. B. King, Theor. Chim. Acta 63, 323 (1983).
38. R. B. King, Theor. Chim. Acta 64, 439 (1984).
39. R. B. King, Theor. Chim. Acta 64, 453 (1984).
40. W. G. Klemperer, J. Chem. Phys. 56, 5478 (1972).
41. W. G. Klemperer, J. Am. Chem. Soc. 94, 6940 (1972).
42. W. G. Klemperer, J. Am. Chem. Soc. 94, 8360 (1972).
43. W. G. Klemperer, J. Am. Chem. Soc. 95, 380 (1973).
44. W. G. Klemperer, J. Am. Chem. Soc. 95, 2105 (1973).
45. P. G. Mezey, Chem. Phys. Lett. 82, 100 (1981).
46. P. G. Mezey, Int. J. Quantum Chem., Quantum Biol. Symp. 8, 185 (1981).
47. P. G. Mezey, Int. J. Quantum Chem., Quantum Biol. Symp. 10, 153 (1983).
48. P. G. Mezey, Theor. Chim. Acta 63, 9 (1983).
49. P. G. Mezey, Can. J. Chem. 62, 1356 (1984).
50. M. Randić, J. Chem. Info. Comput. Sci. 15, 105 (1975).
51. M. Randić, J. Chem. Info. Comput. Sci. 17, 171 (1977).
52. M. Randić, Int. J. Quantum Chem. 15, 663 (1979).
53. M. Randić, J. Comput. Chem. 4, 73 (1983).
54. M. Randić and M. I. Davis, Int. J. Quantum Chem. 26, 69 (1984).
55. D. H. Rouvray, R. I. C. Rev. 4, 173 (1971).
56. D. H. Rouvray, Am. Scientist 61, 729 (1973).
57. D. H. Rouvray, Chem. Tech. 3, 379 (1973).
58. D. H. Rouvray, Chem. Brit. 10, 11 (1974).
59. D. H. Rouvray, Chem. Brit. 13, 52 (1977).
60. O. Sinanoğlu, J. Am. Chem. Soc. 97, 2309 (1975).
61. O. Sinanoğlu, Chim. Acta Turc. 3, 155 (1975).
62. O. Sinanoğlu and L.-S. Lee, Theor. Chim. Acta 51, 1 (1979).
63. O. Sinanoğlu, J. Math. Phys. 22, 1504 (1981).
64. O. Sinanoğlu, Chem. Phys. Lett. 103, 315 (1984).
65. A. Graovac, I. Gutman, and N. Trinajstić, "Topological Approach to the Chemistry of Conjugated Molecules", Springer-Verlag, Berlin 1977.
66. D. Bonchev and N. Trinajstić, J. Chem. Phys. 67, 4517 (1977).
67. O. Mekenyan, D. Bonchev, and N. Trinajstić, Int. J. Quantum Chem. 18, 369 (1980).
68. D. Bonchev, Ov. Mekenyan, and N. Trinajstić, J. Comput. Chem. 2, 127 (1981).
69. J. V. Knop, W. R. Müller, Ž. Jeričević, and N. Trinajstić, J. Chem. Info. Comput. Sci. 21, 91 (1981).
70. B. Džonova-Jerman-Blažič and N. Trinajstić, Comput. Chem. 6, 121 (1982).
71. J. V. Knop, K. Szymanski, Ž. Jeričević, and N. Trinajstić, J. Comput. Chem. 4, 23 (1983).
72. M. Barysz, N. Trinajstić, and J. V. Knop, Int. J. Quantum Chem., Quantum Chem. Symp. 17, 441 (1983).
73. N. Trinajstić, Ž. Jeričević, J. V. Knop, W. R. Müller, and K. Szymanski, Pure Appl. Chem. 55, 379 (1983).
74. S. El-Basil, P. Křivka, and N. Trinajstić, Croat. Chem. Acta 57, 339 (1984).

75. R. F. W. Bader, J. Chem. Phys. 73, 2871 (1980).
76. I. Ugi, D. Marquarding, H. Klusacek, G. Gokel, and P. Gillespie, Angew. Chem., Int. Ed. Engl. 9, 703 (1970).
77. G. Ege, Naturwissenschaften 58, 247 (1971).
78. W.Hässelbarth and E. Ruch, Theor. Chim. Acta 29, 259 (1973).
79. K. Mislow, Bull. Soc. Chim. Belg. 86, 595 (1977).
80. J. G. Nourse, J. Am. Chem. Soc. 102, 4883 (1980).
81. I. Ugi, P. Gillespie, and C. Gillespie, Trans. N. Y. Acad. Sci. 34, 416 (1972).
82. J. Dugundji and I. Ugi, Fortschr. Chem. Forsch. 39, 19 (1973).
83. I. Ugi, J. Bauer, J. Brandt, J. Friedrich, J. Gasteiger, C. Jochum, and W. Schubert, Angew. Chem., Int. Ed. Engl. 18, 111 (1979).
84. V. Kvasnička, Collect. Czech. Chem. Commun. 48, 2097, 2118 (1983).
85. V. Kvasnička, M. Kratochvíl, and J. Koča, Collect. Czech. Chem. Commun. 48, 2284 (1983).
86. J. W. McIver, Jr. and A. Komornicki, J. Am. Chem. Soc. 94, 2625 (1972).
87. O. Ermer, Structure and Bonding 27, 161 (1976).
88. J. N. Murrell, Structure and Bonding 32, 93 (1977).
89. J. N. Murrell, in "Quantum Theory of Chemical Reactions", Vol. 1, Eds. R. Daudel, A. Pullman, L. Salem, and A. Veillard, D. Reidel, Dordrecht 1980.
90. J. N. Murrell, S. Carter, S.C. Farantos, P. Huxley, and A. J. C. Varandas, "Molecular Potential Energy Functions", J. Wiley, Chichester 1984.
91. J. Dugundji, P. Gillespie, D. Marquarding, I. Ugi, and F. Ramirez, in "Chemical Applications of Graph Theory", Ed. A. T. Balaban, Academic Press, London.
92. S. Beran, P. Čársky, P. Hobza, J. Pancíř, R. Polák, Z. Slanina, and R. Zahradník, Usp. Khim. 47, 1905 (1978).
93. Z. Slanina, Advan. Quantum Chem. 13, 89 (1981).
94. Z. Slanina, Radiochem. Radioanal. Lett. 22, 291 (1975).
95. E. Clementi, Bull. Soc. Chim. Belg. 85, 969 (1976).
96. Z. Slanina, P. Berák, and R. Zahradník, Collect. Czech. Chem. Commun. 42, 1 (1977).
97. M. J. S. Dewar and G. P. Ford, J. Am. Chem. Soc. 99, 7822 (1977).
98. Z. Slanina, Collect. Czech. Chem. Commun. 40, 1977 (1975).
99. Z. Slanina, Collect.Czech. Chem. Commun. 42, 1914 (1977).
100. Z. Slanina, Chem. Phys. Lett. 50, 418 (1977).
101. Z. Slanina, Z. Phys. Chem. (Wiesbaden) 132, 41 (1982).
102. Z. Slanina, Int. J. Quantum Chem. 23, 1553 (1983).
103. Z. Slanina and R. Zahradník, J. Phys. Chem. 81, 2252 (1977).
104. Z. Slanina, Chem. Phys. Lett. 52, 117 (1977).
105. Z. Slanina, Advan. Mol. Relax. Interact. Process. 14, 133 (1979).
106. Z. Slanina, Int. J. Quantum Chem. 16, 79 (1979).
107. Z. Slanina, Chem. Phys. Lett. 82, 33 (1981).
108. Z. Slanina, Thermochim. Acta 78, 47 (1984).

Chapter 28

ON THE REDUCED GRAPH MODEL

Nenad Trinajstić* and Pavel Křivka
*The Rugjer Bošković Institute, P.O.B. 1016, 41001 Zagreb, Yugoslavia and Department of Mathematics, Higher School of Chemical Technology, 53210 Pardubice, Czechoslovakia

ABSTRACT

The reduced graph model is discussed and applied to the enumeration of Kekulé structures for several classes of benzenoid hydrocarbons.

INTRODUCTION

The reduced graph model has been introduced as an alternative way to represent benzenoid-type networks (1,2). This model has been shown to be very useful in the combinatorial problems of benzenoid systems (1-5) such as the enumeration and generation of Kekulé structures, the enumeration and generation of conjugated circuits, the counting of all benzenoid hydrocarbons for a given number of benzene rings, the construction of the sextet polynomials, etc.

Here we wish to clarify the graph theoretical aspects of the reduced graph model and give all necessary definitions. We will test this model on the enumeration of Kekulé structures of benzenoid hydrocarbons. The problems of enumeration (i.e. the production of the total number) and display (i. e. the construction of all the perfect matchings) of Kekulé structures are continuously being discussed in the literature (1,3,4,6-23). In addition, recent interest in Kekulé structures has been generated by their use in structure-resonance theory (24), in the conjugated circuits model (25), in valence bond resonance energy calculations (26), in the valence bond model using only significant valence structures (27,28), in molecular orbital resonance theory (29), in various valence bond calculations (30-33), in the unified valence bond theory (34), etc. Finally, Kekulé structures are important in understanding the mathematical basis for the intimate connection between Pauling's VB model and Hückel's MO model (29,35), and in the history of modern chemistry, in which they play a significant role (36-38).

REDUCED GRAPH MODEL

Let G be a connected graph (structure) in an infinite hexagonal planar lattice H. The infinite hexagonal lattice H is a planar bipartite infinite 3-regular graph. The number of vertices in a graph which is 3-regular is always even (39). Three disjunctive sets of parallel edges arranged in rows are present in the lattice H. We can arbitrarily choose one of these three sets and call it vertical and the remaining two we can denote (in two different ways) as left or right diagonal. The hexagonal lattice H with edges denoted as vertical, left diagonal, and right diagonal is called the oriented lattice H. Horizontal rows of hexagons in the oriented lattice H are called levels of lattice H.

The oriented lattice H may be transformed into the trigonal planar lattice T according to the following transformation:

$$V(T) = \{\text{vertical edges of } H\}$$

$$E(T) = \{(v_1, v_2)| \quad \text{either}$$

v_1, v_2 belong to the same ring in H or v_1, v_2 are connected by a diagonal edge $\}$.

In accordance with the orientation of H, we can distinguish three disjunctive sets of edges in T: horizontal, left diagonal, and right diagonal. Horizontal rows of vertices in T are called horizontal levels of T. By this transformation each graph G in H is transformed into a graph R(G) that is a part of T and is called a reduced graph.

One of the uses of the reduced graph model is in the enumeration of the 1-factors of G. 1-factor of G is a graph F such that:

(a) F is a spanning subgraph of G

(b) $V(F) = V(G)$

(c) The components of F are only K_2 graphs, i.e. F is a 1-regular graph.

1-factorization is construction of 1-factors. 1-factors are isomorphic to Kekulé graphs (40) which are employed to depict Kekulé structures of benzenoid hydrocarbons.

The carbon skeletons of benzenoid hydrocarbons are graph-theoretically represented by benzenoid graphs (41-43) and are denoted here by G. A benzenoid graph is a bipartite planar graph which can be constructed in the plane by assembling h regular hexagons in such a way that two hexagons have exactly one common edge or are disjoint.

Below we demonstrate the transformation of a hexagonal network H into a trigonal network T, and simultaneous changing the representation of a given benzenoid hydrocarbon from the benzenoid graph to the reduced graph (see Example 1).

Example 1

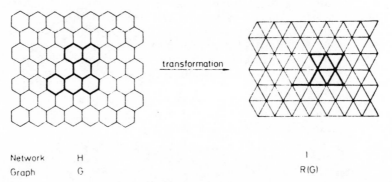

transformation

Network   H |
Graph   G   R(G)

We will now investigate the 1-factors of graph G in the oriented lattice H. It is clear that whenever we determine which of the vertical edges belong to the 1-factors and which do not, this completely determines the assignment of all other (diagonal) edges. We need therefore work only with the vertices of the reduced graph R(G). If we have a horizontal edge in R(G) with no triangle above (and/or below), we can purely formally add above (and/or below) a vertex, and thus create an upper (and/or lower) triangle. After doing the above whenever necessary, we obtain the <u>complete reduced graph</u> of G, CR(G). In CR(G), the <u>set of all added vertices</u> to R(G) to form upper triangles is denoted by UT and the set of all added vertices to R(G) to form lower triangles is denoted by LT, respectively (see <u>Example 2</u>).

Example 2

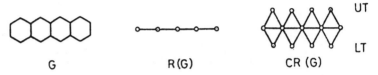

$$G \qquad\qquad R(G) \qquad\qquad CR\ (G)$$

A sequence of adjacent vertices in T, connected only with diagonal edges, is called the <u>vertical path</u>. We will not require any other paths in T.

The following basic theorem gives the necessary and sufficient condition for graph G to have a 1-factor.

THEOREM

Let G be a graph in the oriented lattice H, CR(G) the complete reduced graph of G, and UT (and LT) the set of all added vertices to R(G) to form upper (and lower) triangles. Then G has a 1-factor if, and only if, there exists a 1-1 mapping f: UT $\leftrightarrow$ LT such that for each $u \in$ UT there exists a path to $f(u) \in$ LT and these paths are pairwise disjunctive, i.e. no two of them have a common vertex. (A similar statement can be found in Ref. 8 supported only by pictorial representation for case UT = 1).

Proof

Let graph G have a 1-factor K and let $S = \{ v \in R(G) \mid v \notin E(K) \}$. We will prove that the set $P = S \cup UT \cup LT$ uniquely determines a collection of paths in CR(G) with the required properties. The main part of the proof is formulated by the following proposition: In a given row in CR(G) let there be n_1 vertices from UT, n_2 from S, and n_3 from LT. In the row below it let there be $\bar{n}_1$ vertices from UT, $\bar{n}_2$ from S, and $\bar{n}_3$ from LT. Then, $n_1 + n_2 = \bar{n}_2 + \bar{n}_3$, and each vertex of UT $\cup$ S from the row above is connected with exactly one vertex of S $\cup$ LT from the row below.

Proof of the proposition: We do not need to consider vertices of LT from the row above, since these are clearly terminal ones. Thus, we have several sequences of vertices from UT $\cup$ S, each

separated by vertices belonging to K or totally
disconnected. Such a sequence corresponds to the
following situation in G:

where some edges in the first row may be missing.
Point A_1 (A_2) is either peripheral (sequences are
disconnected) or the third edge of A_1 belongs to
S (sequences are separated by vertices belonging to
K). Let us consider point B_1 - vertical edge is
either missing (UT) or does not belong to K(S) ,
i.e. one of the two adjacent diagonal edges must
belong to K. This is true for all vertices B_1, B_2,.
...,B_n and as two diagonal edges belonging to K
cannot meet at one point, we have only one possibility for the change of direction, namely

All left diagonal edges before B_i and all right
diagonal edges after B_{i+1} must belong to K, i.e. in
the row below there can be only one vertical edge
belonging to K, namely the edge between B_i and B_{i+1}.
If either A_1 or A_2 (but not both) is missing (i.
e. B_1 or B_n are peripheral), then all the vertical
edges in the second row must belong to S$\cup$LT. This
completes the proof of the proposition.
    The proposition shows that we can construct the
desired paths, row after row, starting at the up-
permost level. Since we can reverse the direction
(topsy-turvy position), it is clear that |UT | =|LT |.
    Now, let there be in CR(G) an n-tuple of paths
of the desired properties. Let us denote by S the
set of all vertices of this n-tuple, and let
K' = V(CR(G))\S. We will complete K' up to the 1-
factor K of G: A diagonal edge belongs to K if it
connects two vertical edges that are vertices of
some path in CR(G). Then it is clear that the set
K is a 1-factor of G, q.e.d.

Corollary 1

If G has a 1-factor, then $|UT| = |LT|$.

Corollary 2

If G has a 1-factor, then $R(G)$ has on each level at least n vertices, where $n = |UT|$.

Corollary 3

Let G be a graph in H having 1-factors and let $|CR(G)|_i$ (i=1,2,3) be the corresponding complete reduced graphs representing three different orientations of H. Then:

(a) $(UT)_i = (LT)_i = n_i$; i=1,2,3

(b) The number of n_i-tuples of pairwise disjunctive paths going from $(UT)_i$ to $(LT)_i$ equals the number of 1-factors of G, i.e. is the same for i=1,2,3.

Corolloary 4

A chain of n-rings has n+1 1-factors (8).

DISCUSSION

The procedure, based on the reduced graph model, for the enumeration of Kekulé structures (1-factors) of benzenoid hydrocarbons consists of three steps:

(a) The presentation of a corresponding benzenoid graph,

(b) Its transformation to a complete reduced graph, and

(c) The counting of the vertical paths over the complete reduced graph. (The count corresponds to the number of Kekulé structures).

We will apply this procedure to enumerate the Kekulé structures of anthanthrene.

EXAMPLE 3

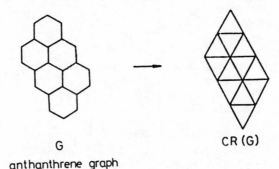

G

anthanthrene graph

CR (G)

Path count

```
a b d g j l
a b e g j l
a b e h j l
a b e h k l
a c e g j l
a c e h j l
a c e h k l
a c f h j l
a c f h k l
a c f i k l
```

The path count over the framework of CR(G) produces
10 paths. Thus, the total number of 1-factors
(Kekulé structures) of G (anthanthrene) is 10, i.e.
K(G) = the path count.

Since each vertical path in CR(G) corresponds to
one 1-factor, these paths can easily be transformed
into Kekulé structures. The generation procedure is
based on the following simple rule: The points in
the vertical path in CR(G) correspond to single
bonds in a given Kekulé structure of a benzenoid

hydrocarbon. We will transform ten paths of CR
(anthanthrene graph) into 10 Kekulé structures of
anthanthrene in three steps: (a) we will first
present a given vertical path in CR(G), (b) CR(G),
with a given vertical path will then be transformed
into a structure with allocated single bonds corresponding to the position of the particular vertical path in CR(G), and (c) this structure will
finally be transformed into a corresponding Kekulé
structure.

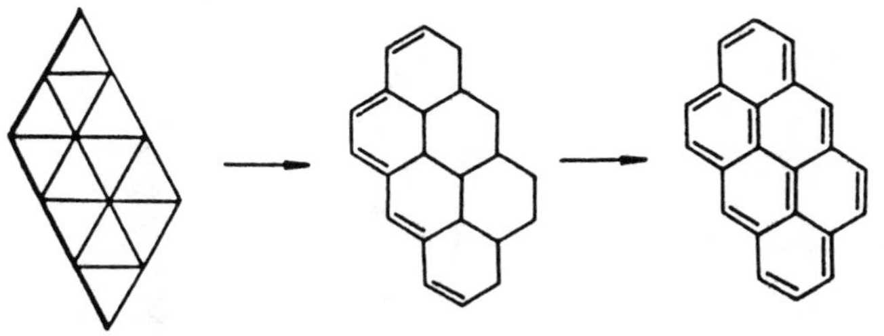

(abdgjl)

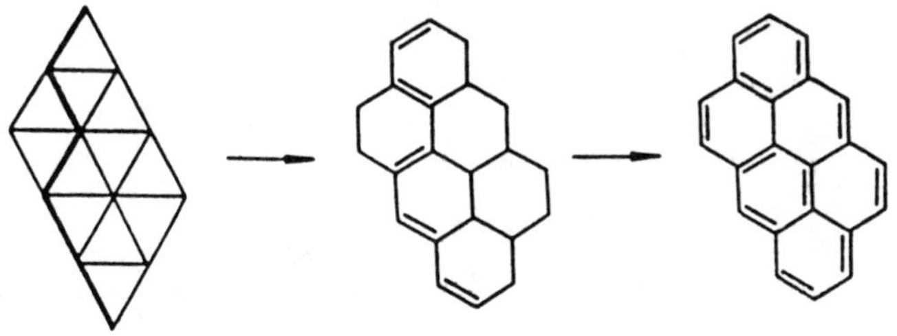

(abegjl)

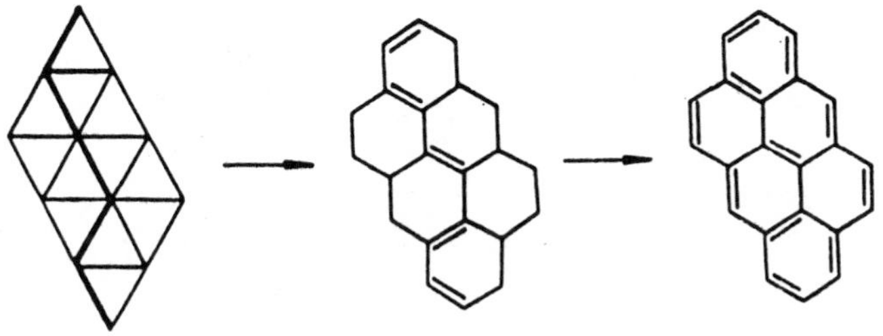

(abehjl)

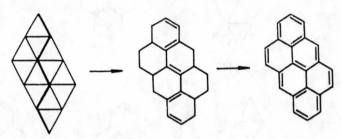

(abehkl)

(acegjl)

(acehjk)

(a c e h k l)

(a c f h j l)

(a c f h k l)

(a c f i k l)

Naturally, when we try to enumerate 1-factors of a given complex graph G, we may use previously known values of the 1-factors for its constituent graphs. Hence, we split a given complete reduced graph corresponding to a complex graph along some convenient vertical path (we can assume that a single vertex represents a path of length zero) to two (or more) fragments with a known number of 1-factors. Then, the total value of 1-factors of G represents the combination of known values of 1-factors for constituting fragments from which are excluded those values which correspond to two paths having one point in common. The procedure can be carried out, of course, only in such cases when there is no path (in any n-tuple) going from one fragment into another.

EXAMPLE 4

G   CR (G)

CR(G) in the above example can neither be broken along path AB_1C nor along path AB_2C (because there always exists a path going from one fragment to another). The only case when we can perform the fragmentation of CR(G) without any problem is when one of the constituting fragments is a chain (and, of course, the fragmentation point is one vertex). Therefore, in the above example it suffices only to change the orientation of G:

G   CR (G)

We can easily find that there are 20 pairs of paths in the first (larger) fragment:

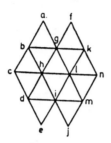

I path	II path
a b c d e	f g h i j
a b c d e	f g l i j
a b c d e	f g l m j
a b c d e	f k l i j
a b c d e	f k l m j
a b c d e	f k n m j ←
a b h d e	f g l i j
a b h d e	f g l m j
a b h d e	f k l i j
a b h d e	f k l m j
a b h d e	f k n m j ←
a b h i e	f g l m j
a b h i e	f k l m j
a b h i e	f k n m j ←
a g h d e	f k l i j
a g h d e	f k l m j
a g h d e	f k n m j ←
a g h i e	f k l m j
a g h i e	f k n m j ←
a g l i j	f k n m j ←

Six pairs among them (denoted by an arrow) contain a path going through the fragmentation point (denoted by a black dot in CR(G)). The smaller fragment contains two paths, and one of them also goes through the fragmentation point. Each of twenty 1-factors belonging to the larger fragment may be combined with both possibilities of the second fragment. However, the obtained value must be corrected for those paths belonging to both fragments which pass through the fragmentation point. Then, the total number of 1-factors for G, K(G), for the above example is equal to K(G) = 20·2 - 6·1 = 34.

If we denote by K(A) and K(B) the total numbers of paths belonging to fragments A and B making up G, and by P(A) and P(B) paths passing through the fragmentation point in both fragments, then the expression for calculating $K(G)$ is given by

$$K(G) = K(A) \cdot K(B) - P(A) \cdot P(B).$$

In some cases the task of calculating 1-factors by this procedure is considerably simpler.

EXAMPLE 5

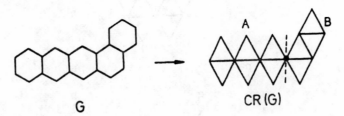

Chain A has four 1-factors, three of them containing the fragmentation point. Fragment B has three 1-factors, one of them containing the fragmentation point. Graph G has, therefore, altogether nine 1-factors, i.e.: $K(G) = 4 \cdot 3 - 3 \cdot 1 = 9$.

Below we give several more examples. Data will be arranged in the following order: (a) benzenoid graph G, (b) complete reduced graph of G CR(G), and (c) the number of 1-factors of G K(G), respectively.

$$K(G) = 6 \cdot 6 - 1 \cdot 5 = 31$$

$$K(G) = 14 \cdot 7 - 8 \cdot 6 = 50$$

G   CR(G)

$$K(G) = 5{\cdot}5 - 1{\cdot}1 = 24$$

G   CR(G)

$$K(G) = 6{\cdot}6 - 3{\cdot}5 = 21$$

The question arises as to whether there is a sufficient condition for structure G in H to have a 1-factor; i.e. can we find some general class of graphs having 1-factors ? In the first place, of course, there is a class of chains of fused hexagons. All members of this class possess 1-factors.

Earlier it was stated (1) that structures called whole lattices always have 1-factors (whole lattices are defined by conditions in Corollary 1, Corollary 2, and the following: Every vertex in the reduced graph, with the exception of vertices belonging to the first and the last level of R(G) is linked with at least two diagonal and one horizontal edge to the adjacent vertices). Unfortunately this is not true. An example of a whole lattice without 1-factors is the structure given below.

EXAMPLE 6

CR (G)

G

Furthermore, it appears that the property of being
a whole lattice is not a property of G, because it
is not invariant with regard to the change of
orientation of H:

G   CR(G)

G   CR(G)

In the first case a combination of two whole lattices is obtained, whilst in the second case a single whole lattice is generated. Thus, there is no
simple sufficient condition available for a graph
to have a 1-factor.

CONCLUDING REMARKS

The reduced graph of a benzenoid graph is defined
and some of its properties investigated. It is applied to the enumeration of Kekulé structures of
benzenoid hydrocarbons. The range of applicability
of the reduced graph model at present is not fully
explored. A lot more work is needed before the usefulness of this model is established. Work is in
progress in this direction (44).

ACKNOWLEDGEMENTS

We are thankful to Professor A.T.Balaban
(Bucharest), Professor S. El-Basil (Cairo), Profes-
sor B.M.Gimarc (Columbia, SC), Dr A.Graovac (Zagreb),
Professor W.C-Herndon (El Paso), Professor D.J.Klein
(Galveston), Professor O.E.Polansky (Mülheim/Ruhr),
Professor M.Randić (Ames), Dr B.Ruščić(Zagreb),
Professor P.Seybold (Dayton), and Dr T.Živković
(Zagreb) for helpful discussions and correspondence
on the reduced graph model, the enumeration of
Kekulé structures and their role in chemistry.

Part of this work was carried out during the visit of one of us (NT) to the Department of Chemistry, Wright State University, Dayton, Ohio, USA. NT is grateful to his host Professor Vladimir Katović for the support, hospitality, and many discussions on the uses of graph theory in chemistry.

REFERENCES

1. B. Džonova-Jerman-Blažič and N. Trinajstić, Comput. Chem. 6, 1211 (1982).
2. B. Džonova-Jerman-Blažič and N. Trinajstić, Croat. Chem. Acta 55, 347 (1982).
3. S. El-Basil, G. Jashari, J.V. Knop, and N. Trinajstić, Monat. Chem. 115, 1299 (1984).
4. S. El-Basil, P. Křivka, and N. Trinajstić, Croat. Chem. Acta 57, 339 (1984).
5. S. El-Basil and N. Trinajstić, J. Mol. Struct. (Theochem) 110, 1 (1984).
6. G. Rumer, Göttinger Nach., 377 (1932).
7. G.W. Wheland, J. Chem. Phys. 3, 356 (1935).
8. M. Gordon and W.H.T. Davidson, J. Chem. Phys. 20, 428 (1952).
9. C.F.Wilcox, Jr., Tetrahedron Lett., 795 (1968).
10. T.F. Yen, Theoret. Chim. Acta 20, 399 (1971).
11. D. Cvetković, I. Gutman, and N. Trinajstić, Theoret. Chim. Acta 34, 129 (1974).
12. W.C. Herndon, Tetrahedron 29, 3 (1973).
13. M. Randić, J.C.S. Faraday Trans. II, 232 (1975).
14. O.E. Polansky and I. Gutman, Math. Chem. (Mülheim/Ruhr) 8, 269 (1980).
15. L. Pauling, Acta Cryst. B36, 898 (1980).
16. J.V. Knop and N. Trinajstić, Int. J. Quantum Chem.: Quantum Chem. Symp. 14, 503 (1980).
17. I. Gutman, Math. Chem. (Mülheim/Ruhr) 11, 1 (1981); Croat. Chem. Acta 55, 371 (1982); Math. Chem. (Mülheim/Ruhr) 17, 3 (1985).
18. C.J. Cyvin, Math. Chem. (Mülheim/Ruhr) 13, 167 (1982); Monat. Chem. 113, 167 (1982); ibid. 114, 13 (1983); Acta Chim. Hung. 112, 281 (1983).
19. N. Trinajstić, Chemical Graph Theory, CRC Press, Boca Raton, FL, 1983, Vol. II, Chapter 2.
20. R.L. Brown, J. Comput. Chem. 4, 556 (1983).
21. A.T. Balaban and I. Tomescu, Math. Chem. (Mülheim/Ruhr) 14, 155 (1983); Croat. Chem. Acta 57, 391 (1984); Math. Chem. (Mülheim/Ruhr) 17, 91 (1985).
22. J.R.Dias, Can. J. Chem. 62, 2914 (1984).
23. J.V. Knop, P. Křivka, K. Szymanski, and N. Trinajstić, Comput. Math. Appls. 10, 369 (1984); J.V. Knop, W.R. Müller, K. Szymanski, and N. Trinajstić, Computer Generation of Certain Classes of Molecules, SKTH, Zagreb, 1985.

24. W.C. Herndon, J. Am. Chem. Soc. 98, 887 (1976);
    Israel J. Chem. 20, (1980); Tetrahedron 38,
    1389 (1982).
25. M. Randić, Chem. Phys. Lett. 38, 68 (1976);
    J. Am. Chem. Soc. 99, 444 (1977); Tetrahedron
    33, 1905 (1977); Mol. Phys. 34, 849 (1977);
    Int. J. Quantum Chem. 17, 549 (1980); J. Phys.
    Chem. 86, 3870 (1982); M. Randić and N. Trinajstić
    J. Am. Chem. Soc. 106, 4428 (1984); M. Randić,
    N. Trinajstić, J.V. Knop, and Ž. Jeričević,
    ibid. 107, 849 (1985).
26. S. Kuwajima, J. Am. Chem. Soc. 106, 6496 (1984).
27. W. Gründler, Z. Chem. 19, 236 (1979); ibid. 19,
    236 (1979); ibid. 19, 391 (1979); ibid. 20,
    391 (1980); ibid. 20, 425 (1980); ibid. 21,
    198 (1981); ibid. 22, 63 (1982); ibid. 22, 63
    (1982); ibid. 22, 235 (1982); ibid. 23, 157
    (1983).
28. W. Gründler, Tetrahedron 38, 125 (1982); Monat.
    Chem. 113, 15 (1982); Theoret. Chim. Acta 63,
    439 (1983).
29. T.P. Živković, Theoret. Chim. Acta 61, 363
    (1982); ibid. 62, 335 (1983); ibid. 63, 445
    (1983); Int. J. Quantum Chem. 23, 679 (1983);
    Croat. Chem. Acta 56, 29 (1983); ibid. 56, 525
    (1983); Chem. Phys. Lett. 107, 272 (1984); J.
    Math. Phys. 25, 2743 (1984).
30. M. Randić, B. Ruščić, and N. Trinajstić, Croat.
    Chem. Acta 54, 295 (1981); B. Ruščić and N.
    Trinajstić, Theoret. Chim. Acta, submitted.
31. W.C. Herndon and H. Hosoya, Tetrahedron 40,
    3987 (1984).
32. P.C. Hiberty and G. Ohanessian, Int. J. Quantum
    Chem. 27, 245 (1985); ibid. 27, 259 (1985).
33. D.J. Klein, T.G. Schmalz, W.A. Seitz, and G.E.
    Hite, Int. J. Quantum Chem.: Quantum Chem. Symp.,
    in press.
34. N.D. Epiotis, Unified Valence Bond Theory of
    Electronic Structure, Springer, Lecture Notes
    in Chemistry, Berlin, 1982, Vol. 21; Unified
    Valence Bond Theory of Electronic Structure:
    Applications, Lecture Notes in Chemistry,
    Springer, Berlin, 1983, Vol. 34.
35. M.J.S. Dewar and H.C. Longuet-Higgins, Proc.
    Roy. Soc. (London) A 214, 482 (1952).
36. L. Pauling, The Nature of Chemical Bond, Cornell
    University Press, Ithaca, New York, 1958,
    second edition, twelfth printing.
37. G.W. Wheland, The Theory of Resonance, Wiley,
    New York, 1952, sixth printing.
38. D.J. Klein, Pure Appl. Chem. 55, 299 (1983).
39. F. Harary, Graph Theory, Addison-Wesley,

Reading, Mass., 1971, second printing.
40. D. Cvetković, I. Gutman, and N. Trinajstić, Chem. Phys. Lett. 16, 614 (1972).
41. D. Cvetković, I. Gutman, and N. Trinajstić, J. Chem. Phys. 61, 2700 (1974).
42. A. Graovac, I. Gutman, and N. Trinajstić, Topological Approach to the Chemistry of Conjugated Molecules, Lecture Notes in Chemistry, Springer, Berlin, 1977, Vol. 4.
43. I. Gutman, Croat. Chem. Acta 56, 365 (1983).
44. N. Trinajstić, work in progress.

Chapter 29

MOLECULAR ORBITAL RESONANCE THEORY APPROACH: APPLICATION AND DEVELOPMENT

T.P. Živković
The Rugjer Bošković Institute, 41001 Zagreb, Croatia, Yugoslavia

ABSTRACT

Molecular orbital resonance theory (MORT) combines the intuitively appealing chemists picture of the molecule, as exemplified by the simple resonance theory (RT), with the numerical advantages of the MO theory. It gives much better description of the SCF ground states of conjugated molecules than the VB approach. It also conceptually enriches the molecular quantum theory, as illustrated by the splitting and the expansion theorems, and their implications.

INTRODUCTION

Historically, two most important methods in the treatment of the quantum chemical problems are the molecular orbital (MO) and the valence bond (VB) theory. The MO theory originated conceptually in physics. Despite undeniable mathematical advantages, this theory lacks chemical intuition, and it has no simple connection with the basic concepts of chemistry, for example, with the very important notion of the chemical bond. The VB theory originated conceptually in chemistry, and it explicitly incorporates via different VB resonance structures the chemists' picture of the molecule. This approach is hence extensively used by the chemists, especially in its simplest form, the resonance theory (RT). However, if applied quantitatively and more rigorously, the VB approach becomes relatively inefficient and numerically quite inferior to the MO approach. This drawback of the VB approach is mainly due to the inadequate treatment of one- and two-particle energy contributions [1]. Physically, the main stabilising force in the molecule is the attractive force between electrons and the nuclei, while the electron-electron interaction, being repulsive, is destabilising. However, in the VB formalism the bonding is attributed mainly to the exchange integral K which is a part of a two-particle electron-electron interaction (provided the basis atomic orbitals are orthonormalized, what is usually assumed), while the one-particle contributions are taken into account only in the next step, as a correction. In the MO approach the bonding is already on the simplest level, the Hückel theory, attributed to the one-particle resonance integral β, and the two-particle contributions are taken into account as a correction in the more sophisticated approaches. This hierarchy of approximations is physically natural, and it explains the computational superiority of the MO over the VB approach [1].

In view of the relative advantages and disadvantages of the MO and VB theory, it is desirable to formulate such a theory which would combine only their advantages. The above discussion suggests the way how this should be done. In order to conserve the close connection with the bond picture, and to retain all the conceptual advantages of the simple RT picture, one has to retain the notion of the resonance and of the resonance structure from the VB approach. In order to reestablish the correct hierarchy of one- and two-particle energy contributions, one has to interpret each particular bond in the MO and not in the VB sense [1,2]. This is the main idea of the Molecular Orbital Resonance Theory (MORT) approach. In this paper we intend to give a simple account of this method, its applications and the most important results.

THE CONFIGURATION INTERACTION SPACE X_n AND RESONANCE STRUCTURES IN THE MORT APPROACH

We shall consider the configuration interaction (CI) space X_n build upon n electrons moving over

2n orthonormalized orbitals X_i $(i=1,...,2n)$. In the MORT approach this space is spanned by regular resonance structures (RRS). The set $R(n)$ of all n-particle RRSs is defined in the following way [3]:

i) Partition the set $B \equiv \{i\}$ containing 2n vertices (i) into subsets B^O and B^* containing n vertices each. By definition, each vertex $(i) \in B^O$ is "source" while each vertex $(i) \in B^O$ is "sink".

ii) Form excited and non-excited bond orbitals (BO):

$$\varphi_s = \varphi_{ij} = \frac{1}{\sqrt{2}} (X_i + X_j) \quad \text{non-excited BO}$$

$$\varphi_s^* = \varphi_{ij}^* = \frac{1}{\sqrt{2}} (X_i - X_j) \quad \text{excited BO}$$

(1)

satisfying the condition

$$(i) \in B^O \quad \text{and} \quad (j) \in B^* \tag{1'}$$

iii) Each normalized determinant containing n mutually disconnected excited and/or nonexcited BOs satisfying the condition (1') is by definition regular resonance structure (RRS) [3].

Graphically, each orbital X_i is represented as a vertex (i). Excited and non-excited BOs are hence represented as oriented and non-oriented bonds, respectively. By convention, in the case of the oriented bond the end of the arrow coincides with the sink vertex

This uniquely defines graphical representation of RRSs. For example, in the case n=2 one has four orbitals X_i, and one can chose the partition $B^O \equiv \{2,4\}$, $B^* \equiv \{1,3\}$ of the set $B \equiv \{1,2,3,4\}$:

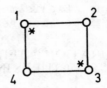

The corresponding set of RRSs is given in Fig. 1.

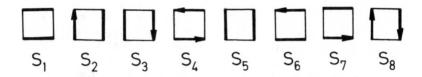

Figure 1. Regular resonance structures in the case n=2.

One can show that non-excited and singly excited RRSs alone span the space X_n [4]. Accordingly, doubly excited RRSs S_4 and S_8 in Fig.1 are linear combinations of other RRSs in this Figure, and hence they need not be considered. Further, the dimension of the space X_n equals $d(n)=(2n)!/(n!)^2$ [3]. In particular, the space X_2 has $d(2)=6$ dimensions. In Fig.1 there are exactly six non-excited and singly excited RRSs, and hence these RRSs are linearly independent. For higher n this is not the case. Thus, if n=3 there are 24 non-excited and singly excited RRSs

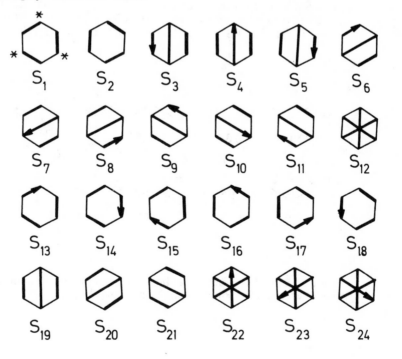

Figure 2. Non-excited and singly excited RRSs in the case n=3. These structures span the CI space X_3.

However, d(3)=20. The set given in Fig.2 is hence overcomplete by four structures. This is due to the fact that RRSs are in general not orthogonal to each other. The nonorthogonality of RRSs is a slight drawback of these structures. However, the some drawback is shared by VB resonance structures as well.

MORT VERSUS VB DESCRIPTION OF MOLECULES. SOME EXAMPLES

In the MORT approach the space X_n is spanned by RRSs. In the VB approach this space is spanned by VB resonance structures. Formally, the two approaches differ only in the choice of the basis set in X_n. However, physically (and computationally) one choice of the basis set can be highly adventageous over another.

We shall now compare the VB and MORT descriptions of the ground states of conjugated molecules. As an example consider the butadiene molecule. The corresponding CI space is the space X_4. In the VB approach the ground state Ψ of the butadiene molecule is a linear combination of VB resonance structures. Some typical VB structures are shown in Fig.3.

0.2210 0.0100 0.2264 0.2320 0.0207

1 2 3 4 5

0.0207 0.1251 0.0538 0.0513 0.0009

6 7 8 9 10

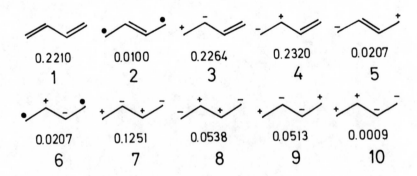

Figure 3. Some VB structures of butadiene and the corresponding structural weights according to Hiberty and Ohanessian [5].

Structure 1 is the Kekulé structure and it corresponds to the chemists simplified description of
the butadiene molecule. Hence, one expects this
structure to be dominant and all other structures
to contribute only slightly to the ground state.
This is however not the case. Hiberty and Ohanessian
[5] have recently calculated structural weights
of structures 1 to 10 in the SCF ground state
$\psi \equiv \psi_{SCF}$ of butadiene. These structural weights are
also given in Fig.3. The Kekulé structure 1 contributes only about 22% to the SCF ground state!
In addition, ionic structures 4 (there are two such
structures due to the degeneracy) contribute about
23% to the SCF ground state, i.e. more than the
Kekulé structure!! The contribution of some other
ionic structures is also significant. Accordingly,
the VB Kekulé structure is not a very good approximation to the SCF ground state of butadiene.

Consider now this molecule in the MORT approach.
The ground state is here a linear combination of
RRSs. To a very good approximation the π-electron
ground state of an even conjugated hydrocarbon can
be written in the spin-separated form $[1,3]$.

$$\psi = |\Phi \; \bar{\Phi} > \tag{2}$$

where Φ and $\bar{\Phi}$ are normalized spin-α and spin-β
substates, respectively. In particular, the SCF ground
state ψ_{SCF} is <u>exactly</u> of the form (2). One can now
represent the state Φ (and equally the state $\bar{\Phi}$) as
a linear combination of RRSs

$$\Phi = \sum_i \lambda_i S_i \tag{3}$$

The approximation (2) substantionally simplifies the
MORT approach. In particular, in the butadiene case
instead to consider all four-particle RRSs spanning
the space X_4 (which has $d(4)=70$ dimensions), one
has to consider only all two-particle RRSs spanning
the space X_2 (which has only $d(2)=6$ dimensions).
These latter structures are given in Fig.4 together
with the corresponding structural weights W_i.

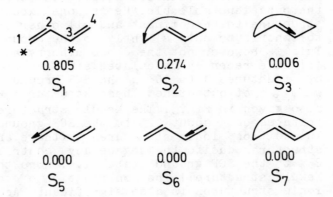

0.805	0.274	0.006
S_1	S_2	S_3
0.000	0.000	0.000
S_5	S_6	S_7

Figure 4. Linearly independent RRSs of butadiene and the corresponding structural weights. See Fig. 1.

Structural weights are defined according to [6]

$$W_i = |< \Psi_{SCF} | S_i \bar{S}_i >|^2 \tag{4}$$

The MORT Kekulé structure S_1 contributes now about 80% to the butadiene ground state. The contribution of this structure to the ground state is thus dominant, in accord with the chemical intuition. Another striking feature of the MORT description is that structures S_5, S_6 and S_7 do not contribute at all to the ground state. We shall return to this point in the next section.

As another example consider the benzene molecule. The structural weight of the two VB Kekulé structures of benzene is (joinly) W=0.0486 [5], to be compared with the structural weight W=0.8100 of the

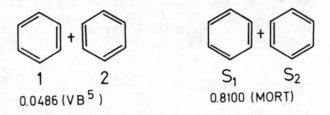

1	2	S_1	S_2
0.0486 (VB⁵)		0.8100 (MORT)	

Figure 5. Benzene VB and MORT Kekulé structures and the corresponding structural weights.

two MORT Kekulé structures of benzene (see Fig.5).
The latter structural weight is calculated according
to [6,7]

$$W = |< \psi_{SCF} | \phi \overline{\phi} >|^2 \qquad\qquad (5)$$

where

$$\phi = (S_1 + S_2)/\sqrt{2.5} \qquad\qquad (5')$$

is the normalized spin-α substate containing the
two Kekulé RRSs [7]. The linear combination of the
two VB Kekulé structures contains only about 5%,
while the linear combination of the two MORT Kekulé
structures contains 81% of the SCF ground state of
benzene! In summary, the MORT Kekulé structures
represent much better the ground state, and if more
accurate description of the ground state is needed,
the progressive inclusion of energetically higher
resonance structures is relatively (in comparation
to the VB approach) fast convergent.

This is a general result which one obtains for
other molecules as well. The poor performance of
the VB Kekulé structures is due to the fact that
the VB approach treats one- and two-particle energy
contributions in an unnatural order [1]. We have
raised this argument in the introduction, and we
now see that the choice of RRSs is fully justified:
these structures conserve the close connection with
the bond picture retaining the intuitively appealing
notion of the resonance, and at the same time they
reestablish the correct hierarchy of one- and two-
particle energy contributions. As a result, the
description of the ground states of conjugated compounds is mathematically quite successful.

Beside the above theoretical argument based on
the hierarchy of energy contributions, there is
another less formal way to explain the superiority
of the MORT approach. Each VB Kekulé structure is
purely covalent, while MORT Kekulé structures contain covalent as well as ionic contributions. The
"true" ground state should contain both, covalent
as well as ionic contributions. One can even show
that the ionic contribution is significantly pre-
ponderant [5]. Accordingly, in the VB approach one
can never expect to obtain a good description of
the ground state, unless one explicitly introduces
ionic structures. In the MORT approach this is not
necessary since on each level of the approximation
the ionic contributions are automatically included.

SPACES X_n^+ AND X_n^- AND THE CHARGE POLARIZATION

In order to apply efficiently the MORT approach to various quantum chemical problems, one has to derive simple and fast algorithms for the evaluation of overlaps and matrix elements of one- and two-particle operators between different RRSs. This is done elsewhere [8]. One finds that the crucial notion in the evaluation of these matrix elements is the notion of the superposition of two RRSs, and the notion of active and passive cycles [8]. The superposition of RRSs S_a and S_b is a graph G_{ab} which is obtained by superimposing graphical representations of these two structures, and it consists of disconnected even cycles $C_\mu \in G_{ab}$ [2,8]. Each cycle $C_\mu \in G_{ab}$ is characterized by two numbers, n_μ and m_μ, where $2n_\mu$ is the number of bonds in C_μ, while m_μ is the number of oriented bonds in C_μ. Cycle C_μ is "passive" if $(n_\mu + m_\mu)$ is even and "active" otherwise. Some examples are given in Fig.6. Thus the

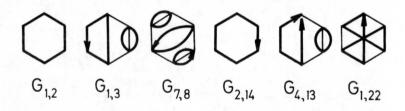

$$G_{1,2} \qquad G_{1,3} \qquad G_{7,8} \qquad G_{2,14} \qquad G_{4,13} \qquad G_{1,22}$$

Figure 6. Some superpositions of RRSs in the case n=3. The corresponding RRSs are shown in Figure 2.

superposition $G_{1,2}$ contains one active cycle C_1 $(n_1=3, m_1=0)$, the superposition $G_{1,3}$ contains two active cycles, C_1 $(n_1=2, m_1=1)$ and C_2 $(n_2=1, m_2=0)$, etc. After the examination of overlaps and matrix elements between different RRSs, one finds that the set $R(n)$ of all these structures splits into two clearly distinguishable subsets $R^+(n)$ and $R^-(n)$ containing "positive" and "negative" RRSs, respectively [3,8]. The algorithm to perform this splitting is simple: if the superposition G_{ab} of RRSs S_a and S_b contains an even number of passive cycles, these two structures are contained in the same subset, if however G_{ab} contains an odd number of passive cycles, these two structures are contained in the opposite subsets [3]. For example, in Fig. 6 the superposition $G_{1,2}$ contains no passive cycle (i. e. zero - an even number). Hence, RRSs S_1 and S_2 in Fig.2 are of the same parity. Applying this criterion to all RRSs in this Figure, one finds that structures S_1 to S_{12} are of one parity (conventionally taken to be positive), while structures S_{13} to S_{24} are of another parity (conventionally taken to be negative). Similarly, in Fig.1 structures S_1 to S_4 are of one parity (positive), while structures S_5 to S_8 are of another parity (negative). The partition on positive and negative structures has many important consequences. Thus one can show that the overlap $S_{ab} = \langle S_a | S_b \rangle$ between RRSs S_a and S_b vanishes if the superposition G_{ab} contains at least one passive cycle [8]. This implies $S_{ab} = 0$ whenever S_a and S_b are of the opposite parity, i.e.

$$S_a \in R^+(n), \quad S_b \in R^-(n) \implies S_{ab} = 0 \qquad (6)$$

Spaces X_n^+ and X_n^- spanned by positive and negative RRSs, respectively, are hence orthogonal to each other.

$$\psi_a \in X_n^+ , \quad \psi_b \in X_n^- \implies \langle \psi_a | \hat{I} | \psi_b \rangle = 0 \qquad (6')$$

where, for the further convenience, we have explicitly written the unit operator $\hat{I}$, $\langle \psi_a | \hat{I} | \psi_b \rangle \equiv \langle \psi_a | \psi_b \rangle$. For example, the space X_3^+ spanned by structures S_1 to S_{12} in Fig.2 is orthogonal to the space X_3^- spanned by remaining twelve structures S_{13} to S_{24}, etc. In general, we call each state $\psi \in X_n^+$ (which is a linear combination of positive RRSs), as well as each state $\psi \in X_n^-$ (which is a linear combination of negative RRSs) an "alternant-like" (AL) state [3,8].

Consider now the operator $\hat{R}_{ii}$

$$\hat{R}_{ii} = 2 \mu_i^+ \mu_i - 1 \qquad (7)$$

where μ_i^+ and μ_i are fermion creation and annihilation operators, respectively, of the orbital X_i. One can show that the matrix element of this operator between RRSs S_a and S_b vanishes whenever these RRSs are of the same parity [8].

$$S_a, S_b \in R^+(n) \quad \underline{or} \quad S_a, S_b \in R^-(n) \implies$$
$$\langle S_a | \hat{R}_{ii} | S_b \rangle = 0 \tag{8}$$

Relation (8) implies

$$\Psi_a, \Psi_b \in X_n^+ \quad \underline{or} \quad \Psi_a, \Psi_b \in X_n^- \implies$$
$$\langle \Psi_a | \hat{R}_{ii} | \Psi_b \rangle = 0 \tag{8'}$$

and in particular

$$\psi \in X_n^+ \quad \underline{or} \quad \psi \in X_n^- \implies \langle \psi | \hat{R}_{ii} | \psi \rangle = 0 \tag{8''}$$

The expectation value of the operator $\hat{R}_{ii}$ vanishes over each AL state $\psi^\pm \in X_n^\pm$. However, the operator $\hat{O}_i = \hat{R}_{ii}/2$ is the effective charge density operator (at the vertex (i)). Accordingly, each AL state has vanishing effective charge at all vertices (i), i. e. it is completely nonpolarized. As a consequence, if the state $\psi \in X_n$ is polarized, then it is neccessarily a nontrivial linear combination of a positive and a negative AL state

$$\psi = \psi^+ + \psi^- \tag{9}$$

where $\psi^+ \in X_n^+$, $\psi^- \in X_n^-$ and $\psi^+ \neq 0$, $\psi^- \neq 0$. The charge polarization is hence due to the interference between subspaces X_n^+ and X_n^- [3,8]. This is quite a remarkable property, and it is completely unlike the VB picture where one has to introduce explicitly the ionic structures. Let us give some examples.

Consider the butadiene molecule. Structures S_1, S_2 and S_3 in Fig.4. are positive, while structures S_5, S_6 and S_7 are negative. Hence each state ϕ which is a linear combination of structures S_1, S_2 and S_3 alone, has the effective charge zero at all four vertices (i)=1,2,3,4. This can be upset only if the state ϕ beside positive structures contains also some negative structures. However, we have found that structural weights of structures S_5, S_6 and S_7 are zero. Hence, the state ϕ corresponding to the butadiene ground state contains only positive RRSs, and it has the effective charge zero at all four vertices. This is spin-α effective charge since the state ϕ is the spin-α substate of the state $\psi = |\phi \; \overrightarrow{\phi}\rangle$. In the same token spin-β effective charge vanishes, and hence the total charge vanishes as well.

The result obtained is by itself not surprising.
Butadiene is an alternant hydrocarbon, and hence
according ot the pairing theorem the effective π-
electron charge should vanish at all carbon atoms.
What we gained is the insight into the mechanism of
the charge polarization: the reason why butadiene is
not polarized is that all RRSs contained in its
ground state are of the same parity. The polarization,
if any, can be due only to the interference between
RRSs of the opposite parity. This is conceptually
a new picture. Another gain is numerical: one has to
consider only positive structures S_1, S_2 and S_3, and
this significantly simplifies the evaluation of the
ground state. Compare this with the VB approach:
there one has to consider all structures in Fig.3.
(there are 16 such structures, since the degeneracy
of these structures should be also taken into account).

In the same way can be treated the benzene molecule. In the spin-separation approximation (2) one
has to consider the space X_3. The corresponding RRSs
are given in Fig.2. One finds that structural weights
of all negative structures S_{13} to S_{24} vanish [4].
Accordingly, the ground state contains only positive
structures S_1 to S_{12}. This substantially simplifies
the evaluation of the ground state (in addition, the
set S_1 to S_{12} is overcomplete by two structures, and
it is sufficient to consider only the first ten
structures S_1 to S_{10} [4]).

One should note the qualitative difference between
the emergence of the charge polarization in the VB
and in the MORT picture. In the VB picture one dis-
tinguishes covalent and ionic structures. Covalent
structures (e.g. structures 1 and 2 in Fig.3.) have
the effective charge zero at each vertex. Ionic
structures (e.g. structures 3 to 10 in Fig.3.) have
the effective charge different from zero, at least
at two vertices. Covalent structures alone can never
produce charge polarization. The charge polarization
can be obtained only if one explicitly includes ionic
structures. This is conceptually a classical picture.
In the MORT approach all structures have the effective charge zero at each vertex. The charge polarization is due to the interference (or, if one prefers,
to the resonance) between structures of the opposite
parity. This is conceptually a quantum picture.

ALTERNANT SYSTEMS AND THEIR CHARACTERISTIC PROPERTIES

The charge polarization is only a special case of many other remarkable properties of spaces X^+ and X^-. All these properties are the consequence of the splitting theorem [3,8,9]. In short, this theorem states the following:

Each operator $\hat{O}$ is a unique linear combination of an "alternant" operator $\hat{O}_{al}$ and an "antialternant" operator $\hat{O}_{nal}$

$$\hat{O} = \hat{O}_{al} + \hat{O}_{nal} \tag{10}$$

where all alternant operators satisfy (6'), while all antialternant operators satisfy (8'). Further, each alternant operator is a unique linear combination of "reduced" alternant operators, while each antialternant operator is a unique linear combination of "reduced" antialternant operators. Reduced alternant and reduced antialternant operators thus form a basis in the space of all alternant and in the space of all antialternant operators, respectively [8,9]. The set of all reduced operators is given elsewhere [9], and the identification of this set is the most important part of the splitting theorem. In particular, the unit operator $\hat{I}$ is a reduced alternant operator, the operator $\hat{R}_{ii}$ is a reduced antialternant operator, etc. [8,9].

Due to the splitting theorem, one can generalize the results obtained in the preceding section to all alternant and to all antialternant operators. One thus finds that alternant and antialternant operators complement each other in the following way:

Alternant operators define alternant systems. These operators satisfy relation (6'), and hence they can be block-diagonalized in subspaces X_n^+ and X_n^-. As a consequence, each hermitian alternant operator has in X_n the complete set of AL eigenstates, i.e. it describes an alternant system [9]. This implies a constructive and exhaustive definition of alternant systems. The definition is constructive since all alternant Hamiltonians can be easily constructed as linear combinations of reduced alternant operators [8,9]. The definition is exhaustive since each operator $\hat{O}$ having the complete set of AL eigenstates can be represented as a linear combination of reduced alternant operators [9]. (Operator $\hat{O}$ is not necessarily an alternant operator, but rather a linear combination of an alternant operator and a "vanishing" operator $\hat{Z}$, $\hat{O} = \hat{O}_{al} + \hat{Z}$. However, as far as the space X_n is considered, operators $\hat{O}$ and $\hat{O}_{al}$ are identical, since $\hat{Z}$ vanishes over X_n. Hence the operator $\hat{Z}$ can be omitted [9]).

Antialternant operators define characteristic properties of alternant systems. These operators satisfy relation (8') and in particular relation (8"). One thus obtains the whole set of properties characteristic to AL states $\psi^{\pm} \in X_n^{\pm}$, i.e. to complementary subspaces X_n^+ and X_n^-. The vanishing of the effective charge density is only a particular example. Another example is the vanishing of the bond orders between vertices of the same parity, etc. The complete set of these properties is given elsewhere [9].

Alternant systems and their properties have been investigated also by other authors using the MO theory [10]. All these approaches are based on the pairing theorem [10]. The most general constructive definition of an alternant system was given by McLachlan [10]. He has shown that the eigenstates of the PPP Hamiltonian describing an alternant hydrocarbon satisfy the pairing theorem [10]. Koutecký has defined some more general alternant Hamiltonians satisfying the pairing theorem, but unfortunately in a rather implicite way [10]. In all these approaches it is cruical to prove the pairing theorem, since in the case of neutral alternant systems this theorem guaranties the vanishing of bond orders between atoms of the same parity, for some eigenstates of the Hamiltonian [8-10]. The results obtained in the MORT approach are however much more general:

i) Each alternant Hamiltonian can be explicitly constructed. No further generalization of the notion of alternant systems is possible [9].

ii) All characteristic linear properties of alternant systems are obtained [9]. Beside charge density and bond orders this includes many other properties.

iii) These properties are shown to be characteristic to all AL states, i.e. to entire spaces X_n^+ and X_n^-, and not only to particular eigenstates of alternant Hamiltonians (which vary form case to case).

ANTIALTERNANT PERTURBATION OF ALTERNANT SYSTEMS

According to (10) each Hamiltonian $\hat{H}$ can be written in the form

$$\hat{H} = \hat{H}_{al} + \lambda \hat{V}_{nal} \qquad (11)$$

where H_{al} is an alternant operator, $\hat{V}_{nal}$ is an antialternant operator, and λ is a parameter. Hence, each system can be considered to be a perturbed alternant system: $\hat{H}_{al}$ is the Hamiltonian of the unperturbed alternant system, while $\lambda \hat{V}_{nal}$ is an antialternant perturbation. This leads to a particular kind of the perturbation expansion with many interesting properties [4,11]. In general, the eigenstate $\psi(\lambda) \in X_n$ of the Hamiltonian $\hat{H} \equiv \hat{H}(\lambda)$ is a function of a parameter λ. All the properties of a system, expressed as expectation values $\langle \hat{O} \rangle_\lambda = \langle \psi(\lambda) | \hat{O} | \psi(\lambda) \rangle / \langle \psi(\lambda) | \psi(\lambda) \rangle$ of various operators $\hat{O}$ over the eigenstate $\psi(\lambda)$, are hence also functions of λ. Due to the particular form (11) of the perturbation, these expectation values obey some regularities which are expressed by the expansion theorem [11]. In the case when $\psi(0)$ is nondegenerate, this theorem states the following: each alternant property is an even function of λ, while each antialternant property is an odd function of λ, i.e.

$$\langle \hat{O}_{al} \rangle_\lambda = \langle \hat{O}_{al} \rangle_{-\lambda} \quad , \quad \langle \hat{O}_{nal} \rangle_\lambda = -\langle \hat{O}_{nal} \rangle_{-\lambda} \qquad (12)$$

In addition, these functions are analytic in λ (for all real λ) [4].

Relations (12) in particular imply

$$\langle \hat{H}(\lambda) \rangle_\lambda = \langle \hat{H}(-\lambda) \rangle_{-\lambda} \qquad (13)$$

i.e. the energy $E(\lambda) \equiv \langle \hat{H}(\lambda) \rangle_\lambda$ is an even function of λ.

In order to illustrate the significance of the expansion theorem, we shall give here only one among many interesting consequences of this theorem. Relations (12) imply

$$\left[\frac{d}{d\lambda} \langle \hat{O}_{al} \rangle_\lambda \right]_{\lambda=0} = 0 \qquad (14)$$

i.e., the first derivative of an alternant property vanishes in the point $\lambda=0$. Physically, this derivative is the rate of change of the property represented by the operator $\hat{O}_{al}$ as a function of an infinitesimal perturbation $\lambda \hat{V}_{nal}$. This quantity is recognized to be a polarizability, in a generalized sense. For example, the operator $\hat{V}_{nal}$ can be chosen to be the effective charge density operator $\hat{O}_i = \hat{R}_{ii}/2$, while the operator $\hat{O}_{al}$ can be chosen to be a bond-order operator connecting vertices of the opposite parity (this is, as required, an alternant operator [3,8]). In this case the expression (14) represents the bond-charge polarizability. One thus obtains the result that in an alternant system (point $\lambda=0$) bond-charge polarizability vanishes, provided the bond connects vertices of the opposite parity. This particular result has been obtained independently by Coulson and Longuet-Higgins within the Hückel theory [12]. The present result is completely general, since it applies to each alternant system, i.e. Hamiltonian $\hat{H}_{al}$ can be any hermitian alternant operator.

It is now obvious how other analogous results can be obtained. One has only to combine all possible antialternant perturbations $\lambda \hat{V}_{nal}$, with all possible alternant operators $\hat{O}_{al}$. Each such combination leads to a particular relation involving various polarizabilities. Due to the expansion theorem, one thus generates a large set of linear relations which involve different polarizabilities, and which are satisfied by all nondegenerate eigenstates of alternant systems [4].

REFERENCES

1. Živković, T.P., Theoret. Chim. Acta, 62, 335 (1983)
2. Živković, T.P., Croat. Chem. Acta, 56, 29 (1983); 56, 525 (1983); 57, 1553 (1984); Theor. Chim. Acta, 61, 363 (1982); Int. J. Quant. Chem. 23, 679 (1983)
3. Živković, T.P., Croat. Chem. Acta, 57, 367 (1984)
4. Živković, T.P., unpublished results
5. Hiberty, P.C., Ohanessian, G., Int. J. Quant. Chem. 27, 245 (1985)
6. This definition slightly differs from the one applied by Hiberty and Ohanessian. However, if their definition is applied, the difference between the VB and MORT results is even more striking [4]
7. Živković, T.P., Chem. Phys. Letters, 107, 272 (1984)
8. Živković, T.P., J. Math. Phys., 25, 2749 (1984)
9. Živković, T.P., J. Math. Phys., 26, 1626 (1985)
10. Coulson, C.A., Rushbrooke, G.S., Proc. Cambridge Philos. Soc. 36, 193 (1940); Pople, J.A., Trans. Faraday Soc., 49, 1375 (1953); Brickstock, A., Pople, J.A., ibid., 50, 90 (1954); McLachlan, A. D., Mol. Phys., 4, 49 (1961); 2, 271 (1959); Koutecký, J., J. Chem. Phys., 44 3702 (1966)
11. Živković, T.P., Croat. Chem. Acta, 57, 1575 (1984)
12. Coulson, C.A., Longuet-Higgins, H.C., Proc. Roy. Soc., A191, 39 (1947); A192, 16 (1947); A193, 447, 456 (1948); A195, 188 (1948)

INDEX